16th International Environmental Specialty Conference 2018

Held as Part of the Canadian Society for Civil Engineering Annual Conference 2018

Fredericton, Canada
13-16 June 2018

Editors:

Kripa Singh
Catherine Mulligan

ISBN: 978-1-5108-8972-9

Printed from e-media with permission by:

Curran Associates, Inc.
57 Morehouse Lane
Red Hook, NY 12571

Some format issues inherent in the e-media version may also appear in this print version.

Printed by Curran Associates, Inc. (2019)

For permission requests, please contact Canadian Society for Civil Engineering at the address below.

Canadian Society for Civil Engineering
521-300 rue St-Sacrement
Montréal, QC
H2Y 1x4

Phone: (514) 933-2634
Fax: (514) 933-3504

info@csce.ca

Additional copies of this publication are available from:

Curran Associates, Inc.
57 Morehouse Lane
Red Hook, NY 12571 USA
Phone: 845-758-0400
Fax: 845-758-2633
Email: curran@proceedings.com
Web: www.proceedings.com

TABLE OF CONTENTS

Building Tomorrow's Society
Bâtir la Société de Demain

Fredericton, Canada
June 13 to 16, 2018 / *13 au 16 juin, 2018*

TREATMENT OF 17B-ESTRADIOL AND ITS METABOLITES IN WASTEWATER

Bennett, Jessica L.[1,2], Mackie, Allison L.[1], Park, Yuri[1] and Gagnon, Graham A.[1]
[1] Department of Civil and Resource Engineering, Dalhousie University
[2] js344308@dal.ca

Abstract: Recently, estrogens have been used to feminize and increase growth in American eels (*Anguilla rostrata)* being grown in a recirculating aquaculture system (RAS). The main objective of this study was to reduce concentrations of estrogens in aquaculture wastewater (17β-estradiol (E2) and its metabolites estrone (E1) and estriol (E3)) to below detection limits in order to assure stakeholders that no detrimental environmental effects will come from discharging this waste stream. Advanced oxidation processes (AOPs), including UV and UV with addition of hydrogen peroxide (H_2O_2), are known to effectively degrade trace organic contaminants. This study tested these AOPs for the removal of low concentrations (~10 µg/L) of E2 and its metabolites. Estrogen concentrations were analyzed using liquid chromatography-tandem mass spectrometry (LC-MS/MS). Increasing UV dose from 100 to 1000mJ/cm^2 increased the degradation of estrogens to near or below detection limits. Analytes were more effectively degraded by UV in aquaculture wastewater than in a pure water matrix. Adding H_2O_2 did not considerably or consistently improve the efficacy of the UV treatments. The results of this study indicate that these AOP water treatment processes are effective at degrading E1, E2, and E3 in aquaculture wastewater.

1 INTRODUCTION

The use of RAS (recirculating aquaculture systems) recirculation systems in the rearing of *Anguilla* species has been studied extensively (Dalsgaard 2013) *A. rostrata,* along with other freshwater eels, do not differentiate their sex until late in life, and will develop based on their environmental conditions (Colombo and Grandidr 1996). Estrogens have been used to encourage female sex differentiation (Degani and Kushnirov 1992) and increase growth in freshwater eels to great success (Degani et al. 2003); However, estrogens distributed into the aquatic environment are known to cause health concerns for freshwater organisms (Folmar et al. 1996). As such, it is crucial to control estrogen release into the environment from aquaculture settings.

Advanced oxidation processes (AOPs), specifically UV and UV/H_2O_2, have been used to remove estrogens from pure water and wastewater matrices. UV AOPs have been largely successful in degrading estrogenic compounds from pure water and wastewater (Cédat et al. 2016) but have not been tested for this purpose in an aquaculture wastewater matrix.

The aim of this study was to determine the efficacy of two AOPs (UV and UV/H_2O_2) at degrading E2 and its metabolites (E1, E3) in both pure water and aquaculture wastewater from American eel production. The results of this study will help to determine the best treatment techniques to be further studied for aquaculture wastewater treatment at pilot-scale.

2 MATERIAL AND METHODS

2.1 Chemicals Reagents

E1, E2, E3, and $^{13}C_6$-Estradiol (internal standard stock solution of estrogen compounds), were purchased from Fisher Scientific (Ontario, Canada) and Cambridge Isotope Laboratories, Inc. (Massachusetts, USA) at a concentration of 1mg/mL in methanol and acetonitrile. Reference working stock solutions were prepared at a concentration of 1mg/L in methanol and stored at -20 °C. Methanol and acetonitrile (LC-MS grade) were purchased from Fisher Scientific (Ontario, Canada). Deionised water was generated by Milli-Q system (Reference A^+, Millipore). To test the efficacy of different treatments for the removal/degradation of E1, E2, and E3 in an aqueous solution, each analyte was prepared at a concentration of 10µg/L in ultrapure water and aquaculture effluent water.

2.2 Collection of Aquaculture Effluent Samples

Effluent was collected from a small RAS located at Dalhousie University's Aquatron Facility with a daily system volume of 7323L. Makeup flow was calculated to be approximately 10% of system volume and was maintained at approximately 0.5L/minute. Wastewater samples were collected from the top of a swirl separator effluent outflow and refrigerated at 4°C until use.

2.3 AOP Batch Experiments

A bench scale experiment of UV and UV/H_2O_2 photo-oxidation was carried out as follows: 125mL water samples containing 10µg/L of E2 or E2 and its metabolites were treated using a bench-scale collimated beam unit (PS1-1-120, Cargon Carbon) equipped with a 1kW medium-pressure mercury UV lamp in the UVB + UVC range (200 – 300nm). Samples were exposed to UV fluences of 100, 500, and 1000mJ/cm^2. Hydrogen peroxide (H_2O_2) was dosed from a 3mg/mL solution to achieve 1 or 10 mg/L concentrations and quenched with 2,950units/mg bovine catalase. For the UV/H_2O_2 experiment, 1 or 10mg/L of H_2O_2 was immediately spiked into the water samples as they were exposed to UV irradiation.

2.4 Analytical Methods

Liquid chromatography-tandem mass spectrometry (LC-MS/MS) analysis was performed by a combination of high performance liquid chromatography (HPLC) with mass spectrometry (MS). The HPLC system (from Agilent 1260) binary pump was used for all estrone (E1), 17β-estradiol (E2), estriol (E3) and $^{13}C_6$ estradiol (ISTD) detections. An Agilent Poroshell EC C-18 (3.0 X 50 mm, 2.7 µm) column was used and was maintained at 30°C throughout the run. The mobile phase comprising of ultrapure water (A) and the mixture of acetonitrile and methanol, 30%:70%: v/v (B) at 0.6mL/min was used for LC condition run. 80µL of sample was injected through the column. The gradient at 30% organic solvent (B) was initially held and the proportion of organic solvent was programmed to linearly increase to 100% over 10 min. A post-time of 5-min was added to make sure all the analytes were eluted within the run time. An additional 10 minutes of post time was allowed the column to re-equilibrate before next injection. This resulted in a total cycle time of 26 minutes. Mass spectrometry was performed on an Agilent 6460 Triple quadrupole mass spectrometry (QqQ). Source parameter settings were optimized and two transitions: a quantifier (most abundant product) and qualifier for target analyte were used for most of the compounds to increase specificity for method. Data acquisition and analysis was performed using Agilent MassHunter software (Version Rev B.08.00).

3 RESULTS

3.1 The Effect of UV and UV + H2O2 on 10µg/L of E2 Alone

[redacted]Pure Water Matrix

UV fluence had the greatest impact on E2 removal, with a high dose of UV (1000mJ/cm^2) resulting in greatest removal using UV alone and UV + 1mg/L H_2O_2. The addition of H_2O_2 did not seem to markedly improve analyte removal in comparison to UV alone. E2 was more readily removed alone than in a mixture of its metabolites over all fluences and treatments, with the exception of a high dose of UV in concurrence with UV + 10mg/L H_2O_2 treatment, which showed similar removal of E2 both alone and in a mixture.

[redacted]Aquaculture Wastewater Matrix

UV fluence had the greatest impact on E2 removal in an aquaculture wastewater matrix E2 was removed to below the detection limit (BDL) of our LCMS-MS (1µg/L in aquaculture wastewater) at medium and high UV doses. We achieved consistently better removal of E2 in an aquaculture wastewater matrix in comparison to the pure water matrix. E2 removal was not considerably different alone or in a mixture of metabolites at any treatment or UV dose.

3.2 The Effect of UV and UV + H2O2 on 10µg/L of E1, E2, and E3 in a Mixture

[redacted]Pure Water Matrix

UV fluence had the greatest impact on the removal of E1, E2, and E3 in a solution. E1 was readily removed in every instance, reaching complete degradation to BDL of the LCMS-MS (0.4ug/L in a pure water matrix) after a medium dose of UV over all three treatments. E2 and E3 were removed at a similar rate. The addition of H_2O_2 did not seem to considerably improve the removal of E1, E2, or E3 in comparison to UV alone.

[redacted]Aquaculture Wastewater Matrix

UV fluence had the greatest impact on removal of E1, E2, and E3 in an aquaculture wastewater matrix (Figure 2). All metabolites were removed to BDLs at both medium and high doses, with the exception of E2 treated with a medium dose of UV and 10mg/L H_2O_2, where E2 was removed to just above the detection limit. Analytes were removed considerably better in an aquaculture wastewater matrix in comparison to a pure water matrix

4 DISCUSSION

Our study has shown that AOPs (i.e., UV, and UV/H_2O_2) are effective at degrading E2 and its metabolites (E1, E3) in both pure water and an aquaculture wastewater matrix. UV was more effective at degrading analytes in a wastewater matrix than in a pure water matrix, and the addition of H_2O_2 did not considerably improve analyte removal. In some instances, H_2O_2 addition did result in higher degradation, but not consistently (i.e., a higher dose did not always result in more degradation).

Other studies have discerned that UV radiation alone is not largely successful at degrading estrogens (Li et al. 2016) but is much more effective in combination with H_2O_2 (Rosenfeldt and Linden 2004). We did not find that UV alone, or in tandem with H_2O_2, degraded E2 and its metabolites to BDLs (below detection limits) up to a dose of 1000mJ/cm^2 (with the exception of E1, which was readily degraded to BDL at a dose of 500mJ/cm^2) in pure water.

We did not find that the addition of H_2O_2 improved analyte removal. Theoretically, H_2O_2 should increase removal efficiency of most compounds, as the interaction between H_2O_2 and UV light creates strong oxidant hydroxyl radicals (OH·), which effectively break down chemical bonds. The literature leads us to conduct more research would need to be completed to observe more definitive results.

The results of this research are important both economically and environmentally for the aquaculture industry. These preliminary results indicate that UV is effective at removing these analytes from wastewater.

Acknowledgements

We would like to thank the Nova Scotia Department of Fisheries and Aquaculture Research and Development program for providing the funding for this research, and NovaEel Inc. for providing the wastewater to carry out this work. We would also like to thank the staff and students at the Centre for Water Resource Studies at Dalhousie University for their help and support throughout this project.

References

Cédat, B., de Brauer, C., Métivier, H., Dumont, N. and Tutundjan, R. 2016. Are UV photolysis and UV/H2O2 process efficient to treat estrogens in waters: Chemical and biological assessment at pilot scale. Water research, 100: 357-366.

Colombo, G. and Grandidr, G. 1996. Histological study of the development and sex differentiation of the gonad in the European eel. Journal of Fish Biology, 48(3): 493-512.

Dalsgaard, J., Lund, I., Thorarinsdottir, R., Drengstig, A., Arvonen, K. and Pedersen, P. B. 2013. Farming different species in RAS in Nordic countries: Current status and future perspectives. Aquacultural engineering, 53: 2-13.

Degani, G. and Kushnirov, D. 1992. Effects of 17β-estradiol and grouping on sex determination of European eels. The Progressive Fish-Culturist, 54(2): 88-91.

Degani, G., Tzchori, I., Yom-Din, S., Goldberg, D. and Jackson, K. 2003. Growth differences and growth hormone expression in male and female European eels [Anguilla anguilla (L.)]. General and comparative endocrinology, 134(1): 88-93.

Folmar, L. C., Denslow, N. D., Rao, V., Chow, M., Crain, D. A., Enblom, J., ... & Guillette Jr, L. J. (1996). Vitellogenin induction and reduced serum testosterone concentrations in feral male carp (Cyprinus carpio) captured near a major metropolitan sewage treatment plant. Environmental Health Perspectives, 104(10): 1096.

Li, M., Xu, B., Liungai, Z., Hu, H. Y., Chen, C., Qiao, J. and Lu, Y. 2016. The removal of estrogenic activity with UV/chlorine technology and identification of novel estrogenic disinfection by-products. Journal of Hazardous Materials, 307: 119-126.

Rosenfeldt, E. J. and Linden, K. G. 2004. Degradation of endocrine disrupting chemicals bisphenol A, ethinyl estradiol, and estradiol during UV photolysis and advanced oxidation processes. Environmental Science & Technology, 38(20): 5476-5483.

Building Tomorrow's Society
Bâtir la Société de Demain

Fredericton, Canada
June 13 to 16, 2018 / *13 au 16 juin, 2018*

APPLICATION OF ELECTROCHEMICAL ADVANCED OXIDATION PROCESSES (EAOPS) FOR TREATMENT OF CONCENTRATE FROM PSYCHROPHILIC ANAEROBIC DIGESTER UNITS

Choudhury, Mahbuboor[1,2], Anwar, Nawrin[1], Rajagopal, Rajinikanth[1] and Rahaman, Md. Saifur[1]
[1] Building, Civil and Environmental Engineering, Concordia University, Montreal, Quebec, Canada
[2] mahbuboor.choudhury@concordia.ca

Abstract: This study applied electrochemical advanced oxidation processes (EAOPs) like electro-oxidation and electrocoagulation to treat the concentrate deriving from a psycrophilic anaerobic municipal organic solid waste digester units. The concentrate or effluent from anaerobic digestion units contains high concentrations of organics and suspended materials. As a result, such concentrate poses a formidable challenge for treatment in conventional wastewater treatment facilities. Recently, EAOPs have demonstrated a promising ability to reduce a variety of pollutants from wastewater efficiently. The EAOPs have the advantage of not requiring the addition of chemical reagents in the treatment process, and these can also be easily implemented by integration in an automated treatment system. Hence, a quick and efficient treatment for high strength wastewater is possible with EAOPs. This study used an open cylindrical glass electrolysis cell with Boron-Doped Diamond (BDD) as the anode and stainless steel as the cathode for the electrooxidation process. The electrolysis cell also contained an aluminum plate in bipolar arrangement for the electrocoagulation process. The electrooxidation and electrocoagulation process stabilized the organics, both dissolved and particulate, in the wastewater. Preliminary tests were conducted to assess the effects of different operational parameters (e.g., current density, treatment time) of the electrolytic cell on the overall treatment efficiency. Based on the preliminary results, combined electro-oxidation and electrocoagulation processes demonstrated great potential for treating the concentrate from anaerobic municipal organic solid waste digester units.

1 INTRODUCTION

Anaerobic digestion (AD) is an engineered method designed to decompose organic solid waste through the synergistic action of a consortium of microbes under oxygen-free condition (Kalysuzhnyi, S;Veeken, A; Hamelers 2000, Li et al. 2011). It offers an appealing treatment option for agricultural wastes, food wastes, and wastewater sludge as it can simultaneously be applied to reduce the pollution load from wastes and recover renewable energy (Chen et al. 2008). AD processes are classified by key operating parameters and design featuresT such as continuity (batch vs. continuous), operating temperature (psychrophilic, mesophilic, and thermophilic), reactor design (plug-flow, complete-mix, and covered lagoons), and solid content (wet or dry) (Li et al. 2011, Environment Canada 2013). The dry/high solids AD process usually treats solid waste with less than 80% moisture content and the feedstock have a total solids content higher that 15-20% (Rapport, J.; Zhang , R.; Jenkins, B M; Williams 2008, Environment Canada 2013). The high solids AD processes are advantageous than wet AD in terms of use of smaller sized reactors, low energy requirements for heating, minimal material handling, lower parasitic energy loss, and shows comparable biogas production to wet AD systems (Luning et al. 2003, Guendouz et al. 2008).

The fresh feedstock in a high-solid AD process is inoculated in the reactor to speed up the reaction processes. It is a much faster process than degradation of solid wastes in a landfill, but slower than in wet AD (Martin et al. 2003). The concentrate or leachate from AD reactor is usually recycled to inoculate the fresh feedstock (Rapport, J.; Zhang , R.; Jenkins, B M; Williams 2008, Environment Canada 2013). Leachate recycling enables the colonization of the bacteria throughout the digester by providing a passive transport mechanism for microbial communities. This provides the operational benefit of reduced handling costs and higher reactor volume utilization (Li et al. 2011). Any surplus of concentrate from AD process is directed to a wastewater treatment plant for further treatment before discharge (Environment Canada 2013). The concentrate produced from anaerobic digestion of organic solid waste contains high concentrations of organics, nutrients, and total solids. Its quality is inferior to usual municipal wastewater quality received in most wastewater treatment plants. Hence, there is a scope for developing effective treatment schemes for concentrates deriving from anaerobic digesters treating municipal organic solid wastes.

The present study considered the use of electrochemical advanced oxidation processes (EAOPs) like electrooxidation (EO) and electrocoagulation (EC) as a quick and efficient means to treat the concentrate from a psychrophilic dry AD system managing municipal organic solid wastes. Application of psychrophilic dry AD systems provides low-cost solid waste management options in cold climates (Massé et al. 2014, Rajagopal et al. 2017). These electrochemical processes avoid any addition of external chemicals to purify the wastewater and hence can be considered a "cleaner" technology than many conventional treatments. Also, these processes require less skilful supervision and can be easily integrated as parts of a more extensive treatment operation. EO process use derivatives of oxygen as an oxidizer. The direct action of electric current is used to oxidize or reduce species exclusively at the electrodes, with the indirect action generating an oxidant, which reacts electrochemically with the species in solution. The possible dominant species of oxidants include superoxide ($O_2^{\circ-}$), hydrogen peroxide (H_2O_2), ozone (O_3), atomic oxygen (O°), hydroxyl radical (OH°), hydroperoxyl radical (HO_2°), molecular oxygen (O_2) and singular oxygen ($^1O_2$) (Ferro 2006, Sultana et al. 2018). On the other hand, EC involves the generation of coagulants in situ by electrically dissolving metal ions (aluminum or iron) at the anodes (Chen 2004). The nascent metal ions are very efficient coagulants and quickly form large particulate flocs, which in turn can chemically adsorb pollutants from wastewater (Shen et al. 2003).

Preliminary results of the combined application of EO and EC for treatment of concentrate from a psychrophilic dry AD reactor have been reported in this paper. Treatment performances of EO and EC regarding chemical oxygen demand (COD) and electrical conductivity removal from concentrate originating from a psychrophilic AD system treating municipal organic solid wastes demonstrated excellent potential for application of these EAOPs.

2 MATERIALS AND METHODS

2.1 Characterization of AD Concentrate Sample

The concentrate from a dry psychrophilic AD system treating mixed food waste with cow manure was used in the present study. The AD system was installed near Sherbrook, Quebec, Canada and was operated at a low temperature (20 – 25 °C). This temperature range has been reported to be psychrophilic in the literature for anaerobic digestion systems (Massé et al. 2014, Rajagopal et al. 2017). The characteristics of the concentrate are presented in Table 1. It is evident from the data that the concentrate contained high concentration of organics and nutrients.

Table 1: Physicochemical characteristics of the concentrate from a dry psychrophilic AD system

Parameter (Units)	Mean Value (± Standard Deviation)
pH	7.63 ± 0.17

Parameter (Units)	Mean Value (± Standard Deviation)
Conductivity (mS)	29.30 ± 1.70
Chemical Oxygen Demand (mg/l)	10658 ± 253
Total Nitrogen (mg/l)	994.50 ± 16.24
Total Kjeldahl Nitrogen (mg/l)	964.25 ± 17.89
Ammonia Nitrogen (mg/l)	767.86 ± 26.91

2.2 Experimental Set-up

An open, cylindrical glass cell with one anode and one cathode in a vertical alignment and monopolar parallel connection was used at room temperature (22 ± 2 °C) for electrochemical treatment. BDD (Fraunhofer USA, Inc.) and stainless steel electrodes (with effective surface area of 25 cm2) were used as anode and cathode, respectively. In addition to the monopolar arrangement of anodes and cathodes, an aluminum electrode was placed in a bipolar arrangement between the anode and cathode. The distance between the different electrodes were kept at 5 mm. A sample of 450 ml volume was added to the cell. A DC power supply (E364xA, Agilent Technologies, Rockaway, NJ, USA) was used for all experiments and the experiments were performed in galvanostatic mode. The electrochemical experiments were conducted for 12 hours at an applied current intensity of 0.6 A and 0.8 A. Samples were collected at different intervals (0, 2, 4, 6, 8, 12 hours) to measure the studied parameters. The BDD, aluminum, and stainless steel electrodes were cleaned by immersion in 1 M H2SO4 for 30 minutes after each experiment to remove any adsorbed molecules at the electrode surface, if any, and after that the electrodes were rinsed with deionized (DI) water until the pH of the water on the electrodes was observed to be neutral (Abdessamad et al. 2013). Rinsing the electrodes with sufficient DI water ensured removal of acid on the electrode surfaces and prevented the electrodes from becoming polarized.

2.3 Analytical Measurements

This paper reports preliminary data of COD and conductivity removal from the concentrates. COD was measured by reactor digestion method using Hach kits (London, Ontario, Canada) and the concentration was obtained using a spectrophotometer (DR 2800, HACH) (HACH 2012). The pH and conductivity of the collected samples were measured using a pH meter and a conductivity meter (both from Oakton Instruments, Vernon Hills, IL, USA), respectively.

3 RESULTS AND DISCUSSIONS

3.1 Electrical Conductivity Reduction with Application of Combined EO-EC

Figure 1 shows the electrical conductivity profile of the AD concentrate with application of combined EO-EC treatment at an applied current intensity of 0.6 A. Visual observation of the concentrate during the EO-EC treatment revealed foam generation in the first 8 hours of treatment, which may be due to electrofloatation occurring in the cell. After this initial period visible flocs were observed in the reactor cell, and the samples collected for analysis also had flocs in them, which settled rapidly in the collection container. With increasing treatment time the rate of decrease of electrical conductivity was observed to be higher.

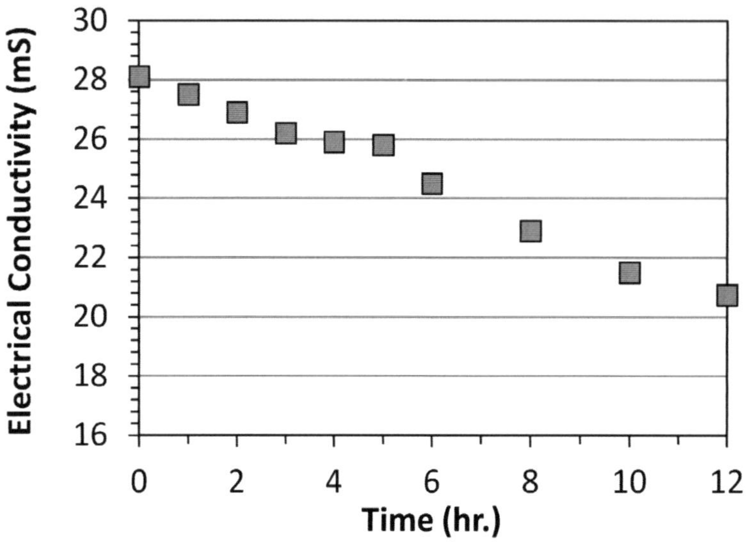

Figure 1: Electrical conductivity profile of Psychrophilic AD concentrate with application of combined EO-EC treatment at an applied current intensity of 0.6 A.

3.2 COD Reduction with Application of Combined EO-EC

COD profiles were obtained for two different applied current intensities: 0.6 A and 0.8 A. Figure 2 indicates the normalized COD reduction profile in the concentrate with time for the two applied current intensities.

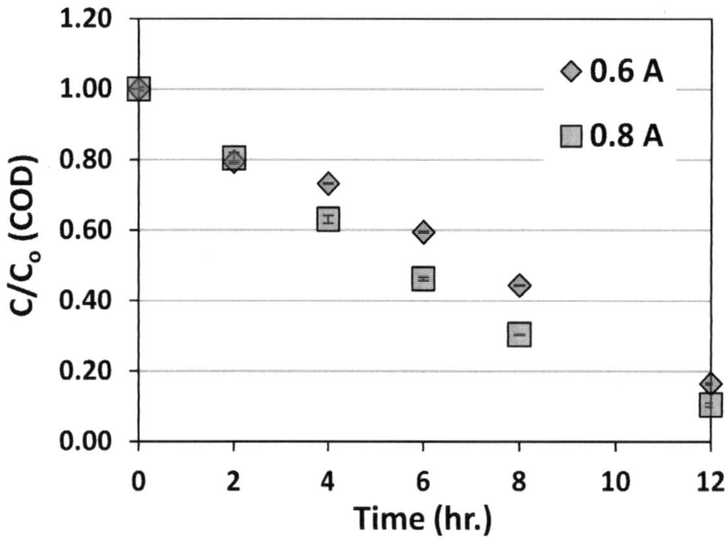

Figure 2: Variation with time of the normalized COD for the combined EO-EC treatment of Psychrophilic AD concentrate at applied current intensities of 0.6 A and 0.8 A.

The COD reduction enhanced with increased applied current intensity. In the initial few hours the effect of increased current intensity on COD reduction was not noticeable. However, the differences were more prominent in the following part, as up to 14% more removal was observed at 8 hours with 0.8 A applied current intensity. Over the extended duration of treatment, the enhanced effect of increased applied current intensity was evident.

3.3 Instantaneous Current Efficiency and Specific Energy Consumption with respect to COD removal

The instantaneous current efficiency (ICE, %) and the specific energy consumption per unit mass of COD removal (KWh kg^{-1}) for different applied current intensities (shown in Figure 3(a) and Figure 3(b), respectively) were calculated using the following equations (Eq. 3-1 and Eq. 3-2, respectively) (Li et al. 2014):

[3-1] $ICE = \frac{[(COD)_0 - (COD)_t]}{8It} FV$

[3-2] $E = \frac{UIt}{1000 V x \Delta COD}$

Where, COD0 and CODt are the solution COD values at time t=0 and t=t, respectively (mg/l), ΔCOD = (COD0 – CODt) (mg/l), V is the solution volume (liter), F is Faraday constant (96,485 C/mole), I is the applied current (A), 8 is the oxygen equivalent mass (g/eq), and U is the applied voltage (V)

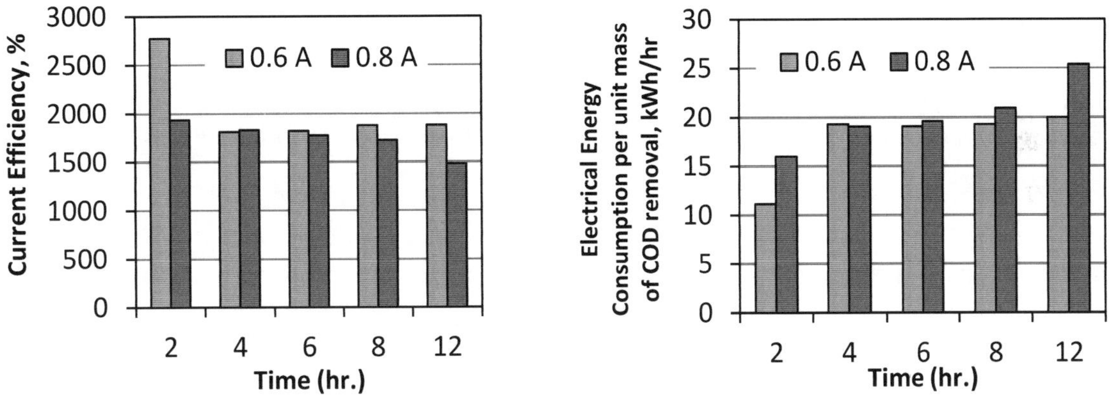

(a) (b)

Figure 3: Current efficiency, % (a) and Specific electrical energy consumption per unit mass of CID removal, kWh/kg (b) for the combined EO-EC treatment of Psychrophilic AD concentrate at applied current intensities of 0.6 A and 0.8 A.

It can be observed that both the specific current consumption and current efficiency values for 0.6 A and 0.8 A applied current intensities remained somewhat similar between 4 – 8 hours of treatment. At 12 hours the current efficiency value decreased and the specific electrical energy consumption value increased for higher applied current (i.e. 0.8 A). This may be due to oxidation of readily electro-oxidizable organics at the beginning part of the treatment; as a result the recalcitrant fraction of COD did not decrease later at the same rate even at increased current intensity.

3.4 Observation of Electrode Fouling in EO-EC

Some fouling was observed in the cathode stainless steel electrode during the treatment experiments. No fouling was observed on the BDD anodes. The bipolar aluminum electrode demonstrated decay on the surface facing the stainless steel electrode and some fouling occurred on the surface facing the anode electrode.

4 CONCLUSIONS

The present study showed great potential for the application of combined EO-EC process for treatment of high strength concentrate derived from psychrophilic dry AD systems. COD decreased rapidly in the first 8 hours of treatment. With applied current intensity of 0.8 A the COD reduction rate decreased after 8 hours of electrolysis. This may be due to conversion of all organics to complexes recalcitrant to electrooxidation, which were not oxidized at the same rate. Overall the COD was reduced to 84% and 90% of initial value after 12 hours of electrolysis with 0.6 A and 0.8 A applied current intensity, respectively. Based on the preliminary observed data, further experimental investigations will be carried out to observe removal of nitrogen and organic loads from AD concentrates. Scaling up of the combined EO-EC process for treating large volume of wastewater will be a critical issue. Such scale up would require proper electrode arrangement throughout the reactor and sufficient mixing in the reactor to circulate the wastewater around the electrodes. Failure to ensure proper circulation of wastewater in the electrochemical reactor might result localized treatment of wastewater in the reactor, which will not remove overall organic load from the wastewater.

Acknowledgements

The authors gratefully acknowledge the Natural Sciences and Engineering Research Council (NSERC) of Canada for providing financial support for this project and Concordia University for providing the extensive laboratory facilities required to perform the experimental work reported in this paper.

References

Abdessamad, N., Akrout, H., Hamdaoui, G., Elghniji, K., Ksibi, M., and Bousselmi, L. 2013. Evaluation of the efficiency of monopolar and bipolar BDD electrodes for electrochemical oxidation of anthraquinone textile synthetic effluent for reuse. Chemosphere, **93**(7): 1309–1316. doi:https://doi.org/10.1016/j.chemosphere.2013.07.011.

Chen, G. 2004. Electrochemical technologies in wastewater treatment. Separation and Purification Technology, **38**(1): 11–41. doi:10.1016/j.seppur.2003.10.006.

Chen, Y., Cheng, J.J., and Creamer, K.S. 2008. Inhibition of anaerobic digestion process: A review. Bioresource Technology, **99**(10): 4044–4064. doi:10.1016/j.biortech.2007.01.057.

Environment Canada. 2013. Technical Document on Municipal Solid Waste Organics Processing. *In* Environment Canada. Available from www.ec.gc.ca.

Ferro, C.A.M.-H.& S. 2006. Electrochemical oxidation of organic pollutants for the wastewater treatment: direct and indirect processes. Chemical Society reviews, **35**(12): 1324–1340. doi:10.1039/b517632h.

Guendouz, J., Buffiere, P., Cacho, J., Carrere, M., and Delgenes, J.P. 2008. High-solids anaerobic digestion: Comparison of three pilot scales. Water Science and Technology, **58**(9): 1757–1763. doi:10.2166/wst.2008.521.

HACH. 2012. Water Analysis Handbook. *In* 7th Editio. HACH Company, Loveland, Colorado, U.S.A., HACH Company, Loveland, Colorado, U.S.A.

Kalysuzhnyi, S;Veeken, A; Hamelers, B. 2000. Two - particle model of anaerobic solid state fermentation. Water Science & Technology, **41**(2): 43–50.

Li, H., Yu, Q., Yang, B., Li, Z., and Lei, L. 2014. Electrochemical treatment of artificial humidity condensate by large-scale boron doped diamond electrode. Separation and Purification Technology, **138**(Supplement C): 13–20. doi:https://doi.org/10.1016/j.seppur.2014.10.004.

Li, Y., Park, S.Y., and Zhu, J. 2011. Solid-state anaerobic digestion for methane production from organic waste. Renewable and Sustainable Energy Reviews, **15**(1): 821–826. doi:10.1016/j.rser.2010.07.042.

Luning, L., Van Zundert, E.H.M., and Brinkmann, A.J.F. 2003. Comparison of dry and wet digestion for solid waste. Water Science and Technology, **48**(4): 15–20.

Martin, D.J., Potts, L.G. a., and Heslop, V. a. 2003. Reaction Mechanisms in Solid-State Anaerobic Digestion. Process Safety and Environmental Protection, **81**(3): 171–179. doi:10.1205/095758203765639870.

Massé, D.I., Rajagopal, R., and Singh, G. 2014. Technical and operational feasibility of psychrophilic anaerobic digestion biotechnology for processing ammonia-rich waste. Applied Energy, **120**: 49–55. doi:10.1016/j.apenergy.2014.01.034.

Rajagopal, R., Bellavance, D., and Rahaman, M.S. 2017. Psychrophilic anaerobic digestion of semi-dry mixed municipal food waste: For North American context. Process Safety and Environmental Protection, **105**: 101–108. Institution of Chemical Engineers. doi:10.1016/j.psep.2016.10.014.

Rapport, J.; Zhang , R.; Jenkins, B M; Williams, R.B. 2008. Current Anaerobic Digestion Technologies Used for Treatment of Municipal Organic Solid Waste. California Integrated Waste Management Board, (March): 90. doi:10.1016/j.biortech.2006.11.048.

Shen, F., Chen, X., Gao, P., and Chen, G. 2003. Electrochemical removal of fluoride ions from industrial wastewater. Chemical Engineering Science, **58**(3–6): 987–993. doi:10.1016/S0009-2509(02)00639-5.

Sultana, S., Rahman, M., Ahmed, C., Bakr, R., Anwar, N., and Rahaman, S. 2018. Effectiveness of electro-oxidation and electro-Fenton processes in removal of organic matter from high-strength brewery wastewater. Journal of Applied Electrochemistry, **48**(5): 519–528. Springer Netherlands. doi:10.1007/s10800-018-1185-3.

Building Tomorrow's Society
Bâtir la Société de Demain

Fredericton, Canada
June 13 to 16, 2018 / *13 au 16 juin, 2018*

ASSESSMENT OF COSTS OF CENTRALIZED AND DECENTRALIZED WATER SYSTEMS ON FIRST NATION COMMUNITIES

Vogel, Timothy M.[1,4], Zagozewski, Rebecca [3], Fonstad, Terrance A.[2], and McPhedran, Kerry N.[1]
[1] Department of Civil, Geological, and Environmental Engineering, University of Saskatchewan, Canada
[2] College of Engineering, University of Saskatchewan, Canada
[3] School of Public Health, University of Saskatchewan, Canada
[4] tim.vogel@usask.ca

Abstract: Water systems on First Nations fail to provide safe drinking water throughout Canada. Many of these communities have individual, household cisterns as part of a decentralized water system that are prone to contamination causing health issues. It has been suggested that there is insufficient funding provided by Indigenous and Northern Affairs Canada (INAC) for water infrastructure projects on First Nations which is preventing communities from addressing these issues that can be mitigated by moving to a safer, centralized distribution water system. Currently, the funding formula used by INAC focusses too simply on the construction costs of water infrastructure projects without providing any consideration for human health related costs. This research presents a proposed expansion to the funding formula to include the potential costs associated with health and social impacts that a decentralized distribution system has on a community. After creating an updated formula, a comparison study was performed on an example Saskatchewan First Nation where the capital cost of a centralized and decentralized system was compared using INAC's funding formula. Additional annual costs were added to an updated funding formula showing a 'true' cost of decentralized systems based on human health related costs. Taking these added costs into consideration, it was found that the less safe decentralized system actual cost was higher than the safer centralized system for this example community. Overall, the methods of funding allocations for communities in need of upgraded water systems must be expanded to reflect long term impacts as justification for greater capital investment.

1 BACKGROUND AND INTRODUCTION

1.1 Drinking Water in Indigenous Canada

Human health in First Nations communities has historically been negatively impacted by poor access to clean drinking water (Waldner et al. 2017). Although the issue has been recognized, communities still struggle for solutions to drinking water challenges that must address technical, social, and political aspects specific to each community. The responsibility for providing water infrastructure on First Nations is shared by Indigenous and Northern Affairs Canada (INAC), Health Canada (HC), and Environment Canada (EC) (Bradford et al. 2016). This structure of multiple overlapping Federal departments has historically complicated the implementation of water, and other infrastructure, projects on First Nations. In addition, these departments generally only provide 80% of water infrastructure costs with the nation's Chief and Council being responsible for covering the remaining 20% of the infrastructure costs (typically collected via user fees in non-First Nations communities) (Bradford et al. 2016). This lack of full funding is an ongoing issue since First Nations distribute water without charge to their communities given water is considered a human right and has a spiritual resource beyond its utility as a physical necessity.

Funding issues directly impact the quality of water treatment and distribution to the community leading to the creation of health challenges in First Nations that have existed for decades. In Canada, it is more likely for a First Nations community to experience waterborne illnesses and to have at risk water infrastructure as compared to the national average (Boyd 2006). Overall, there are 70 First Nations in Saskatchewan and 617 in Canada (Statistics Canada 2011). Saskatchewan is home to 11.7% of Canada's First Nations population making up 10.7% of the total provincial population (Statistics Canada 2011). Of these, 30% of water systems on First Nation communities are described as high risk (Black and McBean 2017) which is defined by Burnside (2011) as a water distribution system that has major deficiencies that could impact the health of a community. In Saskatchewan, 26% of First Nations have high risk water systems (Burnside 2011). There are a wide range of health issues that are prevalent on First Nations as a result of high risk water systems including gastrointestinal illnesses, skin disease, and kidney disease. Additionally, limited or restricted access to safe drinking water can also have an impact on mental health by causing anxiety and stress (Bradford et al. 2016). These health and social costs are not currently considered in the funding formula used to determine water infrastructure costs including the choice to install either centralized (piped-to-homes) versus decentralized (truck-to-cistern) systems. Generally, centralized water systems delivering water directly to household taps are less prone to producing human health issues versus decentralized systems that produce numerous potential contamination sources discussed in detail below.

1.2 Water Systems and Health

Despite technological advances that have led to improved drinking water treatment systems, there are still cases of failing Canadian water systems negatively impacting human health. Possibly the most prominent case in Canada's recent history was in 2000 with an *E. coli* outbreak in Walkerton, Ontario, where over 2,000 people were impacted by tainted water including 7 known fatalities (Lebel and Reed 2010). In Saskatchewan, the 2001 outbreak of gastroenteritis in North Battleford affected approximately 6000 people (Hrudey and Hrudey 2004). Given their populations, it is easy to track these large outbreaks in municipal supplies of larger communities. In contrast, health issues arising from small community and private water supply systems such as individual households' cisterns or wells are harder to assess and, in general, regulatory frameworks focus less on their protection (Kreutzwiser et al. 2011). This is a major issue given that Statistics Canada (2007) estimated that 5 million Canadians receive drinking water from small, decentralized systems each serving less than 300 people. Additionally, the Burnside report (2011) estimated that 26% of homes on First Nations in Saskatchewan are on a decentralized system. This value agrees with Duncan and Bowden (2009) that estimated 25% of prairie reserve lands are dependent on hauling water, which they call the "prairie problem".

There are several components of a community's water distribution system that can lead to the creation of high risk to human health due to poor water quality. For example, even if a community has an effective water treatment plant that produces safe drinking water, the lack of a centralized distribution via high-pressure piping leads to individual homes lacking safe water. Overall, the majority of First Nations in Saskatchewan have homes on a decentralized water distribution system (Burnside 2011) that has various stages involved in transporting treated water from the water treatment plant to household taps that can result in contamination. Firstly, water is collected in a truck from the plant and then delivered to holding tanks (cisterns) to individual homes. Contamination of the clean water can occur during the filling of the truck or the filling of the households' holding tanks. Secondly, if the cistern is not properly maintained or replaced at the end of its lifecycle, it can fail in keeping out contamination such as pathogens, fecal matter, and chemicals. Additionally, when water delivery trucks are down for maintenance the community has to ration their water. Further, low-pressure centralized distribution overcomes the truck-related issues but still has the cistern issues given the water is distributed to the homes at low-pressure and generally stored in cisterns prior to use.

1.3 Economic Burden of Illness/Cost of Intervention

A lack of access to safe drinking water has the potential for creating negative health effects that have intrinsic economic consequences. The most common illnesses associated with waterborne pathogens are gastrointestinal issues with associated direct or indirect costs that have been determined previously. Direct costs may include: prescription medication, over the counter treatments, and provision of alternative water (usually bottled water). Indirect costs can include: lost time from work for the sick and their caregivers, lost business, and travel costs to healthcare providers (WHO 2012). A 2006 study of these costs estimated the economic burden to be $1,089 per case (all dollar values in CDN unless otherwise stated) with an annual cost per capita of $115 (Majowicz 2006). Comparatively, a 2008 study of a British Columbia town found similar results for cost per case and annual cost per capita as $1,342 and $130, respectively (Henson et al. 2008). These costs are typical for a household, but the economic burden is greater when an outbreak occurs impacting an entire water system. For example, a Cryptosporidium outbreak in Ireland was reported to have a cost of USD$142,000/day during a 158-day boil water advisory (Chyzheuskaya et al. 2017). During a 1993 outbreak in Milwaukee, WI, data from 11 hospitals showed that mild, moderate, and severe illnesses had total costs of USD$116, USD$475, and USD$7,808, respectively (Corso 2003). The impacts of these types of outbreaks have been used to exhibit that the costs associated with reactive treatment of illnesses are often higher than costs associated with proactive infrastructure upgrades that would help to eliminate illnesses before they occur.

Intervention with water infrastructure can reduce the economic burden associated with illness. The World Health Organization estimates that with USD$11.3 billion of infrastructure upgrades that the worldwide annual savings would be USD$84 billion for elimination and/or reduction of costs for health-related treatments. However, the savings are region-specific making direct comparisons to Canadian infrastructure potentially incorrect. A more realistic comparison would be with the United States where the projected annual cost to get the entire population onto a piped network of water and sewer is USD$2.32 billion with a benefit of USD$9.01 billion (WHO 2012). Failing and/or inadequate water systems on Canadian First Nations have gained notice by all levels of government with intervention efforts in the form of regulations and budget allocations. For example, the Canadian Federal government allocated $600 million over five years to support the 2003 First Nations Water Management Strategy (SDWF 2008). However, these efforts have arguably fallen short of expectations across Canada. For example, by 2011 25% of water systems in Saskatchewan were still in a state likely to cause health concerns and drinking water advisories (INAC 2011).

1.4 Current Funding Framework

First Nations Infrastructure projects are currently primarily funded by INAC's Capital Facilities Management Plan. The Cost Reference Manual (CRM) (2005) is an updated tool created by INAC in 1978 for the use for/by communities to estimate facility costs and to aid in planning for capital projects. This manual applies costs to components of a project (see Table 1) based on values typical for construction in Toronto, Ontario. Multipliers to these values are applied based on the 'remoteness' of a community that is determined by the distance of the community from the nearest 'city centre'. For communities in Saskatchewan these city centres include Regina, Prince Albert, Saskatoon, and The Pas (located in Manitoba). This remoteness factor allows for a percentage of the cost of up to twice the amount of the Toronto model. Additionally, other site-specific indices are applied based on the nature of the site; required transport methods; and construction schedule, personnel, and administration. An example of this application is shown in Table 1.

The value for each criteria is multiplied by the other criteria to determine the site-specific indices for a project as follows:

[1] Site Specific Index = $\sum(a) * \sum(b) * \sum(c) * \sum(d) * \sum(e) * \sum(f)$

For example, a Saskatchewan community in remoteness zone 2, 50km-350km from nearest city, and with the maximum site-specific indices shown in Table 1 would have its calculated project cost multiplied by a factor of 1.20.

Funds for water infrastructure projects are allocated by this method up to a maximum of $10,000 dollars per home plus 50% of the housing connection price. To qualify for centralized, high pressure piped water, the density of homes must average 3 units per acre and lot frontage cannot be more than 30 m (INAC 2011) with alternative systems considered when density decreases past this point. These alternatives include decentralized systems (trucked water) and more limited centralized low diameter piped and low pressure piped water systems requiring cisterns. Based on INAC's Level of Service Standards, funding decisions are based on the lowest life cycle costing of these alternatives for a 20-year term.

Table 1: Example of the Calculation and Application of Site Specific Multipliers

Criteria		Multiplier %
(a) Materials	Lumber Not Locally Available	2
	Locally Available Aggregates	0
	Building Supplies Not Locally Available	1
(b) Administration	Restricted Lead Time	1
(c) Nature of Site	Normal Soil	0
	Level and Treed	1
(d) Transportation	Road	0
(e) Personnel	Semi-skilled and Unskilled Labour	1
(f) Accommodation	Within Daily Travel	4

2 METHODOLOGY

The overall goal of this research was to compare the funding formula used by INAC to fund water infrastructure to an updated formula meant to reflect the true cost of this infrastructure for a community. The INAC funding formula was taken from the 2005 CRM that provides Class C and Class D cost estimates for indicated infrastructure maintenance and/or upgrades. The item costs used were taken from the 'Water Supply, Treatment, and Distribution' section of the Facility Unit Costs in the Capital Cost Manual. The item costs from the CRM were compared to item costs from other industry sources to determine if there was any association between sources and/or justification of the values given in the CRM given the lack of referencing to real-world costs in this manual. The values from the CRM (2005) were compared to a recently completed water distribution project on a First Nation in Saskatchewan (2012), an analysis of three contractor bids for water service to a subdivision and low density layout on a First Nation in Saskatchewan (2011) and a feasibility study of a low pressure distribution system for a Saskatchewan First Nation (2017).

Only values from the CRM were used to calculate the capital cost of distribution systems to help maintain consistency. The total construction cost was then multiplied by the remoteness index and site-specific indices dependent on the project location. The following equation shows how the total cost was derived using the current funding formula:

[2] CRM Cost = (Construction Cost)*(Remoteness Index)*(Site Specific Indices)

The values derived in comparison to this manual are also Class C and Class D cost estimates. The updated funding formula includes values beyond construction by including economic burden of disease, operation and maintenance of water delivery trucks, and impact on road infrastructure. These added costs are not meant to be exhaustive, they only serve as a first step to full cost accounting that currently excludes a number of relevant costs.

[3] Modified CRM Cost = (Construction Cost) + (Cost of Illness) + (Water Truck Maintenance) + (Increased Road Maintenance)

These formulas were applied to an example community of 100 homes (Figure 1) and using an average of 5 people/household as indicated by Statistics Canada (2011). The density of these homes is typical for rural households and piping lengths are based on similarly sized projects. The values for construction were taken from the CRM and were applied to compare the capital cost of centralized and decentralized systems.

Figure 1 shows what a typical community would look like with homes marked as grey boxes. Typically, the homes in higher density subdivision style layouts near the water treatment plant would receive piped, high pressure water (based on their higher density) and homes in a more rural (i.e., less dense) layout have cisterns. The centralized system calculation determines the cost to pipe water to all homes including the low density rural layouts. This schematic is based on a community similar in layout and density to allow for comparison of the capital cost for centralized and decentralized systems.

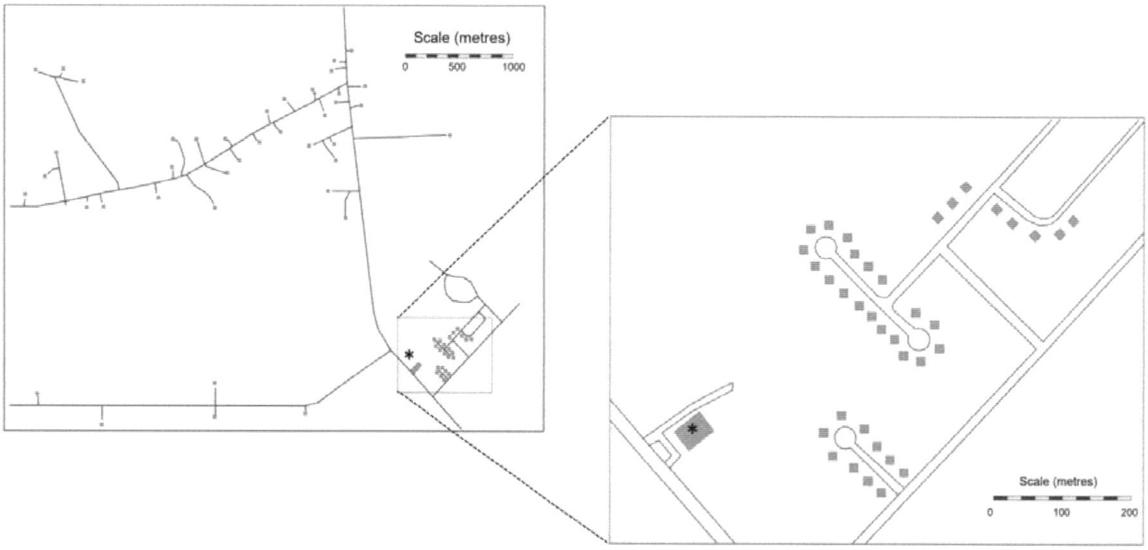

Figure 1: Schematic of Sample Community for Example Density and Layouts
(* Indicates the water treatment plant)

It should be noted that Figure 1, although representative of an actual community, shows a distinct difference between the preferred living/household arrangements (anecdotally, subdivisions are not preferred) and the water distribution systems' usage. The homes near the water treatment plant, typically the only ones on centralized system, are shown in the inset figure. This shows how a community's living arrangements are dictated by water service rather than suiting water service to a more preferred, rural housing layout. The 'subdivision style' layout is used to meet the requirements for funding of a centralized system based on the lowest schedule of services from INAC. However, traditionally people on First Nations choose to live spread out in family units (McLoud 2005).

The updated formula was applied in addition to the capital cost for the decentralized system as shown in eq. 3. The value for the cost of illness ($130) was taken from a study of economic impact of gastrointestinal illness on a Canadian community (Henson 2008). This value was then adjusted to account for increased occurrence, greater travel costs and psychological impacts that translate to a Saskatchewan First Nation community (Patrick 2010, Plummer et al. 2011, Reynolds et al. 2008). The operation and maintenance costs for water delivery trucks are allocated to communities based on the Capital Asset Inventory System. A sample agreement of what a typical heavy hauler truck would pay a municipality for its impact on rural roads was used to estimate the cost of extra maintenance and loss of road life from daily hauling of water in the First Nation community (Ministry of Highways 2013). The updated formula is applied over a proposed lifetime of 20 years (based on INAC's Lifecycle Cost) to show the ongoing impact and justify greater capital.

Please note that the sources of the monetary values being compared are from different years and have therefore been adjusted for inflation using the consumer price index from the Department of Finance's Private Sector Survey (Department of Finance 2017).

3 RESULTS

Values and formulas presented above were examined to help better understand the overall funding process and to assess its validity. Cost of materials and methodology common to water infrastructure projects were compared to determine the basis of the CRM values and see if they were similar to other sources. The results are divided into three sections: updating of the funding formula; application to an example community; and comparison of common values. The first section shows how the funding of water infrastructure projects can be expanded to reflect a truer actual cost of the infrastructure's lifetime. The second section applies this updated funding formula to an example community to compare cost over a lifecycle of 20 years. The last section provides an overview of the values used in the funding formula from the CRM that were compared to costs of methods and materials from other industry sources on water infrastructure projects to see how accurate the CRM is in estimating projects in Saskatchewan.

3.1 Updated CRM Funding Formula

The CRM funding formula was updated to include indices other than construction costs taken from relevant studies (Henson 2008; Ministry of Highways 2013; INAC 2011) for potential health impacts of contaminated water including the cost of gastrointestinal disease treatment, the operation and maintenance of water delivery trucks, and the increased road maintenance costs related to heavy water truck delivery impacts on community roads. These indices are not meant to be exhaustive in their extent of other issues but serve as a first step towards updating the CRM formula to more realistically portray the true costs. Additionally, despite the inadequacies of the CRM indicated previously, its use for comparisons between centralized and decentralized systems was acceptable given it allows for standardization of the various unit costs shown in Table 1.

Clean drinking water can be contaminated in various stages of a decentralized water system including during the filling of the water delivery truck, during the filling of cisterns, and within the cisterns. The contamination during the delivery process can be remedied via proper connections between the various water containers and having properly trained operators making the transfer. Unfortunately, cisterns at the home are prone to contamination (e.g., animals entering the cisterns) as the collars get damaged or the access is not securely restricted. It has been shown that some communities using cisterns that have been improperly installed have drinking water advisories from their installation date (Baird et al. 2013). The most common illnesses from drinking water issues are gastrointestinal with direct costs for medication and visits to health care professionals and indirect costs due to loss of work time and travel to health services. Henson's (2008) study on the direct and indirect cost of gastrointestinal illness in a British Columbia community found these costs to be $130 per capita annually. This value is conservative given that it is estimated that only 5% of gastrointestinal illnesses are reported and treated (Reynolds et al. 2008). However, contamination, and therefore related illness, is 2.5 times more likely on First Nations than other Canadian communities (Patrick 2010). Further, there are other water-related issues such as skin rashes that are not considered (Plummer et al. 2011). Currently, we tripled the costs to $390 per capita annually based on Henson's study community to a Saskatchewan First Nation to better account for the increased frequency of contamination, other types of physical and psychological illness, and remoteness of these communities. This initial increase is arguably subjective in nature needing further validation, however, serves as a reasonable first approximation based on the costs provided by Henson (2008) and the higher risk provided by Patrick (2010).

Other costs added to the funding formula are the operation and maintenance of water delivery trucks and the roads impacted by the heavy trucks. From the CRM, the purchase of a new water delivery truck is approximately $160,000 and the annual funds provided for maintenance is $1,000 per cistern, based on the Capital Asset Inventory System from INAC. Hauling water is a burden on the operation and maintenance capacity of the community requiring about one truck for every 50 cisterns. These trucks are prone to breaking down as they are under heavy demand leaving the community to ration water when they are unavailable and/or or to pay for water from other sources (e.g., bottled water). The impact that these vehicles have travelling daily on gravel roads can increase the road maintenance cost for a community as well as decrease the life of the roads. Saskatchewan Ministry of Highways (2013) suggests compensation of $82.26/km for hauling in the summer and $41.13/km hauling in the winter to municipalities for heavy trucks their impact on rural roads. These values were used in the updated cost formula to account for the impact of the water delivery trucks on community roads.

3.2 Application to a Sample Community

As a case study, a comparison was made using the CRM for centralized (Table 2) and decentralized (Table 3) distribution using the values for a community project of approximately 100 homes and 500 people. The quantities used for these calculations include information from a project brief of a low pressure water distribution system that was completed on a First Nation in Saskatchewan in 2011 (Consultant A 2012). This project serviced 100 homes with density typical to rural residences on First Nations. This simple cost estimate does not include all project expenses but is limited to what the CRM has defined. However, we assumed that components absent from a more complete estimate would be similar in value to what would be absent from a decentralized project. Table 2 shows an initial capital cost of approximately $30,000 per home ($2.8 million total) which is three times the amount ($10,000) that is typically allocated for water infrastructure projects.

Table 2: Application of CRM to Calculate Capital Cost of an Example Centralized Distribution System

Item		Units	Quantity	Unit Cost ($)	Total ($)
Installed Pipe	25mm	m	1,040	57	59,280
	38mm	m	18,300	69	1,262,700
	50mm	m	10,100	80	808,000
	75mm	m	2,700	100	270,000
	100mm	m	8,700	118	1,026,600
Isolation Valves	38mm	ea.	9	853	7,677
	50mm	ea.	11	853	9,383
	75mm	ea.	2	853	1,706
	100mm	ea.	6	853	5,118
Curbstops		ea.	100	278	27,800
Air release valves		ea.	6	5,600	33,600
				Total	3,511,864

In Table 3 the total capital cost is in close agreement ($13,648) with the typical allocation of $10,000 per home considered by INAC. However, this price could vary by community for the size of cistern and number of water delivery trucks.

Table 3: Application of CRM to Calculate Capital Cost of an Example Decentralized System

Item	Units	Quantity	Unit Cost ($)	Total ($)
Cistern (4500L)	ea.	100	5460	546,000
Housing Connection	ea.	100	5000	500,000
Water Delivery Truck	ea.	2	159,420	318,840
			Total	1,364,840

In addition to the initial capital cost of a decentralized system shown in Table 3, other annual costs were added to reflect the true cost of the system to the community (Table 4). Operation and maintenance values do not include routine maintenance of cisterns. The CAIS does not currently provide any operation and maintenance funding to low pressure distribution systems so for equal comparison it was omitted from decentralized systems as well. The amount of funding allocated for operation and maintenance of water delivery trucks is dependent on the number of cisterns and not the number of delivery trucks. This funding includes operator wages, fuel, and maintenance.

Table 4: Extra Annual Costs Associated with Decentralized System

Item	Units	Quantity	Unit Cost ($)	Total ($)
Cost of Illness	per capita	500	457	228, 500
Vehicle Maintenance	per cistern	100	1,000	100,000
Road Maintenance (summer)	km	1825	88	160,600
Road Maintenance (winter)	km	1825	44	80,300
			Total	569,400

Overall, the capital cost of a centralized system is $3.51 million and the capital cost of a decentralized system is $1.36 million. Starting from the capital costs, the annual costs of operation and maintenance can be added to each system over time to more accurately show the infrastructure costs over the 20 year lifetime (Fig. 2). The extra annual cost of a decentralized system is $569,400 with the largest portion ($240,900) attributed to the maintenance of roads which is already an issue for many communities. However, the cost of illness is also significant at $228,500 /year. Overall, the greater capital cost of a centralized system is justified in a short time at roughly four years (point of intersection of the systems in Fig. 2).

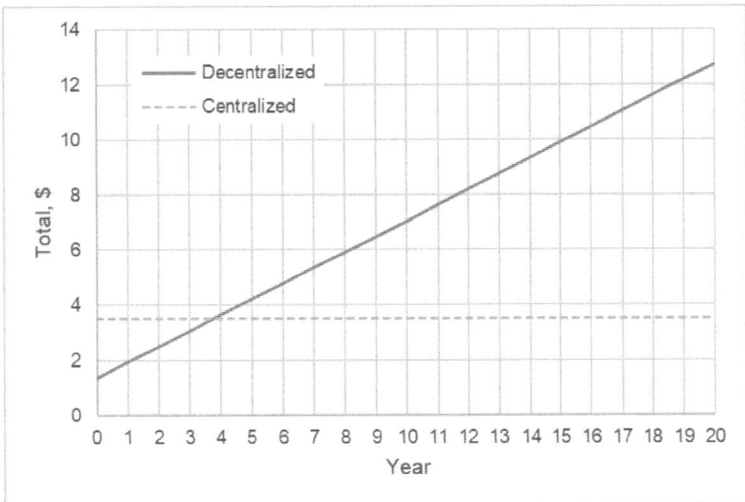

Figure 2: Annual Cost of Centralized and Decentralized Water Distribution (Values in Millions)

3.3 Comparison of Common Values

Common values for construction methods and materials used in water infrastructure projects were compared to see if there was any association or consistency in their determination. The sources that were compared were: INAC's Cost Reference Manual (2005), a completed rural centralized distribution project (2012), a feasibility study for centralized distribution (2017), and a bid analysis of three contractor's bids on a water infrastructure project (2011).

Table 5: Comparison of Values of Common Materials and Methods from Various Industry Sources

Item	Unit	2018 Value (CAD/unit)				
		INAC CRM	Community Project	Consultant	Contractor High	Contractor Low
Trench	m	*	33	96	196	95
Plough	m	*	11	30	N/A	N/A

25mm Pipe**	m	57	34	102	230	112
50mm Pipe**	m	80	35	103	N/A	N/A
75mm Pipe**	m	100	42	106	N/A	N/A
Gate Valve	ea.	852	330	N/A	N/A	N/A

* CRM includes only the cost of installed pipe and does not specify by which method
** Values for piping include the cost of installation

Clearly, general comparisons for each metric indicate no strong association between values for sources. The highest overall values were from contractors but the three contractor bids analyzed varied significantly. The CRM values were lower than the consultant and contractor bids indicating their inaccuracy given these values should align with actual costs. Interestingly, the lowest values were for the community infrastructure project that indicates the possibility of providing a centralized system for less than the CRM or an engineering consultant estimated costs. This lower price appears to be due to a source of inexpensive material that limited costs for the specific project. The consultant's feasibility study relates to the least expensive contractor. However, it is acknowledged that values for water infrastructure projects vary dependent on economic, geographic, political, and social factors making it difficult to generalize for projects on a national level and requires unique examination for every project.

4 CONCLUSIONS

Overall, it was determined that the INAC Capital Facilities Management Plan and CRM inadequately determines the cost of water infrastructure in Saskatchewan (and likely throughout Canada). The fundamental flaw with CRM is indexing all parameters to the urban Toronto example. Altus Group provides a construction cost index every year and is considered to be an industry standard (Altus Group 2017). Using the values from this report and looking at data from Statistics Canada, it is shown that the cost of labour and materials for new construction varies year to year and by geography. Clearly, there is no simple, linear method to relate values from one city to another year by year as is done with the CRM. In general, the determination of index values from the CRM is neither referenced or justified and appears arbitrary lacking these details. Additionally, the multipliers are not justified for use year-over-year and, even if justified, the CRM is out-of-date with the most recent revision done over a decade ago in 2005.

In general, there is insufficient funding for both capital projects and ongoing operation and maintenance of water systems on First Nations. Trucked water and household cisterns are the lowest cost based on capital but these systems pose a greater impact to community health and infrastructure in the years after installation. Estimating for only construction cost has failed to provide technical systems that succeed long term and is more costly long term based on the true costs presented here. The greater capital cost of a centralized system is justified considering the extra costs from failing decentralized systems and that fact that extra capital costs are recovered within approximately 4 years. The benefits to human health and ensuring safe drinking water for a community is worth the extra investment on its own but this research has found it economically justified as well.

Acknowledgements

This research has benefited from insight and information from The School of Public Health at the University of Saskatchewan, The Federation of Sovereign Indigenous Nations, and File Hills Qu'Appelle Tribal Council. This research was funded by RBC's Blue Water Project and The College of Engineering at the University of Saskatchewan.

References

Altus Group. 2017. Canadian Cost Guide 2017. altusgroup.com.

Baird, J.M. Summers, R. and Plummer, R. 2013. "Cisterns and safe drinking water in Canada." Canadian Water Resources Journal 38 (2):121-134.

Black, K and McBean, E. 2017. "First Nations' water sustainability and Security Strategy: Tools and methodologies for community-driven processes for water treatment in Indigenous communities." Technology in Society 50:57-65. doi: 10.1016/j.techsoc.2017.04.004.

Boyd, D.R. 2006. The water we drink : an international comparison of drinking water quality standards and guidelines : a report. Edited by Library Canadian Electronic and Foundation David Suzuki, International comparison of drinking water quality standards and guidelines.

Bradford, L.E. Bharadwaj L.A. Okpalauwaekwe, U. and Waldner, C.L. 2016. "Drinking water quality in Indigenous communities in Canada and health outcomes: a scoping review." Int J Circumpolar Health 75 (1):32336. doi: 10.3402/ijch.v75.32336.

Canada. Indian and Northern Affairs Canada issuing, body, and body Neegan Burnside Ltd. issuing. 2011. *Saskatchewan regional roll-up report : final.*

Chaudhary, D.S. Vigneswaran, S. Ngo, H. Shim W.G. and Moon, H. 2003. "Biofilter in water and wastewater treatment." *Korean Journal of Chemical Engineering* 20 (6):1054. doi: 10.1007/BF02706936.

Consultant A. 2012. Construction Cost Estimate Low Pressure Water Distribution Projects.

Consultant B. 2017. Low Pressure Water Distribution Feasibility Study.

Corso, P.S. Kramer, M.H. Blair, K.A. Addiss, D.G. Davis, J.P. and Haddix, A.C. 2003. "Cost of illness in the 1993 waterborne cryptosporidium outbreak, Milwaukee, Wisconsin. (Research)."*Emerging Infectious Diseases* 9 (4):426.

Department of Finance. 2017. Department of Finance Survey of Private Sector Economic Forecasters. https://www.fin.gc.ca/pub/psf-psp/index-eng.asp: Government of Canada.

Duncan, L.F. 2009. *A legal guide to Aboriginal drinking water : a Prairie province perspective.* Edited by Marie-Ann Bowden. S.l.: S.l. : Alberta Law Foundation ; Walter & Duncan Gordon Foundation.

Government of Saskatchewan. 2013. Road Maintenance Agreement Rates. edited by Ministry of Highways. Roads and Streets Construction and Maintenance.

Henson, S.J., Majowicz, S.E. Masakure, O. Sockett, P.N. Macdougall, L. Edge, V.L. Thomas, M.K. Fyfe, S. Kovacs, J. and Jones, A.Q. 2008. "Estimation of the costs of acute gastrointestinal illness in British Columbia, Canada." *Int. J. Food Microbiol.* 127 (1-2):43-52.

Hrudey, S.E. and Hrudey, E. 2006. "Safe Drinking Water: Lessons from Recent Outbreaks in Affluent Nations." *Water Intelligence Online* 5.

INAC. 2005. Cost Reference Manual. edited by Indigenous and Northern Affairs Canada. Canada: Government of Canada.

INAC. 2011. "Water and Wastewater Policy and Level of Services Standards (Corporate Manual System)." Government of Canada, accessed January 12

Kreutzwiser, R. de Loë, R. Imgrund, K. Conboy, M.J. Simpson, H. and Plummer, R. 2011. "Understanding stewardship behaviour: Factors facilitating and constraining private water well stewardship."*Journal of Environmental Management* 92 (4):1104-1114.

Lebel, P.M. and Reed, M.G. 2010. "The Capacity of Montreal Lake, Saskatchewan to Provide Safe Drinking Water: Applying a Framework for Analysis." *Canadian Water Resources Journal* 35 (3):317-338. doi: 10.4296/cwrj3503317.

Majowicz, S.E. McNab, W.B. Sockett, P. Henson, S. Doré, K. Edge, V.L. Buffett, M.C. Fazil, A. Read, S. McEwen, S. Stacey, D. and Wilson, J.B. 2006. "Burden and cost of gastroenteritis in a Canadian community." *Journal of Food Protection* 69 (3):651-659.

Mcloud, N. 2005. Exploring Cree Narrative Memory. Thesis:University of Regina. 174- 200.

Patrick, R.J. 2011. "Uneven access to safe drinking water for First Nations in Canada: Connecting health and place through source water protection." *Health and Place* 17 (1):386-389.

Safe Drinking Water Foundation (SDWF). 2009. A review of the engagement sessions for the federal action plan on safe drinking for First Nations: Position statement, April 14, 2009.,19pp.

Statistics Canada 2007. "Canada's New Government invests $4.2 million in water treatment plant at Mashteuiatsh." *Canada NewsWire*:1.

Waldner, C.L. Alimezelli, H.T. McLeod, L. Zagozewski, R. Bradford, L.E. and Bharadwaj, L.A. 2017. "Self-reported Effects of Water on Health in First Nations Communities in Saskatchewan, Canada: Results From Community-Based Participatory Research." *Environ Health Insights* 11:1178630217690193. doi:

Building Tomorrow's Society
Bâtir la Société de Demain

Fredericton, Canada
June 13 to 16, 2018 / *13 au 16 juin, 2018*

BENCH- AND FIELD-SCALE STUDY OF DISPOSAL OF GYPSUM WALLBOARD RESIDUALS

Mackie, Allison L.[1,3], Anderson, Robert[2], and Gagnon, Graham A.[1]
[1] Department of Civil and Resource Engineering, Dalhousie University, Canada
[2] Robert Anderson Consulting, Inc., Canada
[3] allisonmackie@dal.ca

Abstract: Waste or residual gypsum wallboard has the potential to generate hydrogen sulphide (H_2S) if disposed in a traditional landfill and most landfills do not accept wallboard for this reason. This study compared five materials in mixtures with residual gypsum wallboard from a wallboard manufacturing plant at bench-scale for the mitigation of sulphide generation: cement kiln dust, pasteurized biosolids, agricultural lime, open hearth slag (OHS), and wood ash. These materials were mixed with gypsum wallboard residuals and placed in vertical columns to which synthetic, de-oxygenated rain water was added weekly. Analysis of leachate from the columns indicated that wood ash and OHS were able to reduce leachate sulphide concentrations most significantly compared to control columns with gypsum residuals only. Wood ash was selected over OHS for field-scale testing due to its proximity. Field test plots were constructed atop a wood ash/pulp mill biosolids landfill, comparing a 1:1 volumetric mixture of wood ash:wallboard residuals with wallboard residuals alone in two separate plots. Both plots were covered with a layer of pulp mill biosolids which effectively prevented any H_2S gas emissions from the landfill. The results of the field study indicate that co-disposal of gypsum wallboard with wood ash can help to reduce generation of H_2S gas and that temperature was a key contributing factor controlling H_2S gas generation.

1 INTRODUCTION

Hydrogen sulphide gas (H_2S) in landfills is generated as a by-product of the respiration of sulphate-reducing bacteria (SRB). Gypsum wallboard contains both the sulphate (SO_4^{2-}) and the organic material (i.e., paper backing) required for biological sulphate reduction by SRB to produce H_2S, which is highly toxic in gaseous form. Landfills provide the anaerobic and moist environment required for SRB to thrive. Environmental conditions favoured by SRB are often found at construction and demolition (C&D) debris landfills, where gypsum wallboard is a primary component (Townsend et al. 1999). H_2S gas has a distinct odour of rotten eggs which is detectable by humans at extremely low concentrations (< 10 ppb) and produces mild toxic and neurological effects with long term exposure at low concentrations (1 to 10 ppm) and coma and death resulting from short term exposures at higher concentrations (> 500 ppm; Ko et al. 2015; Kilburn 2003).

Two main options for the mitigation of H_2S gas emission from gypsum wallboard waste or C&D debris landfills are 1) preventing its production or 2) adsorbing it after it's been generated. For option 1, one or more of the environmental conditions necessary for SRB to thrive must be removed (i.e., moisture, absence of oxygen, neutral pH) to prevent the reduction of sulphate to sulphide and thus the production of H_2S. For option 2, various cover materials, from soil to concrete, have been investigated and several have been shown to be able to adsorb H_2S gas generated from gypsum wallboard in C&D landfills (Yang et al. 2006; Plaza et al. 2007; Bergersen and Haarstad 2008, 2014; Xu, Liu et al. 2010; Xu, Townsend and Reinhart 2010; Sahu et al. 2011; Xu et al. 2011). Specifically, the use of calcium- or iron-based materials, as well as household compost, as covers for landfills containing gypsum wallboard waste have been shown to be effective at adsorbing gaseous H_2S emissions.

The goals of this study were 1) to determine, at bench-scale, which of several additives would best reduce sulphide concentrations and raise pH of leachate when mixed with gypsum wallboard residuals and 2) to monitor field test plots of landfilled materials for sulphide generation.

2 MATERIALS AND METHODS

2.1 Materials

Gypsum residuals were sampled from outside of a wallboard production facility in Atlantic Canada. Samples were taken from the anaerobic interior of the piles to ensure SRB were present and in saturated conditions. Bulk density and moisture content (% M) of the gypsum wallboard residuals were approximately 1.2 ± 0.1 g/cm^3 and 49 ± 2%. Agricultural lime (pulverized limestone; $CaCO_3$, 7.0 ± 0.1% M) was contributed by Antigonish Lime (Antigonish, NS, CA), cement kiln dust (CKD; 25% loss on ignition, 43% CaO, 7% available CaO, 0% M) was provided by LaFarge North America (Brookfield, NS, CA), and wood fly ash (0% M) was sampled from a 60-megawatt biomass power plant. Pasteurized biosolids (50 ± 1% M) were produced from municipal biosolids pasteurized using a mix of CKD and quicklime and obtained from the N-Viro® facility in Halifax, NS, CA. Open hearth slag (OHS; 11.9 ± 0.1% M) was contributed by Portside Aggregates and taken from the former steel plant site in Sydney, NS, CA. Figure 1 illustrates all materials used in this study, which were used as provided.

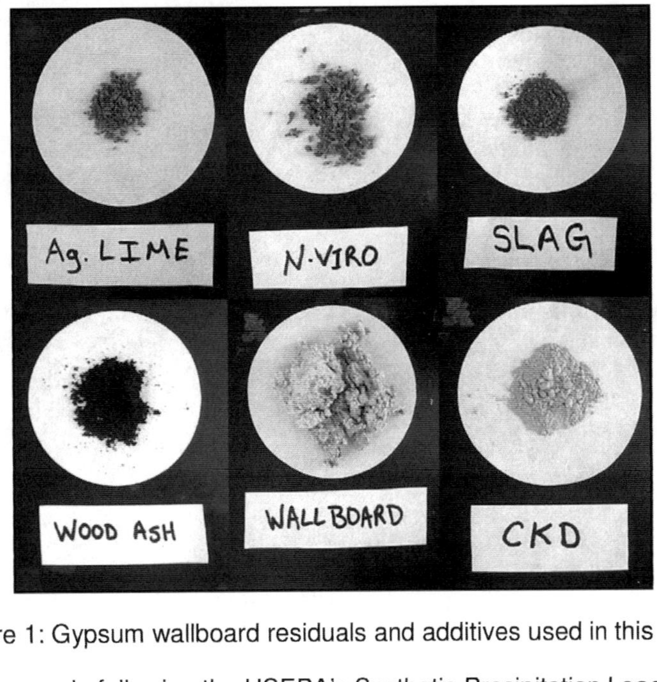

Figure 1: Gypsum wallboard residuals and additives used in this study

Simulated rainwater was made following the USEPA's Synthetic Precipitation Leaching Procedure (SPLP) method. A diluted 60/40 w/w sulphuric/nitric acid mixture was used to reduce the pH of pure (Milli-Q) water to 4.20 ± 0.05 (USEPA 1994).

2.2 Bench-Scale Methods

Leachate test columns were constructed of 10-cm diameter acrylonitrile butadiene styrene (ABS) pipe in 40- or 92-cm long sections, as shown in Figure 2, with 6.35-cm of gravel at the bottom covered with geotextile (landscape fabric), followed by 6.35-cm of rinsed and dried sand, then the gypsum waste mixture (19- or 76.2-cm), then 6.35-cm of rinsed and dried gravel at the top. An initial series of 6 columns was constructed: one with gypsum alone plus one of each of the five alkaline materials (i.e., OHS, CKD, wood ash, agricultural lime, and pasteurized biosolids) mixed with the gypsum waste. Preliminary ratios of additive:wallboard residuals (v:v) for the column trials were selected by mixing varying ratios with simulated rainwater to determine the ratio that would give a pH greater than 9.0 and a low S^{2-} concentration. Duplicates were run using a second set of 6 columns at the same initial ratios. The total depth of the waste mixtures was 19-cm and mixtures were placed in the pipe sections by hand to achieve a comparable level of compaction. Additional sets of columns were tested with 1) a different ratio of additive:gypsum residuals and 2) an increased depth of material (i.e., 76.2-cm).

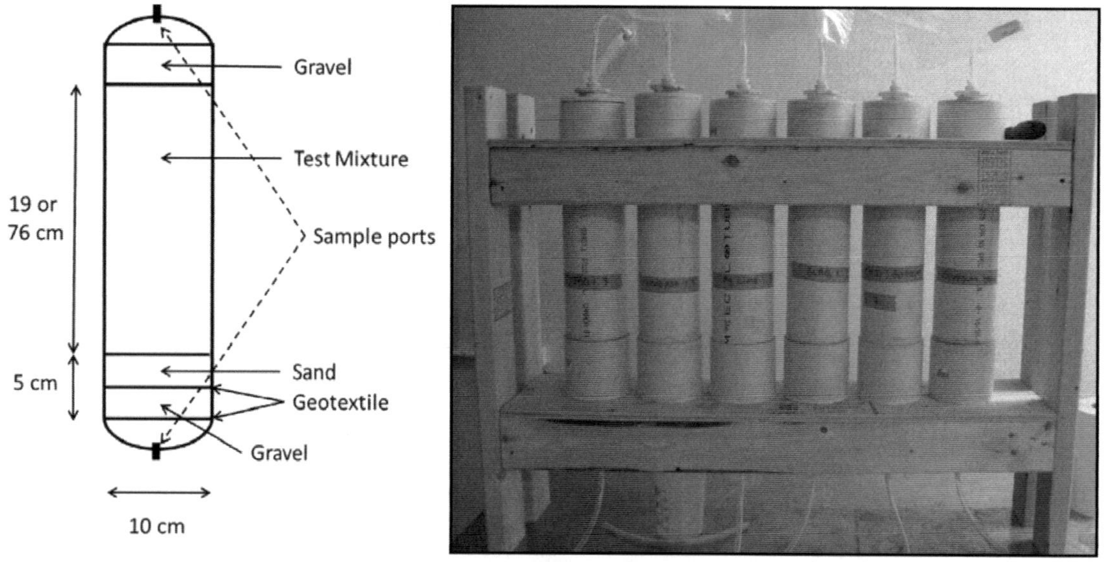

Figure 2: Schematic of column construction (left) and photo of a set of 6 columns (right)

500-mL of simulated rainwater was added to the upper port of each column weekly for 6 weeks followed by a reduction in the added volume to 100-mL per week for a further 6 weeks and intermittently thereafter. Leachate sampling from the bottom sample port was also performed weekly for the first 12 weeks then intermittently thereafter. Leachate samples were collected in amber glass bottles from the lower sample port and analyzed the same day for SO_4^{2-}, S^{2-}, and pH.

2.3 Field-Scale Methods

Two roughly equally-sized field test plots were constructed on top of an estimated 3.2-ha section in the middle of an estimated 15.4-ha wood ash/pulp mill biosolids landfill in eastern NS, CA: one using a 1:1 volumetric mixture of wood ash and wallboard residuals and the other using wallboard residuals alone (i.e., control plot). This landfill has been and continues to be used to dispose of wood ash and pulp mill biosolids in alternating layers. Gypsum residuals and wood ash were mixed using backhoes prior to placement on the landfill in layers totalling approximately 50- to 100-cm. A 30-cm minimum cover of pulp mill biosolids was used over both test plots. Core samples from 3 sites on each test plot were taken on 8 occasions over a 33-week period from May to September. Core samples were taken by first removing the biosolids layer using a shovel then coring into the residuals using a hand auger. Samples were taken from the top and bottom, just below and above the interfaces with the upper and lower biosolids layers. H_2S gas readings were also taken in each borehole as well as above the surface of the landfill. Core samples were processed by mixing a 25-g portion with 100-mL of DI water and letting react for 1-hr, filtering through 0.45-µm cellulose nitrate filters, then measuring SO_4^{2-} and S^{2-}. pH was measured on unfiltered samples. Mixing, filtering, and in some cases dilution of the samples prior to S^{2-} analysis may have oxidized some of the sulphides, resulting in underestimated sulphide concentrations.

2.4 Analytical Methods

pH was measured using an Accumet XL50 meter with an Accumet double junction electrode. SO_4^{2-} (detection limits = 2 to 70 mg/L) and S^{2-} (detection limits = 5 to 800 ug/L) were measured spectrophotometrically on a HACH DR5000 spectrophotometer using EPA-approved methods. The methylene blue S^{2-} test method used determines total sulphides (e.g., H_2S and HS^-) in water; S^{2-} is generally less than 0.1% of a sample even at pH of 14 (Steudel 2000; Yongsiri et al. 2005). H_2S gas was measured using a Crowcon Tetra personal gas monitor calibrated by Avensys Solutions (Montréal, QC) at concentrations of 25 ppm and 255 ppm.

Error bars on graphs and error terms in text represent one standard deviation from the mean of two columns tested using the same additive ratio for the bench-scale trials and from the mean of 3 boreholes for the field-scale trials. Analysis of variance (ANOVA) was used to determine the statistical significance of differences between test conditions. P-values less than 0.05 indicate that there is a significant difference between the means of the two test conditions analyzed at the 95 % confidence interval (Mac Berthouex and Brown 2005).

3 RESULTS AND DISCUSSION

3.1 Bench-Scale Column Trials

3.1.1 Effect of Additive

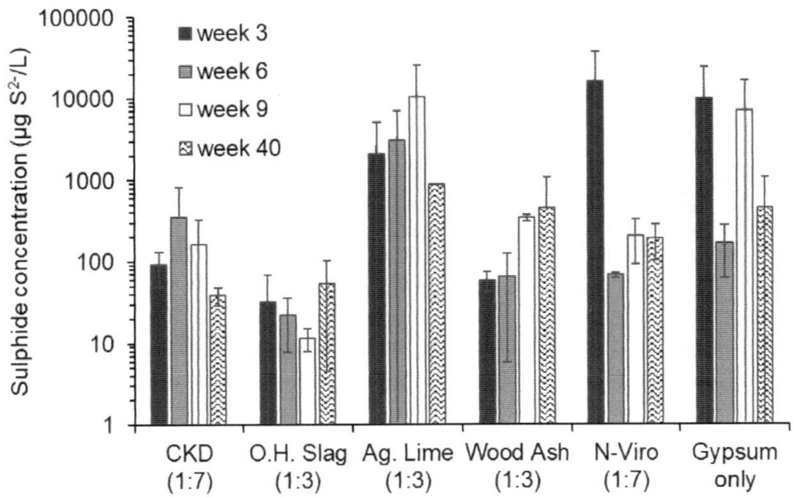

Figure 3: Total sulphides in leachate from columns with varying additives. Note that sulphide concentration is plotted on a logarithmic scale.

Figure 3 shows total sulphide (S^{2-}) concentrations in leachate from columns run with the additives tested in this study. The S^{2-} concentration in leachate samples was highly variable over the course of the study. However, in general, the gypsum-only control and agricultural lime/gypsum mixture resulted in the highest

S^{2-} concentrations while open hearth slag and wood ash resulted in the lowest. No H_2S gas was ever generated from the upper ports of the columns. The S^{2-} analytical method used in this study measures aqueous H_2S and HS^-, along with some metal sulphides. The pK_a for H_2S to HS^- is approximately 7, with H_2S decreasing with increasing pH above 7 (Yongsiri et al. 2005). The leachate from columns in this study had a pH of approximately 7 (Figure 4), therefore, H_2S and HS^- should be approximately in equilibrium in solution. Since columns also remained saturated during testing, it is likely that H_2S remained dissolved in solution rather than being emitted in gaseous form. The Henry's Law volatility constant for H_2S in water is approximately 500 atm per mole fraction at 20 °C (Yongsiri et al. 2005 and references therein) which, when expressed as a solubility constant, is 0.001 mol/m^3 Pa (Sander 2015 and references therein). For comparison, the solubility constant of CO_2 is estimated to be 0.00034 mol/m^3 Pa, while that of methanol is approximately 2.0 mol/m^3 Pa.

CKD and pasteurized biosolids were able to raise the pH more effectively than the other additives, as evidenced by their selected additive:residuals ratios of 1:7, compared to 1:3 for the remainder of the additives. CKD consistently had the highest pH over the course of the study, even at the lower ratio compared to other additives (Figure 4). The pH of the other columns was more consistent over the course of the study, with the gypsum residuals-only columns and the agricultural lime columns having the lowest leachate pH (6.8 ± 0.1 and 6.7 ± 0.1, respectively, at week 10), followed by pasteurized biosolids (7.4 ± 0.9), OHS (7.3 ± 0.1), and wood ash (7.2 ± 0.5) mixtures. None of the columns had pH levels above 9 after week 1 apart from CKD.

Both CKD and pasteurized biosolids contain SO_4^{2-} themselves which would add to the SO_4^{2-} available for reduction by bacterial action. For this reason, CKD and pasteurized biosolids were excluded from further column testing, along with agricultural lime due to its poor performance in terms of increasing pH. Due to the poor pH performance of the remainder of the additives during this experiment (wood ash and OHS), subsequent column tests were run with an additive:wallboard residuals ratio of 1:1.

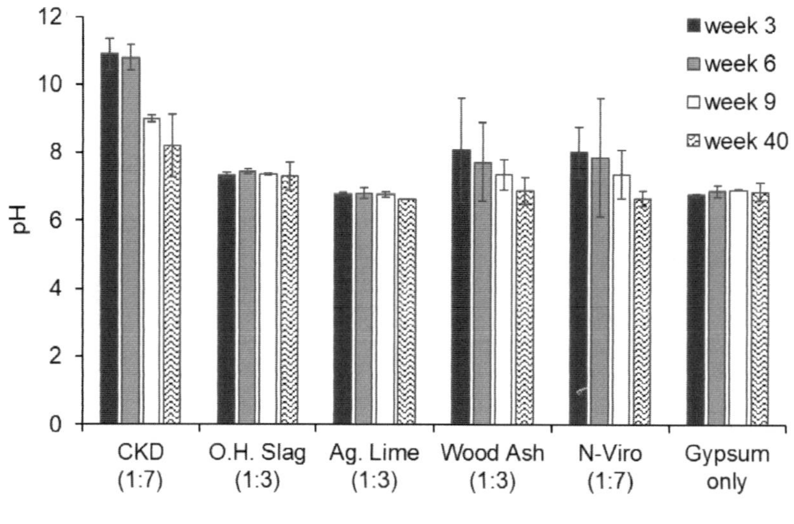

Figure 4: pH of leachate from columns with varying additives

3.1.2 Effect of Column Depth

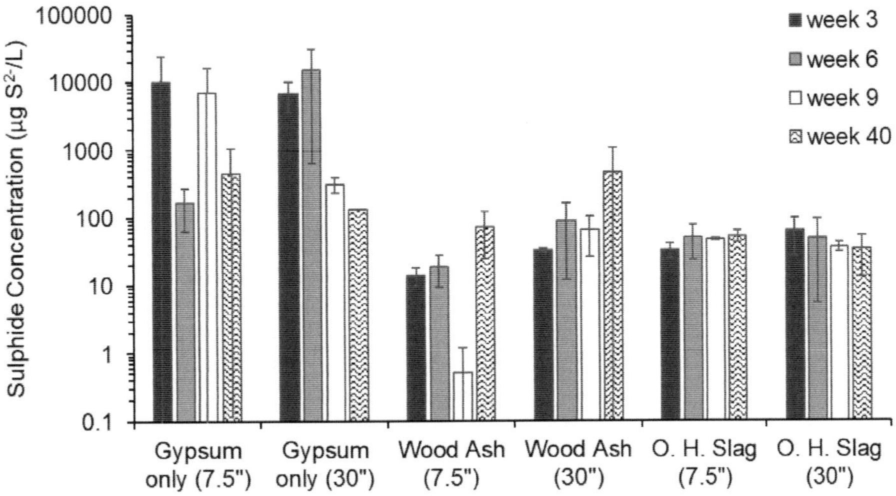

Figure 5: Total sulphides in leachate from columns with varying depths. Note that sulphide concentration is plotted on a log scale.

Increasing depth of material from 7.5" to 30" (19 cm to 76.2 cm) did not appear to have a substantial or consistent influence on S^{2-} generation (Figure 5). Increasing depth also did not result in significantly different pH over the course of the experiment (Figure 6). P-values calculated from ANOVA tests comparing duplicate measurements at each depth for week 2 and week 9 results ranged from 0.15 to 0.43 for S^{2-} and 0.08 to 0.93 for pH. The volume of synthetic rainwater added to the 76.2-cm columns was the same as for shorter columns. The gypsum residuals-only control columns produced much higher S^{2-} concentrations than the 1:1 mixtures with wood ash or OHS, which had comparably low S^{2-} concentrations for the duration of the tests (Figure 5). pH was highest in leachate from wood ash mixture columns and lowest in leachate from gypsum only control columns (Figure 6). This evidence, along with the proximity of the wood ash material to the landfill, led to its selection for field-scale testing.

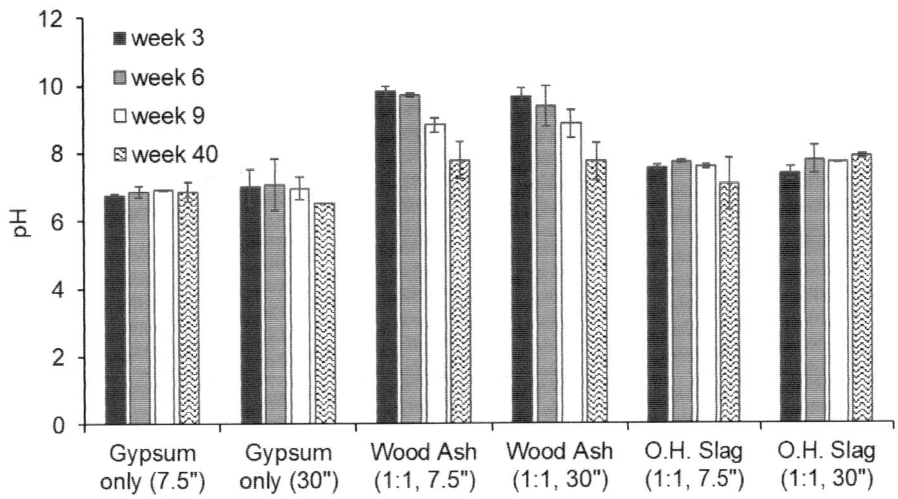

Figure 6: pH of in leachate from columns with varying depths

3.2 Field-Scale Trials

Field test plots were constructed atop an existing wood ash and pulp mill biosolids landfill. Plot A contained a mixture of gypsum wallboard residuals and wood ash at a ratio of 1:1 and plot C contained gypsum residuals only (control). No H_2S gas was ever detected above the surface of the landfill-applied field plots or anywhere near the landfill site once the material was moved and covered with biosolids. H_2S gas within the test plots was not detectable until late June and was detected earlier and in higher concentrations in the gypsum-only plot (C) compared to the gypsum/wood ash mixture plot (A). Readings from both test plots peaked in late August and were above the detection limit of the instrument (>255 ppm) in all boreholes on that sampling visit and the one in late September. H_2S gas diminished to barely detectable levels by November. H_2S becomes less water-soluble with increasing temperature (Yongsiri et al. 2005), therefore increased emission of H_2S gas is expected with increased temperature. Bacterial activity also increases with increasing temperature within the range studied. Both likely contributed to high H_2S gas readings in boreholes during sampling visits when air temperatures were highest. Any H_2S emitted by gypsum residuals applied to the landfill was likely adsorbed by the cover of biosolids.

Figure 7 shows the S^{2-} concentrations found from mixing solid samples from test plot boreholes with pure water were fairly consistent over the course of the study. S^{2-} concentrations spiked during August in both test plots, likely related to the air temperature increasing bacterial activity. Figure 8 shows that, in general, the pH of solid material samples from plot A was slightly higher than that of samples from plot C. However, pH never reached above 9 like it had in the bench-scale trials. This difference was likely due to inadequate mixing at field-scale resulting in pockets of wallboard with no wood ash mixed in as well as the fact that column trials were kept fully saturated while rainfall in field conditions was significantly lower.

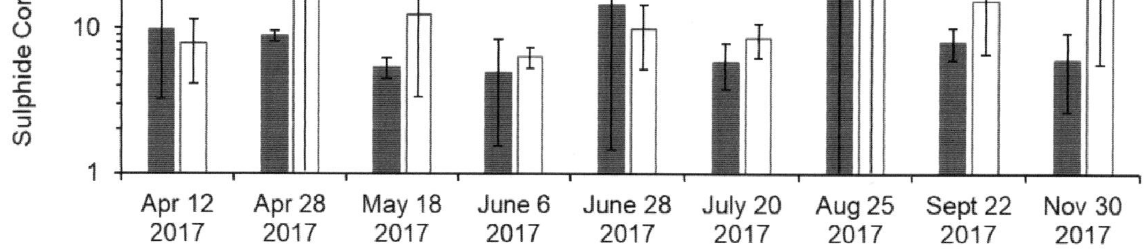

Figure 7: Total sulphides in samples from field test plots. Note that sulphide concentration is plotted on a log scale.

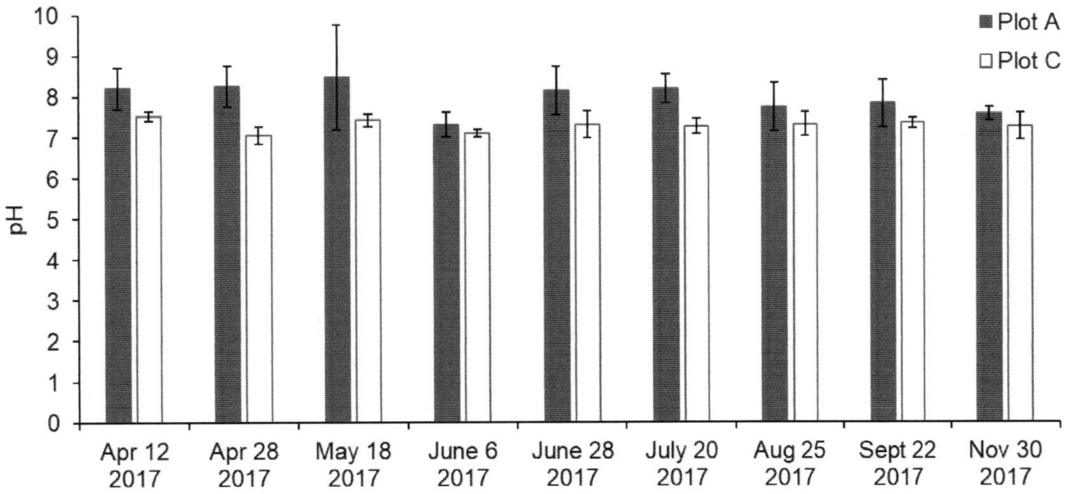

Figure 8: pH of samples from field test plots

4 CONCLUSIONS

5 BENCH-SCALE STUDY:

- No columns produced H_2S gas from upper sample ports during the study.
- Sulphide (S^{2-}) concentrations in leachate were lowest in columns with wood ash or OHS at a 1:1 additive:gypsum wallboard residuals ratio.
- After CKD, which contained SO_4^{2-} and therefore was not tested further, pH was highest with wood ash and generally above 9 in columns with a 1:1 ratio of wood ash to gypsum residuals.
- Increased depth of material from 7.5" to 30" (19 cm to 76.2 cm) did not affect S^{2-} concentrations in or pH of leachate.

Field-scale study:

- Hydrogen sulphide (H_2S) gas was never detected above the surface of the landfill. It is likely that any escaping gas was adsorbed by the covering layer of pulp mill biosolids.
- Warmer temperatures led to generation of H_2S gas within both the control pile, plot C, and the gypsum/wood ash mixture, plot A. Detected concentrations took longer to begin in plot A and were observed to be consistently lower in the wood ash mixture. H_2S concentrations in plot C were above the detection limit of the detector (>255 ppm) from late June to late September and in plot A during August and September sampling events.
- Sulphide concentrations in solids mixtures were consistent throughout the study period and were not significantly different between plots A and C.
- pH was slightly but not significantly higher in test plot A compared to control plot C.

Differences between the field study and bench study results (i.e., lower field pH, no H_2S gas generation in columns) were likely due to inadequate mixing and lower rainfall (i.e., less saturation) in the field compared to bench study. Mechanical mixing in the field resulted in pockets of gypsum without wood ash and saturated conditions in the column trials led to sulphides remaining in dissolved form and not being released as gas.

Acknowledgements

The authors would like to acknowledge Nova Scotia Business Inc., who provided funding for this project. The authors would also like to thank Keith Porter, Brett Penney, Toni Stanhope, and Blair Nickerson from Dalhousie University and Lise LeBlanc of LP Consulting, Ltd. for their assistance with the project.

References

Mac Berthouex, P. and Brown, L.C. 2002. Statistics for Environmental Engineers, 2nd Ed., CRC Press, Boca Raton, FL, USA.
USEPA 1994. Synthetic Precipitation Leaching Procedure. EPA Method 1312. Washington, DC, USA.
Bergersen, O. and Haarstad, K. 2008. Metal Oxides Remove Hydrogen Sulfide from Landfill Gas Produced from Waste Mixed with Plaster Board under Wet Conditions, Journal of the Air & Waste Management Association, 58(8): 1014-1021.
Bergersen, O. and Haarstad, K. 2014. Treating landfill gas hydrogen sulphide with mineral wool waste (MWW) and rod mill waste (RMW). Waste Management, 34(1): 141-147.
Kilburn, K.H. (2003). Effects of hydrogen sulfide on neurobehavioral function. Southern Medical Journal, 96(7): 639-647.
Ko, J.H., Xu, Q., Jang, Y.-C. 2015. Emissions and Control of Hydrogen Sulfide at Landfills: A Review, Critical Reviews in Environmental Science and Technology, 45(19): 2043-2083.
Plaza, C., Xu, Q., Townsend, T., Bitton, G., Booth, M. 2007. Evaluation of alternative landfill cover soils for attenuating hydrogen sulfide from construction and demolition (C&D) debris landfills. Journal of Environmental Management, 84: 314-322.
Sahu, R.C., Patel, R., Ray, B.C. 2011. Removal of hydrogen sulfide using red mud at ambient conditions. Fuel Processing Technology, 92(8): 1587-1592.
Sander, R., 2015. Compilation of Henry's law constants (version 4.0) for water as solvent. Atmospheric Chemistry & Physics, 15(8): 4399-4981.
Steudel, R. 2000. The chemical sulfur cycle. Environmental technology to treat sulfur pollution: Principles and engineering, P.N.L. Lens and L. Hulshoffpol, eds., International Water Association, London, 1-31.
Townsend, T.G., Jang, Y., Thurn, L.G. 1999. Simulation of construction and demolition waste leachate. Journal of Environmental Engineering, 125(11): 1071-1081.
Xu, Q., Townsend, T., Bitton, G. 2011. Inhibition of hydrogen sulfide generation from disposed gypsum drywall using chemical inhibitors. Journal of Hazardous Materials, 191(1): 204-211.
Xu, Q., Townsend, T., Reinhart, D. 2010. Attenuation of hydrogen sulfide at construction and demolition debris landfills using alternative cover materials. Waste Management, 30(4): 660-666.
Xu, Q., Liu, F., Townsend, T.G., Abichou, T., Chanton, J. 2010. Tire-Derived Steel for Hydrogen Sulfide Removal in Landfill Cover. Pract. Period. Hazard. Toxic Radioact. Waste Management, 14(3): 211-214
Yang, K., Xu, Q., Townsend, T.G., Chadik, P., Bitton, G., Booth, M. 2006. Hydrogen sulfide generation in simulated construction and demolition debris landfills: Impact of waste composition. Journal of the Air & Waste Management Association, 56(8): 1130-1138.
Yongsiri, C., Vollertsen, J. Hvitved-Jacobsen, T. 2005. Influence of wastewater constituents on hydrogen sulfide emission in sewer networks. Journal of Environmental Engineering, 131(12): 1676-1683.

Building Tomorrow's Society
Bâtir la Société de Demain

Fredericton, Canada
June 13 to 16, 2018 / *13 au 16 juin, 2018*

DESIGN AND CONSTRUCTION OF A SNOWMELT MANAGEMENT FACILITY IN SOUTHERN ONTARIO

Senior, Matthew [1,3], Scheckenberger, Ron [1] and Moore, Ryan [2]
[1] Amec Foster Wheeler, Canada
[2] City of Hamilton, Canada
[3] matt.senior@amecfw.com

Abstract: Winter conditions, including snow and ice, are a reality for Canadian drivers, including those in Southern Ontario. Local municipalities and other government agencies are typically charged with ensuring safe roadway conditions for motorists, including clearing and plowing snow. In areas with limited space within the roadway right-of-way, plowed snow may need to be transported off-site and disposed of, with little or no environmental or runoff controls. Untreated snowmelt water has been demonstrated to be a significant environmental concern to downstream receiving watercourses, resulting in elevated levels of suspended solids, chloride (salt), as well as numerous other water quality contaminants.

A recent example of the design, and ultimate construction of a snowmelt management facility in Hamilton, Ontario is presented. An integrated "treatment train" approach was employed for the facility, which included the design of a large appropriately graded asphalt melting pad, an oil/grit separator for pre-treatment, and a specifically designed wet pond facility involving a retrofit of an existing stormwater management (SWM) facility for final treatment and dilution. The system was also designed to have a complete SWM function (quantity and quality control) under non-snowmelt (i.e. rainfall-driven) conditions.

Design considerations specific to the management of snowmelt contaminants (salt in particular) are discussed and presented. Perspectives of different stakeholders involved in the management of roadway snow are also considered, including municipalities (engineering and operations and maintenance groups), regulatory authorities, and practitioners.

1 INTRODUCTION

As a northern hemisphere nation, Canada experiences seasonal winter conditions, including snow and ice. Given Canada's geography, the winter season may be extensive. In the City of Hamilton, located in Southern Ontario, the daily average temperature is below zero degrees Celsius for December through March, although snow routinely occurs in April and November (Environment Canada, 2018). Winter conditions may therefore be present for up to one third of the calendar year.

Maintaining roadway transportation links during the winter period is critical, to ensure the uninterrupted movement of goods and services, and to allow citizens safe travel. Government agencies are typically charged with ensuring safe roadway conditions for users; including Provincial agencies for major highways (the Ministry of Transportation in Ontario), and local governments (upper and lower tier municipalities) for arterials, collectors, and local roadways. Winter roadway maintenance typically involves a number of different activities, including monitoring, de-icing (brine spray), plowing, and salting and sanding (Ministry of Transportation, 2018). Salt is most effective at temperatures of -12°C or higher, whereas sand may be used at lower temperatures, or in locations where there is an immediate need for traction, such as hills, curves, bridges and intersections (Ministry of Transportation, 2018). The City of Hamilton typically relies on a blend of road salt and sand, typically referred to as "pickle". This mixture consists of varying ratios of salt to sand (1:3, 1:5, or 1:10) depending on requirements.

In most cases, plowed snow is simply directed to the side of the roadway right-of-way; either the ditch (rural roadway cross-section) or between the curb and the sidewalk (urban roadway cross-section). In areas with limited space within the right-of-way (older urban areas, bridges), or in other areas with space constraints, such as parking lots, plowed snow may be collected by trucks and transported off-site for storage and melting.

The water quality of plowed snow from roadway areas has been recognized as being a significant environmental concern. Snowmelt water from such plowed sources may be contaminated with a number of different pollutants, including petroleum products/additives, corded or abraded metals, de-icing and anti-skid agents (i.e. salt and sand), abrasion of roadway surfaces and litter (Marsalek et al., 2003). Road salt in particular is a primary contaminant of concern in roadway snow. As such, the proper disposal and management of snow is a key consideration for government agencies tasked with winter road maintenance and ultimately those dealing with excess snow management.

This paper briefly reviews considerations and guidance with respect to the design of snowmelt management facilities in Southern Ontario. A recent example of the design and ultimate construction of a snowmelt facility in the City of Hamilton, Ontario is presented. Perspectives of different stakeholders involved in the management of roadway snow are also considered, including municipalities (engineering and operations and maintenance groups), regulatory authorities, and practitioners.

2 SNOWMELT MANAGEMENT

2.1 Snowmelt Contaminants

The water quality of urban snowmelt water has been recognized as being a significant environmental concern. Melt water may be contaminated with a number of different pollutants, including petroleum products/additives, corded or abraded metals, de-icing and anti-skid agents (i.e. salt and sand), abrasion of roadway surfaces and litter (Marsalek et al., 2003). Numerous studies have reviewed the processes involved with pollutant accumulation and release in urban snowmelt (i.e. Exall et al (2010)). A substantial portion of snowmelt pollutants can be attributed to atmospheric sources (combustion, incineration, processing and manufacturing). Notwithstanding, snowmelt pollutants are also clearly linked to urban land uses and vehicle activity and traffic density (Exall et al., 2010). Aged snowpacks, or snowpacks which have been extensively processed and worked may also be expected to have higher levels of contaminants.

Road salt is a particular contaminant of concern in urban snow from roadway and paved areas. It has been estimated that approximately 5 million tonnes of road salt are used per year in Canada (Environment Canada and Health Canada, 2001). Road salt is highly soluble, and the majority of applied road salt will ultimately impact surface water (creeks, rivers, lakes) or groundwater sources. As an example, monitoring studies have shown that Ontario's Lake Simcoe (surface area of approximately 750 km2) has experienced a five-fold increase in salt concentrations from the 1970s to present (McDiarmid, 2017). As part of its 2001 report, Environment Canada considered road salt to be "toxic" and recommended it be listed as such as per the Canadian Environment Protection Act. Ultimately however, this did not occur, reportedly due to concerns from the salt and transportation industries (Chung, 2008). Environment Canada however promotes a voluntary "code of practice" to reduce salt usage, which the majority of Ontario municipalities have adopted. While this practice appears to have contributed to a reduction in the "normalized" application rate, it has not translated in an overall reduction in salt usage (Lembcke et al, 2016).

2.2 Regulatory Guidance and Experience in Southern Ontario

Based on the preceding, urban snowmelt is expected to have an adverse impact on receiving systems. Notwithstanding, based on a review by Environment Canada (2012), only 16% +\- of municipal road organizations have runoff and meltwater collection engineered sites in place (with no change since 2007). The same study however indicated that 77% of municipal road organization implement "good housekeeping practices" at snow disposal sites (Environment Canada, 2012).

Within southern Ontario, the earliest and most well-known snowmelt management facility is operated by the Town of Richmond Hill. This facility, constructed in 2003, was the first of its kind in southern Ontario. The facility uses a three step "treatment train" approach, with a large asphalt melting pad, oil/grit separator for pre-treatment, and a wetland stormwater management facility (Exall et al, 2010). Since that time however, there remains few other reported examples of snowmelt management facilities constructed in Southern Ontario. Some other examples in eastern Ontario include the Carp snow disposal facility outside the City of Ottawa.

The Ontario Ministry of the Environment and Climate Change (MOECC) has generated "Guidelines on Snow Disposal and De-icing Operations in Ontario" (2011). Those guidelines provide general guidance on snow disposal, however do not provide any specific or formal direction on the design of snow disposal sites or snowmelt management facilities, beyond suggested criteria in the selection of a suitable site.

The Transportation Association of Canada (TAC) does provide guidance on snow storage and disposal (2013). Similar to the MOECC guidelines, TAC provides suggestions on the assessment and evaluation of candidate sites, including snow hauling distances, land use (both the site and surrounding area, i.e. zoning), and environmental considerations (surface and sub-surface). The "treatment train" approach is again promoted, with a solid melting pad (able to accommodate trucks and heavy equipment) and meltwater collection pond, fitted with impermeable bases to limit infiltration and protect groundwater resources (TAC, 2013). The guidelines also suggest the potential consideration of an oil/grit separator for pre-treatment.

Some technical guidance in the design of snow storage disposal facilities is also provided in an overarching study on road salt management undertaken by the University of Waterloo (Stone et al, 2010). This document also promotes the "treatment train" approach to snowmelt, with a melting pad and wet pond SWM facility, an oil and grit separator for additional pre-treatment. The document provides similar guidance to the other sources on considerations with respect to the siting and sizing of potential facilities.

In general, the foregoing guidelines promote removal of debris and sediments (and contaminants attached to sediments, such as metals), and advocate a dilution-based approach to chloride management, to avoid shocking receiving systems. As it is a dissolved contaminant, physical processes alone (i.e. settling; typical of stormwater treatment) cannot remove chloride. Wastewater treatment processes would be required, which are typically not considered as part of snowmelt management systems, given cost and the associated operations and maintenance requirements. Consideration can be given connecting snowmelting areas to the sanitary sewer system (and ultimately a wastewater treatment plant), however such a connection may not be feasible in all cases, due to servicing or capacity constraints. Halophytic (salt-tolerant) plantings could potentially be considered, to aid in the uptake of dissolved chloride. However, such plants are typically non-native and considered invasive species, which can result in other negative impacts.

3 PROJECT EXAMPLE

3.1 Project Location and Purpose

As outlined in the preceding section, frequently a planning study is undertaken by the government agency or municipality charged with roadway snow removal, in order to identify preferred sites for snow disposal and management. For the subject project example, this process was not required. The City of Hamilton historically used its Public Works yard on Upper Ottawa Street for winter maintenance activities (ref. Figure 1).

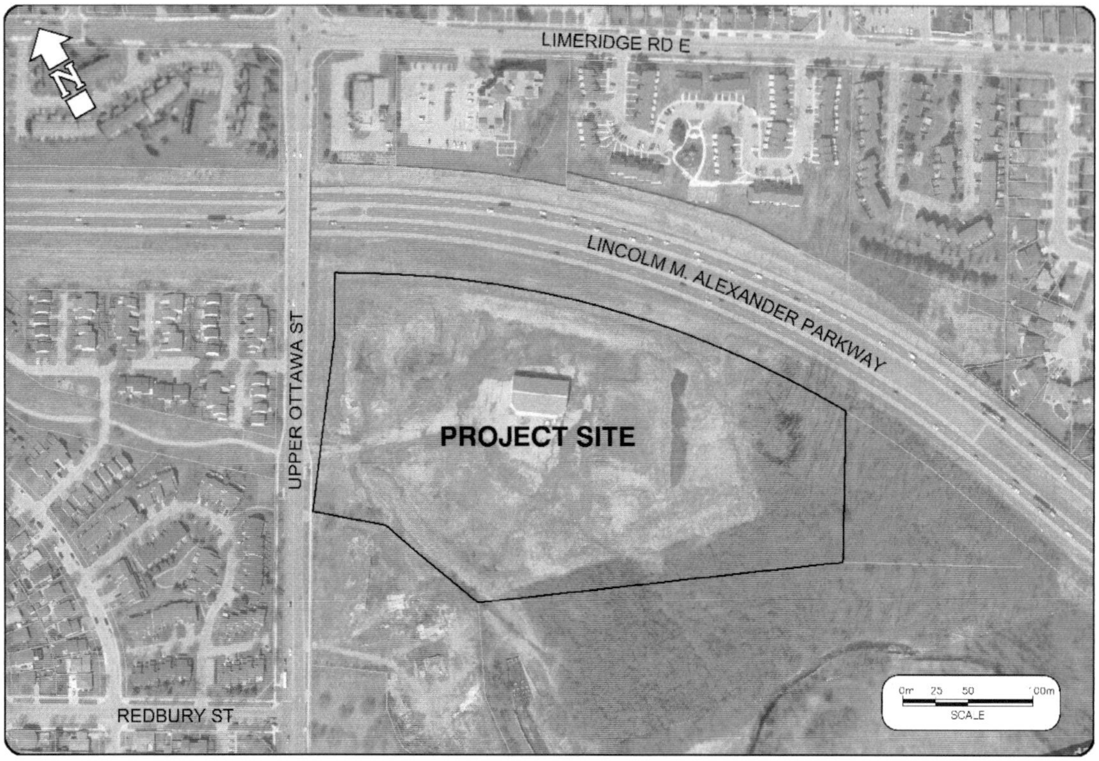

Figure 1: City of Hamilton Upper Ottawa Works Yard Site

The site is bounded by two major roadways, the Lincoln Alexander Parkway to the north, and Upper Ottawa Street to the west. The south and east limits of the site are generally forested/naturalized and include the headwater/outlet of Red Hill Creek, which is enclosed in a large storm sewer structure upstream of the site. The site itself (covering some 6 ha +\-) is generally open space, with a salt storage shed structure in the centre (including a hydro transformer for lights, and a brine storage tank area). In previous years, the City of Hamilton would pile plowed snow informally in available spaces on site. An informal dry pond feature was located at the south limits of the site, receiving sheet flow drainage (including meltwater) from the site. The pond ultimately discharged to the adjacent watercourse receiver, Red Hill Creek.

In 2015, the City of Hamilton engaged Amec Foster Wheeler to support the re-development of the site, to formalize the snowmelt and snow disposal management process, and to make other site improvements, including consideration of privacy measures to limit impacts to existing residents on the west side of Upper Ottawa Street, immediately opposite the site. The City of Hamilton further expressed a desire to address environmental concerns with the project, while also developing a cost-effective, relatively low maintenance system, which also addressed secondary concerns, such as impacts to adjacent residents.

3.2 Snowmelt Management Facility Design Considerations

The snowmelt management facility (SMF) was designed using guidance from available regulatory sources and best practices, as summarized in the preceding sections. It was considered that the "treatment train" approach was the most appropriate, combining an asphalt melting pad, an oil/grit separator, and a wet pond. The layout of the site and SMF are presented in Figure 2.

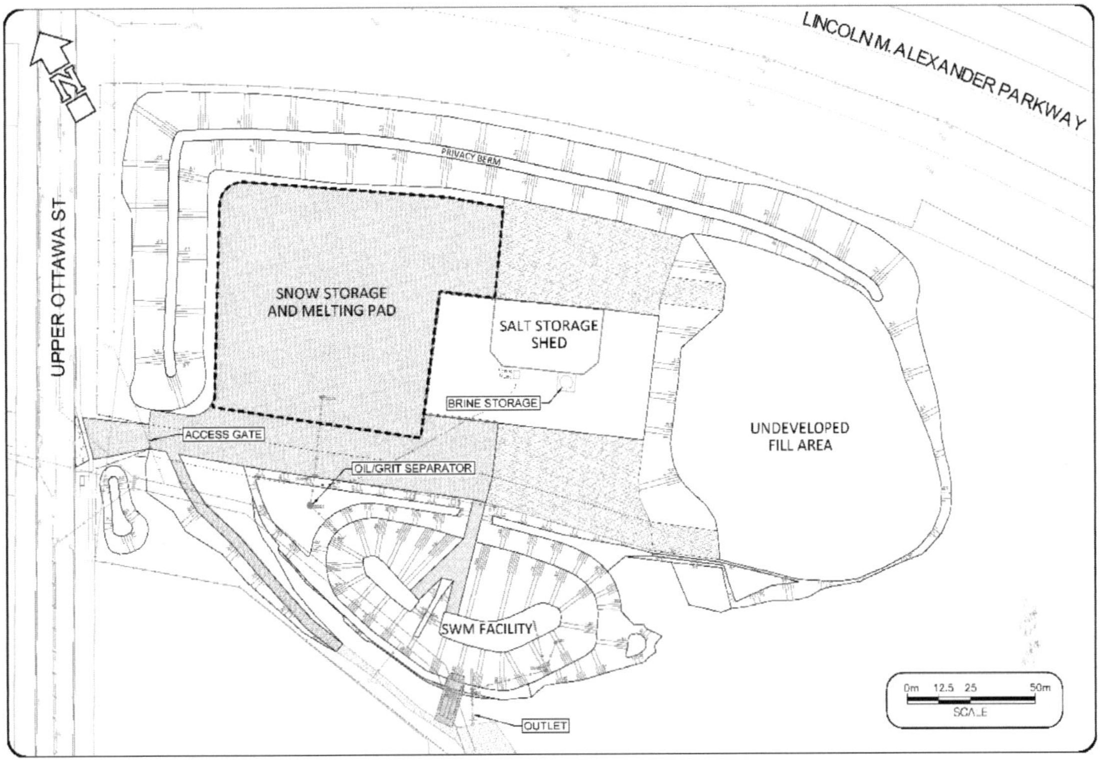

Figure 2: Upper Ottawa Snowmelt Management Facility Layout

As part of maintaining a cost-effective design, a cut/fill balance for earthworks was a design objective to avoid the high cost of off-site disposal of material. Material removed to support the creation of the wet pond and grading/sub-grade for the asphalt melting pad was used to create a privacy berm around the west and northern limits of the site, to reduce light and noise impacts to adjacent residents from operations work. Additional material was also used to grade a uniform fill area at the east limits of the site, which was considered for temporary stockpiling and other maintenance activities by the City.

A large (1.0 ha +\-) asphalt melting pad was the first component of SMF design, intended to promote melting and settling of larger sediment and debris/garbage. The asphalt structure of the pad was carefully designed, accounting for the loads expected from dump trucks full of snow, as well as other snow management equipment. As per available design guidance, relatively gentle melting pad slopes of 1-2% are typically recommended to promote settling of coarser sediments and debris, with a sloping plane or v-shaped pad to allow for the collection of meltwater and associated direction of flow (Stone et al, 2010). Due to site grading constraints, designed grades ranged from 0.5 to 3%, however the majority of the grades remained consistent with the suggested 1-2% range. Concrete curbs were also implemented to direct flow and separate the melting pad area from adjacent uses. The pad was designed to drain from north to south, allowing for south facing exposure for melting.

Meltwater from the pad is collected/captured by a double catchbasin, and directed to an oil/grit separator (the second component of the SMF) for pre-treatment and settling of coarser sediment material. The oil/grit separator is intended to reduce maintenance frequency for the downstream wet pond facility, given that the oil/grit separator can be more easily cleaned and maintained using mechanical means such as a hydrovac truck. In order to size the unit appropriately, a higher event mean concentration (EMC) of sediment (600 mg/L) was assumed as compared to typical stormwater (80 to 125 mg/L). A coarser particle size distribution was also assumed based on the use of sand for winter maintenance. The unit was ultimately sized for an average annual total suspended solids (TSS) removal of 81%. Sizing calculations undertaken by the manufacturer were also used to ensure that the storage sump within the unit was sufficient to require clean-out no more than annually based on sediment concentrations and expected snowmelt volumes.

The final component of the SMF is a wet pond facility. The wet pond facility is intended to provide final settling of suspended sediment material (and attached contaminants such as metals), as well as dilution of chloride to the extent possible, to minimize "shock" discharges. Based on a review of available literature, there is a lack of guidance on specific wet pond design parameters for snowmelt management. As such, the guidance and direction available within the Ontario Ministry of the Environment's Stormwater Management Planning and Design Manual (2003) were applied. Futhermore, an oversized permanent pool volume was provided (approximately 2.7 times the minimum required value), to assist in diluting dissolved chloride concentrations.

Monitoring (i.e. Richmond Hill Facility) has noted that high concentrations of chloride during snowmelt periods can lead to a density-based stratification within the wet pond due to heavier chloride-laden water (Exall et al, 2011). The facility was able to delay and dilute chloride concentrations in the released flows, however releases could still occur at toxic levels given that a wet pond ultimately does not remove the dissolved chloride (Exall et al, 2011). Other studies (Lembcke et al, 2016) have also shown stratification in wet ponds, with higher chloride concentrations noted at the bottom of the facility. By contrast, thermal stratification also occurs in wet ponds particularly during warm weather conditions, where the difference between the top and bottom of the wet pond permanent pool can be substantial (up to 9oC in Lembcke et al, 2016).

Stone et al (2010) note that reverse sloped outlet pipes (which are typical for conventional SWM facilities to draw cooler water from the bottom of the permanent pool) may actually release denser water with elevated chloride concentrations due to stratification. That study suggested that surface outlets (i.e. weirs) or adjustable outlets may be preferable for that reason. For the Upper Ottawa SMF, a compromise approach was sought, to balance the desire to discharge lighter, less chloride laden water during melt periods, but to also discharge cooler water during summer month periods. A reverse sloped pipe outlet was implemented, however the pipe draws water from the mid-point of the permanent pool.

While the wet pond facility was designed primarily for snow melt management, consideration was also given to operation during conventional warm weather conditions as the Public Works Yard operates year-round. The wet pond facility was designed to achieve "post to pre" stormwater management peak flow control for the 2 through 100 year return period events. Extended detention times in excess of the required 24 hour were also designed for warm weather events (i.e. 4 hour, 25 mm storm event) as well as a "snowmelt" event. The latter was developed by using the maximum estimated snowmelt rate (based on a review of both historical data and available empirical snowmelt equations), and an assumption that this rate would occur during primary daytime melting hours (i.e. 10 AM to 4 PM).

Consideration was given during the design phase to the use of halophytic (salt-tolerant) plantings within the wet pond facility, to aid in the uptake of dissolved chloride. Notwithstanding, following consultation with landscape architect and ecology specialists, this approach was not pursued, given that available halophytic plant species are all non-native and in many cases invasive species.

3.3 Operations and Maintenance

Site works were largely constructed over the course of September to November, 2017. Top course asphalt paving and landscaping activities are planned for spring 2018. Notwithstanding, snow management activities have occurred over the course of December 2017 and January 2018. Preliminary observations have noted the need for a clear operations and maintenance procedure, which needs to be communicated with City staff undertaking work on the site. Snowpiling activities initially occurred at the south limits of the site, on top of the melting pad outlet (double catchbasin). This greatly reduced or eliminated settling of coarse sediments and debris, and could have also lead to an outlet blockage and on-site ponding/flooding. In other areas, snow was piled above and beyond the melting pad curb limits and onto pervious areas. This tended to allow for infiltration of contaminated water, and could also have a negative impact on landscaping and soil stability, due to the expected high chloride levels. In general, for the City of Hamilton SMF snowpiling activities should be restricted to the melting pad area, and should proceed from north to south to maximize potential settling. Consideration should also be given to the space required for maintenance equipment, and safe limits of snow piling. While such experiences may initially be discouraging, they provide an opportunity to improve site operations procedures and ensure that best practices are established appropriately.

As part of the conditions of approval from the Ministry of the Environment and Climate Change (MOECC), the City of Hamilton is required to prepare and implement and operations and maintenance manual, which will largely follow the guidance provided as part of the design submissions made by Amec Foster Wheeler (2017). Other references will also be considered for guidance, including Stone et al (2010) and the Transportation Association of Canada (2013). A field monitoring program of the SMF will also be undertaken by the City of Hamilton as part of the Conditions of Approval from the MOECC; the results of this program may yield further insights.

In addition to winter operations, spring time maintenance activities will also be important. It is expected that a streetsweeper will be required to remove coarse sediment and debris from the melting pad area post snowmelt, and prior to spring rainfalls, which would tend to wash this material downstream. The oil/grit separator unit should be inspected during this period, and a clean-out (likely using a hydrovac truck) co-ordinated as required. The wet pond facility should also be visually inspected for any potential issues. Following spring maintenance activities, the asphalt melting pad should not be used for stockpiling loose materials which could unnecessarily consume capacity within the oil/grit separator.

4 SUMMARY AND FUTURE CONSIDERATIONS

Winter road maintenance, including plowing and disposal of excess snow, will continue to be a consideration for public sector agencies tasked with these works. Based on available data, formal snowmelt management systems have not been widely implemented by these agencies. Given the significant environmental concerns associated with uncontrolled snowmelting, it is expected that snowmelt management facilities will become increasingly common in Southern Ontario and beyond.

A recent example of the design and construction of a snowmelt management facility (SMF) in Hamilton, Ontario was presented. Based on a review of best practices, the SMF was designed using an integrated "treatment train" approach, including an asphalt melting pad, an oil/grit separator for pre-treatment, and a specifically designed wet pond facility. Operations and maintenance considerations are considered key to the long-term success of the facility.

While it is considered that this paper has provided an overview of design considerations and best practices, it is noted that there are a number of other related considerations that merit further review; these include:

- A larger discussion is warranted regarding salt management. Ultimately SMFs can only dilute, not remove chloride concentrations. Reductions in chloride/salt can only be achieved through more careful application/use for road management activities. This may include reduced use (and better winter driving education), alternative de-icing products and procedures, and other options. This discussion would necessarily include cost considerations, including the costs of salt and other maintenance work, as well as liability costs for road safety, and the environmental costs.
- The preceding discussion has focused on disposal of plowed/removed snow for public roadways. Consideration should also be given to the impacts of snowmelt from other roadways (i.e. where snow is piled within the right-of-way), as well as removal and disposal from private sites (i.e. large commercial parking lots).
- The project example cited herein was designed using best practices. However, as part of the site selection consideration should also be given to the sensitivity of the drainage system receiver, including the characteristics of the aquatic community. In some cases, alternative locations, strategies or targets may be warranted.
- Current best practices recommend liners or impermeable barriers to eliminate potential infiltration of snowmelt water. While this is a preferred approach in areas with sensitive groundwater resources (i.e. sourcewater protection zones, particularly in areas which source drinking water from groundwater), in other areas infiltration may be preferable to surface water impacts. This should be considered further,

particularly given the current direction towards the promotion of infiltration based techniques for stormwater management (i.e. low impact development best management practices) in the Province of Ontario.

Acknowledgements

The authors would like to acknowledge the City of Hamilton for its contributions to this paper and the project as a whole. The authors would also like to acknowledge those involved with the design and construction of the project example, including Canadian Infrastructure Products (Hydroguard Oil and Grit Separator), Dougan & Associates (landscaping), and the Charlton Group (prime contractor).

References

Amec Foster Wheeler Environment & Infrastructure. 2017. Upper Ottawa Street Public Works Yard Stormwater Management Report. Amec Foster Wheeler, Burlington, Ontario, Canada.

Chung, A.. 2008. Cheap salt outweighs green concerns. The Toronto Star. Available at: https://www.thestar.com/news/gta/2008/03/02/cheap_salt_outweighs_green_concerns.html

Environment Canada and Health Canada. 2001. Road salts: Priority Substances List Assessment Report. Canadian Environmental Protection Act, Ottawa, Ontario, Canada.

Environment Canada. 2012. Five-year Review of Progress: Code of Practice for the Environmental Management of Road Salts. Environment Canada, Ottawa, Ontario, Canada.

Environment Canada. 2018. Canadian Climate Normals 1981-2010 (Hamilton A). Available at: http://climate.weather.gc.ca/climate_normals/

Exall, K., Rochfort, Q, Marsalek, J., Grapentine, L, Kydd, S. and Nemeth, J. 2010. Assessment of Operation of the Town of Richmond Hill's Snow Storage Facility (RHSSF): Final Report. Water Science and Technology Directorate (Environment Canada), Ottawa, Ontario, Canada.

Exall, K., Marsalek, J., Rochfort, Q, and Kydd, S. 2011. Chloride transport and related processes at a municipal snow storage and disposal site. Water Quality Research Journal of Canada, 46.2: 148-156.

Lembcke, D, Thompson, B, and VanSeters, T. 2016. Salt and Stormwater: The Issues, Challenges and Not so Easy Solutions. Presented at TRIECA 2016 Conference, Toronto, Ontario, Canada.

Marsalek, J., Oberts, G., Exall, K., and Vikland, M. 2003. Review of operation of urban drainage systems in cold weather: Water Quality considerations. Water Sci. Tech. 48 (9): 11-20.

McDiarimid, Margo. 2017. Road salt threatening health of freshwater lakes, study finds. CBC. Available at: http://www.cbc.ca/news/technology/lakes-road-salt-chloride-study-1.4064476

Ministry of the Environment. 2003. Stormwater Management Planning and Design Manual. Government of Ontario, Toronto, Ontario, Canada.

Ministry of the Environment. 2011. Guidelines on Snow Disposal and De-icing Operations in Ontario. Government of Ontario, Toronto, Ontario, Canada.

Ministry of Transportation Ontario. 2018. Winter Road Maintenance. Available at: http://www.mto.gov.on.ca/english/ontario-511/winter-highway-materials-technology.shtml

Stone, M, Emelko, M. B., Marsalek, J., Price, J.S., Rudolph, D.L., Saini, H, and Tighe, S.L.. 2010. Assessing the Efficacy of Current Road Salt Management Programs. University of Waterloo and National Water Research Institute. Waterloo, Ontario, Canada.

Transportation Association of Canada. 2013. Syntheses of Best Practices – Road Salt Management: Chapter 8.0 – Snow Storage and Disposal. Transportation Association of Canada, Ottawa, Ontario, Canada.

Building Tomorrow's Society
Bâtir la Société de Demain

Fredericton, Canada
June 13 to 16, 2018 / *13 au 16 juin, 2018*

EFFECT OF TEMPERATURE ON ELECTRICITY PRODUCTION IN A CLOSED-LOOP PRESSURE RETARDED OSMOSIS PROCESS

Adhikary, Suman[1]; Islam, Md. Shahidul[1]; Sultana, Sormin[1]; Ramamurthy, S. Amruthur[1] and Rahaman, Md. Saifur[1, 2]
[1]Department of Building, Civil and Environmental Engineering, Concordia University, Canada.
[2]saifur.rahaman@concordia.ca

Abstract: The search for sustainable and environmentally friendly energy sources has been ramped-up by an ever-increasing global demand for energy as well as the need to reduce dependency on fossil fuels. Closed-loop pressure retarded osmosis is an integrated system of pressure retarded osmosis (PRO) with a thermally driven separation process feasible for recovering draw solutions. Harnessing pressure gradients (i.e. salinity gradient) via PRO is a promising technology for non-greenhouse emitting renewable energy production. Feed and draw solution temperature has an influence on water flux and, therefore, the temperature can play a significant role in overall electricity production via PRO. In order to investigate the effect of temperature on PRO process, this study primarily examines the impact of temperature on electricity production. Using 1 M NaCl as draw solution, an increase in power density of ~34% (from 6.7 Wm^{-2} to 9.0 Wm^{-2}) was observed when the feed solution temperature was raised from 20^{0}C to 40^{0}C in a PRO process using commercial thin-film composite (TFC) membrane.

1 INTRODUCTION

There has been a growing worldwide need for environmental friendly and sustainable sources of energy. Pressure retarded osmosis (PRO) is one emerging technology by which green energy (i.e. energy production without producing greenhouse gasses) is possible. Harnessing the osmotic pressure differences between two solutions of different salinity gradients can produce, energy in the form of electricity, without producing any CO_2 (Achilli and Childress 2010). It has been estimated that global osmotic power potential is equivalent to around 2.6 TW/year, which could be even higher if high salinity RO brine wastewater from reverse osmosis (RO) desalination plants is taken into consideration (Logan and Elimelech 2012; Chung et al. 2012; Kim, Lee, and Kim 2012). Intermittent energy production systems (solar, wind turbines etc.) cannot provide a continuous supply of energy. Therefore, to get energy without interruption, PRO is a potential viable alternative technology. In 1997, Stat Kraft first implemented PRO technology to produce power and achieved a maximum power density of 5.06 Wm^{-2} when concentrated brine (1.03 M NaCl) was used as the draw solution (Achilli, Cath, and Childress 2009). PRO can be considered economically viable if it produces more than 5 Wm^{-2} (Yip et al. 2011).

[1] $W = J_w \Delta P$

Where, W is the power density (Wm^{-2}), J_w is the water flux (LMH) and ΔP is the applied hydraulic pressure. The effect of operating temperature can play a significant role in overall power production in the PRO process since temperature affects solution physicochemical properties like viscosity, density and diffusion (Touati et al. 2015). Therefore, PRO could utilize existing, low-grade waste heat to affect production.

Because of this, PRO has gained more attention in recent times as a reliable, green energy source enhanced by the utilization of low-grade waste heat, In this research, the effect of temperature on the water flux, power density and reverse solute flux achieved by using a commercial TFC FO flat sheet membrane has been studied to understand the temperature-induced interaction between solute, water and membrane.

2 MATERIALS AND METHODS

2.1 Membranes

The commercial thin film composite FO membrane used in this study was provided by Porifera (Hayward, CA). The membrane is mechanically supported by an integrated woven mesh support layer. Table 1 provides the membrane's intrinsic properties at 20, 30 & 40^{0}C.

Table 1: Summary of membrane and draw solution properties at different operating temperatures

Temperature	Membrane	Water Permeability Coefficient, A (LMH/bar)	Salt Permeability Coefficient B (LMH)
20^{0}C	TFC	4.924 ± 0.074	2.11 ± 0.04
30^{0}C		5.940 ± 0.107	3.56 ± 0.07
40^{0}C		7.798 ± 0.125	5.44 ± 0.21

2.2 Solution chemistries

ACS grade sodium chloride, potassium acetate and sodium propionate (provided by Fisher Scientific) was used in this experiment. Ultrapure water with a resistivity of 18.2 MΩ cm was produced via a Milli-Q system (Millipore Integral 10 - Water Purification System). The performance of the membrane was determined via the conventional method (Han, Ge, and Chung 2014). The membrane was been supported by a fabric spacer on the support side. The membrane was soaked for at least 24 hours prior to use.

2.3 PRO experiment

The laboratory bench scale PRO setup is shown in Fig. 1. The membrane has an effective area of 19.94 cm^2. Mesh spacers placed in the feed channel supported the membrane and enhanced the turbulence in the feed stream. A 2 L reservoir was used for both the feed and draw solution. A hydra-cell, high-pressure pump was used to circulate the draw solution at specific velocities (flow velocity for both draw and feed solution was 0.5 L/min). Containers for each of feed and draw solutions were placed on analytical balances to provide weight of water flux. A thermostatic water bath was used to maintain the temperature. In our study, DI water was used as feed water and 1.0 M NaCl as draw solution.

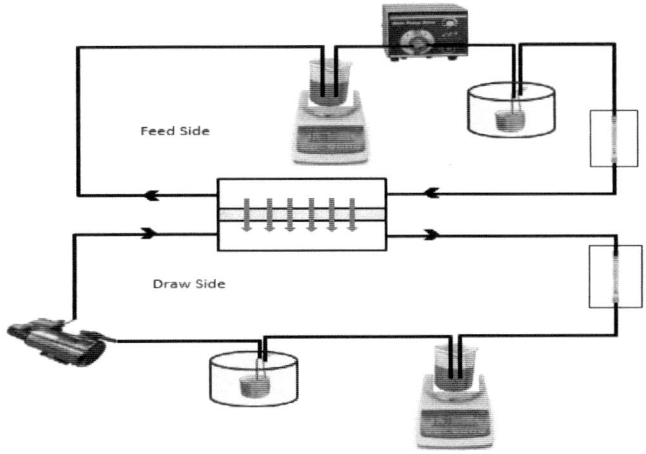

Figure 1: Schematic diagram of bench scale PRO set up

3 RESULTS AND DISCUSSIONS

3.1 Effect of operating temperatures on PRO performance

Fig. 2 summarizes the water flux data for the 1.0 M NaCl draw solution at three different operating temperatures tested under PRO conditions. Water flux increased with the increase in operating temperature of the feed solution. This can be attributed to the higher water permeability coefficient and higher mass transfer coefficient value associated with increased temperature. As seen in table 1, the increase in "A" has a positive impact on higher water flux. Also, with an increase temperature, diffusivity increases, which corresponds to a higher mass transfer coefficient. This higher mass transfer coefficient eventually reduces the internal concentration polarization (ICP) and, therefore, it increases water flux (She, Jin, and Tang 2012). As seen in Fig. 2, water flux increased to ~34% (34.8 LMH to 47.1 LMH) at 100 psi pressure when the feed solution temperature was increased from 20⁰C to 40⁰C. However, from Fig. 2, it is also evident that, as predicted, water flux decreased as the applied pressure increased.

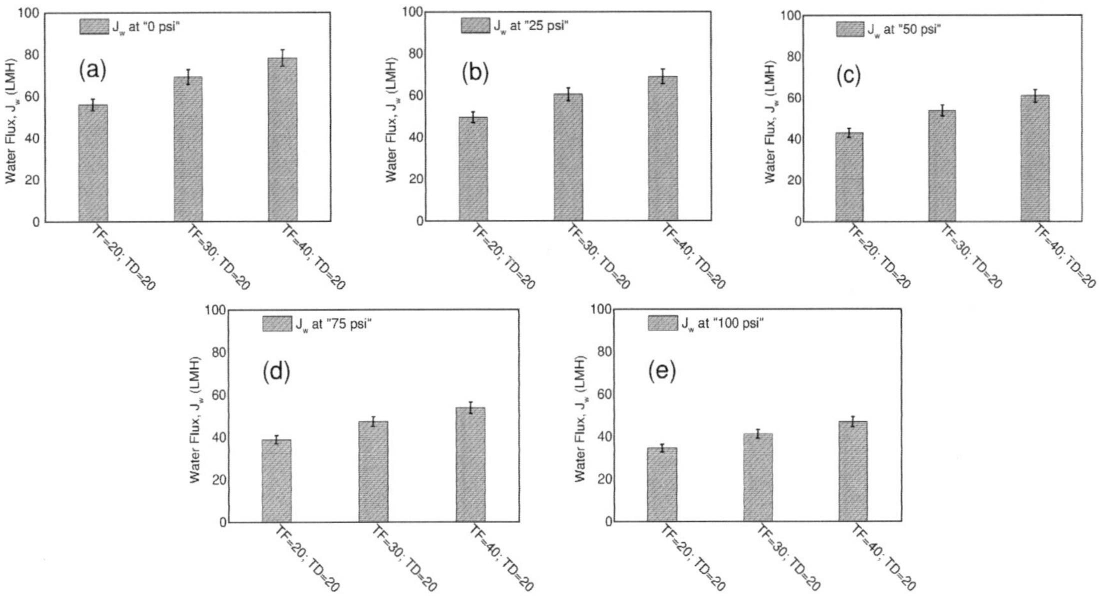

Fig 2. Water flux for 1.0 M NaCl draw solution as a function of working temperatures at the applied pressure of a) 0 psi; b) 25 psi; c) 50 psi; d) 75 psi; e) 100 psi. Membrane used was commercial TFC. Draw and feed flow rate were 0.50 L/min each. Error bars indicate standard deviation among three trials.

3.2 Effect of operating temperatures on power density

The power density illustrated in Fig. 3 has been calculated using equation 1. Fig. 3 shows power density as a function of draw solution pressure for a 1 M NaCl draw solution. This study was conducted for three different feed solution temperatures (20^{0}C, 30^{0}C and 40^{0}C) while draw solution temperature was kept constant (i.e. 20^{0}C). At a constant temperature and concentration, increasing hydraulic pressure caused power density to increase. Power density increased by about ~34% (6.7 Wm^{-2} to 9.0 Wm^{-2}). The highest pressure applied in this study was 100 psi, beyond that pressure membrane internal structure is known to deform. At a constant pressure, conditions that yield high net water flux—corresponding to higher temperature—also yield high power density.

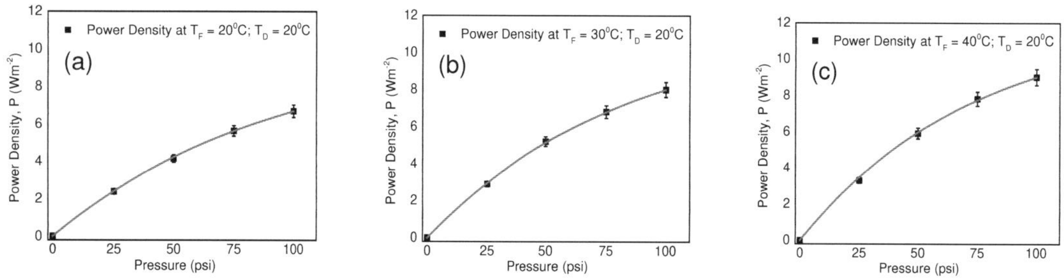

Fig 3. Power density for a 1.0 M NaCl draw solution as a function of draw solution pressure at working temperatures of a) T_F=20^{0}C, T_D=20^{0}C; b) T_F=30^{0}C, T_D=20^{0}C; c) T_F=40^{0}C, T_D=20^{0}C. Membrane used was commercial TFC. Draw and feed flow rate were 0.50 L/min each. Error bars indicate standard deviation over three trials.

3.3 Effect of operating temperatures on reverse solute diffusion on PRO performance

Reverse salt flux into the feed, as a function of temperature, at different applied pressures (0, 25, 50, 75 and 100 psi) is shown in Fig. 4. From this figure, it can be concluded that reverse salt flux shows an upward trend when pressure increases. This upward trend is due to the reduction of dilutive external concentration polarization (ECP) in the draw solution. As water flux decreases with increased pressure, this lowered water flux into the draw solution diminishes the dilutive ECP boundary layer, which causes the concentration at the draw-side membrane interface to increase (Touati et al. 2015). The resulting change in interface concentration causes greater reverse salt flux into the feed solution. Besides this, the increased temperature has a synergistic effect for the reverse salt flux. Since diffusivity increases with increased operating temperatures, the reverse salt flux will be greater for the membrane than at lower temperatures.

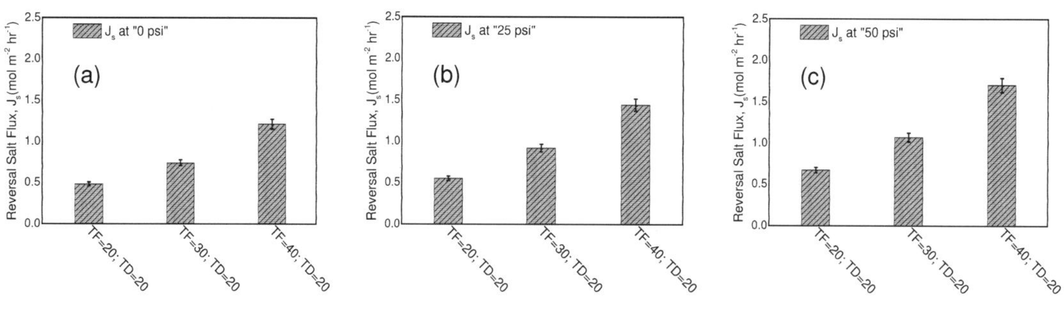

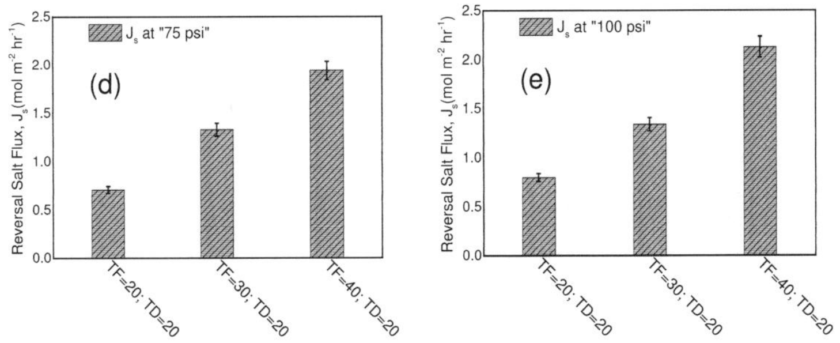

Fig 4. Reverse salt flux for a 1.0 M NaCl draw solution as a function of working temperatures at applied pressures of a) 0 psi; b) 25 psi; c) 50 psi; d) 75 psi; e) 100 psi. The membrane used was a commercial TFC. Draw and feed flow rate were 0.50 L/min each. Error bars indicate standard deviation over three trials.

3.4 Effect of operating temperatures on specific solute flux (Js/Jw) on PRO performance

Specific salt flux (J_s/J_w) is useful to study the effect of reverse salt flux diffusion on PRO performance due to the relationship with water flux and the salt flux. J_s/J_w describes the amount of draw solutes permeating through the membrane, normalized by volumetric water flux (Phillip, Yong, and Elimelech 2010; Tang et al. 2010; Zou et al. 2011). Larger specific solute flux is not desirable, since it may cause enhanced ICP (Lee, Baker, and Lonsdale 1981; Saren, Qiu, and Tang 2011; Yip et al. 2011), and more severe membrane fouling (Zou et al. 2011). The specific solute flux is found to increase along with higher applied pressures and higher operating temperature which can be attributed to the increase of salt flux at higher temperature and pressure.

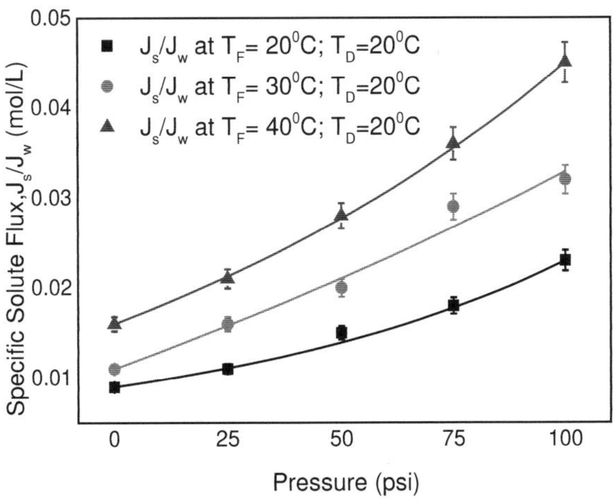

Fig 5. Specific solute flux (J_s/J_w) for a 1.0 M NaCl draw solution as a function of draw solution pressure at different working temperatures. The membrane used was commercial TFC. Draw and feed flow rate were 0.50 L/min each. Error bars indicate standard deviation over three trials.

4 CONCLUSIONS

In the current work, the effect of temperature on PRO performance was investigated. Operating temperature can play a significant role in power production, with an increase powered by waste heat resulting in an ~34% increase in power production. However, higher water flux is associated with higher salt diffusion which may cause severe ICP, which will eventually affects the performance of the membrane used in PRO.

From this study, it has also been seen that with an increase in temperature, specific solute flux (J_s/J_w) was found increased, leading to subsequent draw solute permeation through the membrane.

Acknowledgements

The authors gratefully acknowledge the Fonds de recherché du Québec – Nature et technologies (FRQNT) for providing financial support for this project. The authors also acknowledge the support of Concordia University, Montreal, Canada.

References

Achilli, Andrea, Tzahi Y. Cath, and Amy E. Childress. 2009. "Power Generation with Pressure Retarded Osmosis: An Experimental and Theoretical Investigation." *Journal of Membrane Science* 343 (1): 42–52. https://doi.org/10.1016/j.memsci.2009.07.006.

Achilli, Andrea, and Amy E. Childress. 2010. "Pressure Retarded Osmosis: From the Vision of Sidney Loeb to the First Prototype Installation — Review." *Desalination*, Special Issue in memory of Sidney Loeb (1917-2008), 261 (3): 205–11. https://doi.org/10.1016/j.desal.2010.06.017.

Chung, Tai-Shung, Xue Li, Rui Chin Ong, Qingchun Ge, Honglei Wang, and Gang Han. 2012. "Emerging Forward Osmosis (FO) Technologies and Challenges Ahead for Clean Water and Clean Energy Applications." *Current Opinion in Chemical Engineering*, Energy and environmental engineering / Reaction engineering and catalysis, 1 (3): 246–57. https://doi.org/10.1016/j.coche.2012.07.004.

Han, Gang, Qingchun Ge, and Tai-Shung Chung. 2014. "Conceptual Demonstration of Novel Closed-Loop Pressure Retarded Osmosis Process for Sustainable Osmotic Energy Generation." *Applied Energy* 132 (November): 383–93. https://doi.org/10.1016/j.apenergy.2014.07.029.

Kim, Jihye, Jijung Lee, and Joon Ha Kim. 2012. "Overview of Pressure-Retarded Osmosis (PRO) Process and Hybrid Application to Sea Water Reverse Osmosis Process." *Desalination and Water Treatment* 43 (1–3): 193–200. https://doi.org/10.1080/19443994.2012.672170.

Lee, K. L., R. W. Baker, and H. K. Lonsdale. 1981. "Membranes for Power Generation by Pressure-Retarded Osmosis." *Journal of Membrane Science* 8 (2): 141–71. https://doi.org/10.1016/S0376-7388(00)82088-8.

Logan, Bruce E., and Menachem Elimelech. 2012. "Membrane-Based Processes for Sustainable Power Generation Using Water." *Nature* 488 (7411): 313–19. https://doi.org/10.1038/nature11477.

Phillip, William A., Jui Shan Yong, and Menachem Elimelech. 2010. "Reverse Draw Solute Permeation in Forward Osmosis: Modeling and Experiments." *Environmental Science & Technology* 44 (13): 5170–76. https://doi.org/10.1021/es100901n.

Saren, Qi, Chang Quan Qiu, and Chuyang Y. Tang. 2011. "Synthesis and Characterization of Novel Forward Osmosis Membranes Based on Layer-by-Layer Assembly." *Environmental Science & Technology* 45 (12): 5201–8. https://doi.org/10.1021/es200115w.

She, Qianhong, Xue Jin, and Chuyang Y. Tang. 2012. "Osmotic Power Production from Salinity Gradient Resource by Pressure Retarded Osmosis: Effects of Operating Conditions and Reverse Solute Diffusion." *Journal of Membrane Science* 401–402 (May): 262–73. https://doi.org/10.1016/j.memsci.2012.02.014.

Tang, Chuyang Y., Qianhong She, Winson C. L. Lay, Rong Wang, and Anthony G. Fane. 2010. "Coupled Effects of Internal Concentration Polarization and Fouling on Flux Behavior of Forward Osmosis Membranes during Humic Acid Filtration." *Journal of Membrane Science* 354 (1): 123–33. https://doi.org/10.1016/j.memsci.2010.02.059.

Touati, Khaled, Christopher Hänel, Fernando Tadeo, and Thomas Schiestel. 2015. "Effect of the Feed and Draw Solution Temperatures on PRO Performance: Theoretical and Experimental Study." *Desalination* 365 (June): 182–95. https://doi.org/10.1016/j.desal.2015.02.016.

Yip, Ngai Yin, Alberto Tiraferri, William A. Phillip, Jessica D. Schiffman, Laura A. Hoover, Yu Chang Kim, and Menachem Elimelech. 2011. "Thin-Film Composite Pressure Retarded Osmosis Membranes for Sustainable Power Generation from Salinity Gradients." *Environmental Science & Technology* 45 (10): 4360–69. https://doi.org/10.1021/es104325z.

Zou, Shan, Yangshuo Gu, Dezhong Xiao, and Chuyang Y. Tang. 2011. "The Role of Physical and Chemical Parameters on Forward Osmosis Membrane Fouling during Algae Separation." *Journal of Membrane Science* 366 (1): 356–62. https://doi.org/10.1016/j.memsci.2010.10.030.

Building Tomorrow's Society
Bâtir la Société de Demain

Fredericton, Canada
June 13 to 16, 2018 / *13 au 16 juin, 2018*

ESTIMATING THE ECONOMIC FEASIBILITY OF BIOSOLIDS TREATMENT BY SMOULDERING

Rashwan, Tarek[1,3], Fournie, Taryn[1], Grant, Gavin[2] and Gerhard, Jason[1]
[1] The University of Western Ontario, Canada
[2] Savron, Canada
[3] trashwan@uwo.ca

Abstract: Within wastewater treatment plants (WWTPs), approximately 50% of the operating expenses are associated with managing the solid by-product, biosolids, making it the most expensive component of the entire process. Managing biosolids is a major challenge, demanding complicated processing, expensive transportation, and wasteful disposal. The objective of this research is to investigate the economic feasibility of a new approach to managing biosolids that has potential to be more energy efficient and cost-effective than current disposal methods. Self-sustaining Treatment for Active Remediation applied ex-situ (STARx) was originally developed as a chemical waste management and soil-clean up technology. STARx is based on the concept of smouldering, a flameless form of combustion for solid and liquid fuels. Smouldering is thermally efficient, meaning the process can recycle much of the released reaction energy to destroy fuels containing little energy or significant moisture content (MC) that would otherwise not burn (e.g., via incineration). Previous research has demonstrated that up to 80% MC (wet mass basis) biosolids could be efficiently destroyed via smouldering. This is a key finding, as it suggests that this process can reduce the extensive processing (e.g., dewatering, thickening), which are significant WWTP expenditures. This research uses costs reported by the United States Environmental Protection Agency to estimate "target disposal costs" for retrofitting a STARx system into existing WWTPs that practice either land application or incineration. It was found that a target disposal cost within the range of $341-$823 per dry ton of biosolids is probably required to retrofit a wide range of WWTPs.

1 LITERATURE REVIEW AND PROBLEM STATEMENT

Municipal wastewater treatment plants (WWTPs) are energy intensive and operationally demanding public activities that require significant capital and infrastructure (Shannon et al. 2008). In the United States (U.S.), WWTPs and drinking water services account for 3-4% of all national energy consumption and 30-40% of the total energy consumed by municipalities, costing $4 billion/year (U.S. EPA 2018). The scenario is quite similar in Canada, as both countries experience similar per capita water usage in comparable economies and social norms. Furthermore, as much of North America's WWTP infrastructure approaches the end of its design life, an estimated $271 billion and $26 billion is required in the U.S. and Canada, respectively, to repair and upgrade WWTP infrastructure to fair working condition (CIRC 2016, ASCE 2017). These costs represent 1.4% and 1.5% of the U.S. and Canadian 2017 GDP, respectively (CIA 2018). It is important to note that these costs do not consider upgrades due to evolving regulations (e.g., those that could result from emerging contaminants).

This presents a significant financial burden for North American municipalities, which are already in debt. A 2007 survey conducted collaboratively by The Federation for Canadian Municipalities and McGill

University found that Canada's municipal deficit was $123 billion, where all water supply systems (i.e., drinking water, wastewater, and storm water) constituted 25% of the deficit (Mirza 2007). Optimistically, the resulting benefits from improved WWTP infrastructure are expected to greatly outweigh these costs. Environment Canada (2013) estimated that $5.5 billion was required for Canadian WWTPs over the next 54 years to meet established regulations ($3.0 billion for capital, $1.7 billion for operation and maintenance (O&M), and $0.75 billion for reporting and monitoring). These costs were anticipated to provide $16.5 billion in benefits estimated from numerous projections including increased land value, greater recreational use of water bodies, increased commercial fishing, and reduced water supply costs. In addition, these costs are probably conservative as many environmental and social benefits are extremely difficult to quantify and are often undervalued (Hanley et al. 1998). In another sense, poor WWTP infrastructure could cause insidious environmental damage, where the ramifications and costs from such damage are not well understood or included in most cost-benefit analyses (Abdalla et al. 1992). The above information considers WWTP infrastructure holistically to present justification for the advancement of WWTP technology. It is equally worthwhile to identify the operations that are currently driving the high costs. Generally, the highest costs in WWTPs are associated with biosolids management (Khiari et al. 2004).

Following technical documents on biosolids management from the U.S. Environmental Protection Agency (U.S. EPA), O&M and capital cost ranges for popular biosolids processing and disposal methods have been compiled (U.S. EPA 2000a-g, 2002, 2003a-c, 2006a-b). Table 1 summarizes the high and low ranges of O&M costs for various biosolids processing technologies, and costs for disposal via application or incineration.

Table 1: Estimated Biosolids Processing and Disposal Cost Ranges

Processing | Disposal	Cost Range (2018 USD /Dry Ton of Biosolids)	
	Low	High
Thickening	0.40	4.02
Dewatering	93	299
Polymer Addition	0.0	28.0
Stabilization	15	491
* Land Application	94	456
Incineration	106	341

* Land Application costs may be offset up to $20/ton by selling composted biosolids as fertilizer (U.S. EPA 2002), but these savings were neglected due to infrequent practice.

It is important to caution that these approximations may not accurately represent current technology as the technical documents were published from 2000 to 2006, with some cost information dating as far back as 1987 (U.S. EPA 2000d). However, this information is assumed sufficient to understand the relative cost differences between popular biosolids management technologies. All values have been converted to 2018 USD following historical U.S. inflation rates from The World Bank (2018). It is interesting to highlight the large variability in capital and O&M costs. For example, gravity thickening capital costs vary between $286 500 and $4.60 million, and alkaline stabilization O&M costs vary from $61 to $481 per dry ton of biosolids processed (U.S. EPA 2000a). These costs vary with WWTP scale, inflow conditions, local electricity rates, etc. (King and Painter 1986). Altogether, these costs provide the framework for estimating the economic feasibility for new biosolids management technologies. This is important in identifying target O&M costs and the types of WWTPs these new technologies would best fit into, from a financial perspective. For this study, these costs will be used to discuss the economic feasibility of a novel biosolids disposal technology: Self-sustaining Treatment for Active Remediation applied ex-situ (STARx).

STARx has been in research and development for over ten years and was initially conceived for brownfield soil clean-up and hazardous, organic liquid waste management (Switzer et al. 2009, Pironi et al. 2011, Switzer et al. 2014). Currently, STARx is being implemented as a simple, low cost remediation technology at multiple sites around the world (Scholes et al. 2015, Grant et al. 2016). STARx utilizes self-sustaining smouldering: a flameless form of combustion commonly observed as the glowing, surface reaction surrounding charcoal briquettes in a traditional barbecue (Ohlemiller 1985). By mixing the hazardous wastes into a solid, porous medium (e.g., a bed of sand), a smoulderable mixture can be formed. This is due to: (1) forming permeable pathways to deliver the air (oxidant) to the reaction zone, (2) increasing the surface area of the waste (fuel), and (3) mitigating the heat losses out of the system, thereby recycling energy released from the reaction within the sand bed. This facilitates a thermally efficient process that can destroy a wide range of organic wastes that otherwise face limited disposal options. Furthermore, as the process is self-sustaining, it only requires energy input for ignition, after which the reaction can continue indefinitely with a minimal supply of air. Recently, it was shown that human faeces up to 75% moisture content (MC, wet mass basis) could be robustly smouldered in a simple reactor (Yermán et al. 2015, Yermán et al. 2016, Fabris et al. 2017). This was developed for implementation in a low-cost, simple toilet for improving sanitation in developing countries. Rashwan et al. (2016) demonstrated that biosolids from WWTPs with up to 80% MC could also be successfully smouldered. This is an exciting result as it suggests that biosolids processing (e.g., dewatering and thickening), which require significant O&M costs, may be reduced prior to disposal using a smouldering system, which itself should require minimal O&M costs. This finding has led to the development of a scale-up project that follows the conditions investigated at the lab scale (0.0080 m^3), using reactors at two prototype scales (0.27, 2.4 m^3). This project is ongoing and aims to identify the key benefits and challenges with increasing scale, and better distinguish the operating conditions relevant to the full-scale design of a smouldering system for biosolids.

2 ECONOMIC FEASIBILITY ANALYSIS

An economic feasibility analysis was conducted to estimate the "target disposal costs" (in 2018 US dollars per dry ton of biosolids) required to retrofit eight hypothetical WWTPs under various operating conditions with a new STARx smouldering system for biosolids disposal. A STARx system's O&M costs would need to be less than these target disposal costs to be an economical retrofit. The target disposal costs required for replacing land application or incineration were considered, as these are the most popular disposal methods practiced in Canada (Apedaile 2001).To estimate the target disposal costs, the current O&M costs required for land application and incineration must be understood. Furthermore, capital costs will not be considered, only O&M cost comparisons will be used to estimate the target disposal costs.

Each biosolids disposal option requires a specific sequence of biosolids processing steps, or "processing path", to condition the biosolids prior to disposal. The biosolids processing paths assumed for land application, incineration, and STARx are summarized in Table 2. Land application requires some degree of stabilization to achieve either Class A or B biosolids for safe use and, depending on the desired biosolids MC, may also require rigorous thickening and dewatering with polymer addition (U.S. EPA 1995). Incineration does not usually require stabilization but does require dewatering and thickening (usually with polymer addition) to reach a final MC between 60-85% (U.S. EPA 2003b). As STARx only requires minimal biosolids processing, an economical processing path was chosen to condition the biosolids to 80% MC with a lower heating value greater than 1.6 kJ/g. These properties have been shown suitable for disposal via smouldering (Rashwan 2015). This processing path requires minimal biosolids dewatering and thickening, without polymer addition, where low-end processing costs are assumed sufficient (Table 2). Each processing step has a range of reported costs, and is influenced by many factors including: prior processing, influent characteristics, and plant size (Metcalf and Eddy 2003). For a given processing path, summing the lowest cost estimates for all steps provides a low-end estimate of the processing path cost, while summing the highest cost estimates provides a high-end estimate. Ranges also exist for estimates of the disposal costs, from which the highest and lowest reported values provide the end-points. Taken together, it is possible to define the envelope of expected costs, from the highest possible processing path/disposal costs (high/high), through intermediate cases (high/low, low/high) to the lowest possible processing path/disposal costs (low/low). The range of possibilities in combining processing path/disposal end-point costs is summarized in Table 2.

Table 2: Processing Pathways and Cost End-Points for Considered Disposal Options

Disposal	Expected Processing Path	Processing Path Cost Range	Disposal Cost Range
Land Application	Thickening, Polymer Addition, Stabilization, and Dewatering	High/Low	High/Low
Incineration	Thickening, Polymer Addition, and Dewatering	High/Low	High/Low
STARx	Minimal Thickening and Dewatering	Low	-

Table 3 presents the target disposal costs estimated to replace either incineration or land application with a STARx system in eight hypothetical WWTPs. These target disposal costs are estimated by comparing the expected processing path/disposal costs required for land application and incineration, to the low-cost biosolids processing path assumed for a STARx system (Table 2).

Table 3: STARx Target Disposal Costs for Replacing Incineration and Land Application

(2018 USD/Dry Ton of Biosolids)

PROCESSING PATH COST RANGE	DISPOSAL COST RANGE			
	High		*Low*	
	Incineration	Land Application	Incineration	Land Application
High	$579	$1 185	$344	$823
Low	$341	$471	$106	$109

Table 3 suggests that target disposal costs would vary greatly between WWTPs: from $106 per dry ton of biosolids to replace the most economical WWTP practicing incineration, to $1185 to replace the most expensive WWTP practicing land application. Moreover, replacing land application, which may require a target disposal cost between $109-$1185, may achieve greater financial reward than replacing incineration, which may require a target disposal cost between $106-$579. This mainly results from removing stabilization, which is generally quite expensive (Table 1). This suggests that STARx may be an attractive retrofit for a larger fraction of WWTPs practicing land application compared to WWTPs practicing incineration.

It is important to note that target disposal costs to retrofit the four WWTPs with extreme processing path/disposal costs (i.e., low/low and high/high) are probably least realistic. Presumably, high disposal costs would be incurred from minimally processed biosolids following a low-cost processing path, and vice versa. E.g., low-cost processing generally results in high MC biosolids that would require high transportation costs for land application or extra costs for supplemental fuel for incineration. Therefore, considering these combinations may exaggerate the range of target disposal costs. Instead, it may be more practical to highlight the target disposal costs to retrofit the four WWTPs with intermediate processing path/disposal costs (i.e., high/low and low/high). These target disposal costs vary much less: from $471-$823 and $341-$344 to replace land application and incineration, respectively. This suggests that, if a WWTP intends to retrofit with a new disposal technology that minimizes its biosolids processing path costs, that technology's target disposal cost will probably be near the high-end costs of current disposal practices, i.e., $456 and $341 for land application and incineration, respectively (Table 1).

Currently, STARx contaminated soil treatment system O&M costs can be as low as $25 to $50 per mixture ton, depending on the volume of soils to be treated and the timeframes for remediation. However, these soils do not require mixing with a solid, porous medium (as is the case for biosolids) since the contaminated soil is a smoulderable mixture. Therefore, the presented target disposal costs may be achievable if the biosolids/solid ratio is high and the biosolids processing requirements are low. It is important to note that, as STARx is a new technology, information on O&M costs are very limited.

3 CONCLUSION AND SUMMARY

As WWTPs across North America approach their design life, there is a strong need to retrofit existing WWTPs with low energy, economical technologies. Smouldering combustion represents a promising approach to biosolids disposal that has the potential to greatly reduce a WWTP's energy footprint and serve as an economical retrofit. From the analysis presented, the estimated target disposal cost for a new smouldering system would vary depending on the current WWTP operating conditions, probably between $341-$823 per dry ton of biosolids. Furthermore, WWTPs practicing land application may benefit from greater financial savings by retrofitting with a smouldering system than WWTPs practicing incineration. As the first experimental scale-up study for a biosolids smouldering system is ongoing, the operating conditions are expected to become better understood. This study is anticipated to clarify the degree of biosolids processing and mixture conditions required for disposal, and better approximate the expected O&M costs.

Acknowledgements

This project is funded by the Government of Canada through the Federal Economic Development Agency for Southern Ontario through the Southern Ontario Water Consortium's Advancing Water Technologies Program (Grant SUB02392). Support, in the form of graduate and undergraduate research scholarships, is acknowledged from the Natural Sciences and Engineering Research Council of Canada. In-kind support is provided by Savron, a wholly owned subdivision of Geosyntec Consultants Ltd. The smouldering of biosolids is patent pending under PCT/US2012/035248. Geosyntec/Savron holds an exclusive license to commercialize the STAR technology.

References

Abdalla, C. W., Roach, B. A. and Epp, D. J. 1992. Valuing environmental quality changes using averting expenditures: an application to groundwater contamination. *Land Economics*, 163-169.

Apedaile, E. 2001. A perspective on biosolids management. *The Canadian Journal of Infectious Diseases and Medical Microbiology*, **12**(4), 202.

ASCE. 2017. Wastewater. *2017 Report Card for America's Infrastructure*, American Society of Civil Engineers.

CIA. 2018. The World Factbook. Retrieved February 22, 2018, from https://www.cia.gov/library/publications/resources/the-world-factbook/index.html.

CIRC. 2016. Wastewater. *Canadian Infrastructure Report Card.*

Environment Canada. 2013. Wastewater Systems Effluent Regulations Reporting. *What are the benefits of the Regulations?* Retrieved June 10, 2015, from https://www.ec.gc.ca/eu-ww/default.asp?lang=En&n=0D689118-1.

Fabris, I., Cormier, D., Gerhard, J. I., Bartczak, T., Kortschot, M., Torero, J. L. and Cheng, Y.-L. 2017. Continuous, self-sustaining smouldering destruction of simulated faeces. *Fuel*, **190**, 58-66.

Grant, G. P., Major, D., Scholes, G. C., Horst, J., Hill, S., Klemmer, M. R. and Couch, J. N. 2016. Smoldering Combustion (STAR) for the Treatment of Contaminated Soils: Examining Limitations and Defining Success. *Remediation Journal*, **26**(3), 27-51.

Hanley, N., Wright, R. E. and Adamowicz, V. 1998. Using choice experiments to value the environment. *Environmental and Resource Economics*, **11**(3-4), 413-428.

Khiari, B., Marias, F., Zagrouba, F. and Vaxelaire, J. 2004. Analytical study of the pyrolysis process in a wastewater treatment pilot station. *Desalination*, **167**, 39-47.

King, E. and Painter, H. 1986. Inhibition of respiration of activated sludge: variability and reproducibility of results. *Toxicity Assessment*, **1**(1), 27-39.
Metcalf and Eddy. 2003. *Wastewater Engineering: Treatment and Reuse*, McGraw-Hill, New York.
Mirza, S. 2007. *Danger Ahead: The Coming Collapse of Canadia's Municipal Infrastructure*, Federation for Canadian Municipalities, Ottawa.
Ohlemiller, T. J. 1985. Modeling of smoldering combustion propagation. *Progress in energy and combustion science*, **11**(4), 277-310.
Pironi, P., Switzer, C., Gerhard, J. I., Rein, G. and Torero, J. L. 2011. Self-sustaining smoldering combustion for NAPL remediation: laboratory evaluation of process sensitivity to key parameters. *Environmental Science & Technology*, **45**(7), 2980-2986.
Rashwan, T. L. 2015. Self-Sustaining Smouldering Combution as a Novel Disposal Destruction Method for Waste Water Biosolids. *Civil and Environmental Engineering*. London, The University of Western Ontario. **MESc**.
Rashwan, T. L., Gerhard, J. I. and Grant, G. P. 2016. Application of self-sustaining smouldering combustion for the destruction of wastewater biosolids. *Waste Management*, **50**, 201-212.
Scholes, G. C., Gerhard, J. I., Grant, G. P., Major, D. W., Vidumsky, J. E., Switzer, C. and Torero, J. L. 2015. Smoldering Remediation of Coal-Tar-Contaminated Soil: Pilot Field Tests of STAR. *Environmental Science & Technology*, **49**(24), 14334-14342.
Shannon, M. A., Bohn, P. W., Elimelech, M., Georgiadis, J. G., Mariñas, B. J. and Mayes, A. M. 2008. Science and technology for water purification in the coming decades. *Nature*, **452**(7185), 301-310.
Switzer, C., Pironi, P., Gerhard, J., Rein, G. and Torero, J. 2009. Self-sustaining smoldering combustion: a novel remediation process for non-aqueous-phase liquids in porous media. *Environmental Science & Technology*, **43**(15), 5871-5877.
Switzer, C., Pironi, P., Gerhard, J. I., Rein, G. and Torero, J. L. 2014. Volumetric scale-up of smouldering remediation of contaminated materials. *Journal of Hazardous Materials*, **268**, 51-60.
The World Bank. 2018. World Development Indicators. *Inflation, GDP deflator (annual %)*.
U.S. EPA. 1995. Part 503 Implementation Guidance. Washington, U.S. Environmental Protection Agency. **833-R-95-001**.
U.S. EPA. 2000a. Alkaline Stabilization of Biosolids. Washington, U.S. Environmental Protection Agency. **EPA 832-F-00-052**.
U.S. EPA. 2000b. Belt Filter Press. Washington, U.S. Environmental Protection Agency. **EPA 832-F-00-057**.
U.S. EPA. 2000c. Centrifuge Thickening and Dewatering. Washington, U.S. Environmental Protection Agency. **EPA 832-F-00-053**.
U.S. EPA. 2000d. Filter Press, Recessed Plate. Washington, U.S. Environmental Protection Agency. **EPA 832-F-00-058**.
U.S. EPA. 2000e. In-Vessel Composting of Biosolids. Washington, U.S. Environmental Protection Agency. **EPA 832-F-00-061**.
U.S. EPA. 2000f. Land Application of Biosolids. Washington, U.S. Environmental Protection Agency. **EPA 832-F-00-064**.
U.S. EPA. 2000g. Odor Control in Biosolids Management. Washington, U.S. Environmental Protection Agency. **EPA 832-F-00-067**.
U.S. EPA. 2002. Use of Composting for Biosolids Management. Washington, U.S. Environmental Protection Agency. **EPA 832-F-02-024**.
U.S. EPA. 2003a. Gravity Thickening. Washington, U.S. Environmental Protection Agency. **EPA 832-F-03-022**.
U.S. EPA. 2003b. Use of Incineration for Biosolids Management. Washington, U.S. Environmental Protection Agency. **EPA 832-F-03-013**.
U.S. EPA. 2003c. Use of Landfilling for Biosolids Management. Washington, U.S. Environmental Protection Agency. **EPA 832-F-03-012**.
U.S. EPA. 2006a. Heat Drying. Washington, U.S. Environmental Protection Agency. **EPA 832-F-06-029**.
U.S. EPA. 2006b. Multi-Stage Anaerobic Digestion. Washington, U.S. Environmental Protection Agency. **EPA 832-F-06-031**.
U.S. EPA. 2018. Energy Effeciency for Water and Wastewater Facilities. *Water: Sustainable Infrastructure*. Retrieved February 22, 2018, from http://water.epa.gov/infrastructure/sustain/sustainable_infrastructure.cfm.

Yermán, L., Hadden, R. M., Carrascal, J., Fabris, I., Cormier, D., Torero, J. L., Gerhard, J. I., Krajcovic, M., Pironi, P. and Cheng, Y.-L. 2015. Smouldering combustion as a treatment technology for faeces: Exploring the parameter space. *Fuel*, **147**, 108-116.

Yermán, L., Wall, H., Torero, J., Gerhard, J. I. and Cheng, Y. L. 2016. Smoldering Combustion as a Treatment Technology for Feces: Sensitivity to Key Parameters. *Combustion Science and Technology*, **188**(6), 968-981.

Building Tomorrow's Society
Bâtir la Société de Demain

Fredericton, Canada
June 13 to 16, 2018 / *13 au 16 juin, 2018*

EVALUATION OF A LOW-COST DRONE FOR MONITORING RESTORATION OF WELL PADS

Smith, Kalie[1,2], Tannant, Dwayne D.[1]
[1] School of Engineering, University of British Columbia, Canada
[2] km6smith@mail.ubc.ca

Abstract: Oil and gas development activities disturb the land surface at well pads. These sites must be restored upon decommissioning, and this process takes several years to accomplish. Using drones to assess restoration of well pads can be more efficient and provide better data compared to manual inspection at the ground surface. Fieldwork was conducted in 2017 to evaluate the use of a small, low-cost drone for measuring and assessing reclamation at four well pads on crown land near Hudson's Hope, BC. These sites had an "abandoned" status for one year, were under the same operator, and had not yet received a Certificate of Restoration. This paper presents the results of the investigation and discusses the advantages and limitations of using drones for reclamation assessment purposes.

1 INTRODUCTION

Well pads used by the oil and gas industry must be restored to natural conditions when the sites are abandoned. This paper documents the evaluation of a small, low-cost drone for measuring and assessing the state of reclamation on well pads located on crown land near Hudson's Hope, British Columbia. A drone was flown over well pads to take a series of aerial images. The images were then used to evaluate the state of land disturbance and restoration activities under natural field conditions. The approach used in this research was to use the aerial images supplemented with field observations and photographs taken at the ground surface to characterize as many aspects of restoration as possible.

Four well pads were flown in July and again in September 2017. Three sites (well pads 1, 2, and 3) were clustered close to each other approximately 8 km west of Hudson's Hope. The fourth site (well pad 4) was located 7.4 km northwest of Hudson's Hope. These sites were accessible by road, with no gates. All sites are located in the BWBS biogeoclimatic zone (Boreal White and Black Spruce). Because of its location east of the Rocky Mountains, this region experiences a cold, humid continental climate similar to that of the Canadian Prairies.

2 DRONE AND FLIGHT PLANNING

A small, low-cost drone was used to acquire the aerial images. A DJI Phantom 4 Pro drone (Figure 1) was selected because it can be easily carried to the site and rapidly deployed. Furthermore, this drone model carries a relatively good camera with a mechanical shutter. At the time of its purchase in the summer of 2017, this drone cost $2100 plus taxes. An extra battery and larger capacity microSD card were also purchased for another $300 plus taxes. The selected drone has a practical flight time limit of 20 minutes. One well pad of a size limited to about 6 ha could be safely flown using a single fully charged drone battery.

Several factors affect battery usage, including flight length, average drone speed, temperature, flying in manual control versus automated flight paths, and wind speed and variability.

Figure 1: DJI Phantom 4 Pro drone (DJI 2017)

Different flight plans were used at the well pads. In most cases, the drone was pre-configured to fly a grid pattern under autonomous control. All flight paths specified a drone height of 50 m above ground. This height was selected as a compromise between battery life, number of acquired images, and ground pixel resolution. The images were taken with a 80% overlap. For these cases, PixCapture software was used to set up each flight plan. Details from the September flights are provided in Table 1. An example of a grid flight path flown over well pad 4 is shown in Figure 2.

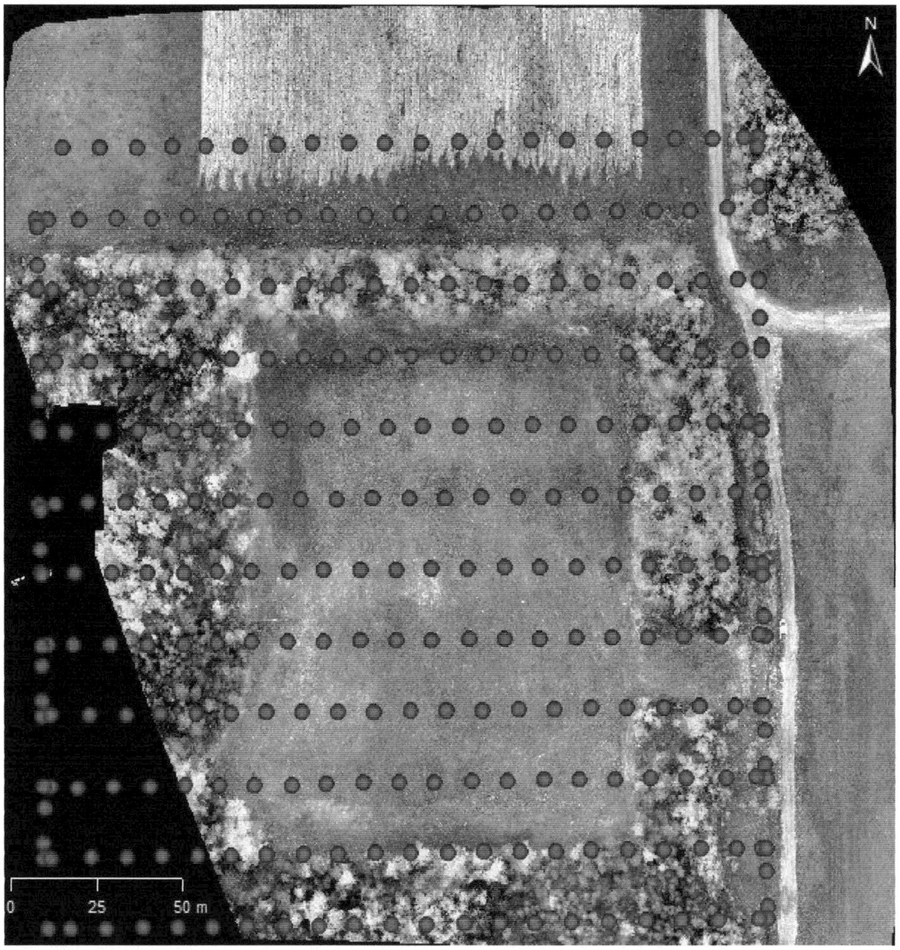

Figure 2: Flight path showing camera location for each photo taken in a grid configuration over well pad 4

Table 1: September flight details

Well Pad	Area Covered (ha)	Average Speed (m/s)	Flight Duration (mm:ss)	Images Taken	Processing Time (h:mm:ss)	Points in Point Cloud (millions)
1	3.26	2.87	13:10	250	4:11:02	25.4
2	3.07	2.92	04:02* 06:19* 01:33*	204	2:04:40	14.5
3	3.83	2.98	09:50	183	1:54:39	13.2
4	5.91	2.43	15:44	296	2:49:25	26.3

*Well pad 2 was flown in 3 separate flights

3 STRUCTURE-FROM-MOTION PHOTOGRAPHIC ANALYSIS

The images from each flight were loaded into Pix4D version 4 (Pix4D 2017) structure-from-motion software for processing to create large point clouds and generate orthomosaics, contour maps, digital surface models, and other outputs. The points were georeferenced using the drone's built-in GPS; no ground control points were used. While the UTM Northing and Easting coordinates are likely within a few metres of accuracy, the elevation values associated with the point cloud could be off by 10s of metres. However, the relative coordinates of the points within the generated point cloud provide a more reliable and useful measure with accuracies estimated to be within a range of 0.1 m.

Depending on what outputs are generated and which settings are selected, processing of the acquired images for one well pad in the structure-from-motion software can take many computer hours. These times were obtained when the computer was dedicated to running only one model. The image processing was performed on a desktop computer with an Intel Xeon® ES 1607 V3, 3.7 GHz quad-core processor with 64GB RAM and Nvidia® Quadro K2200 GPU.

The photographs were taken at an ISO setting of 100. The camera has a 1" CMOS sensor, and the image sizes are 5472x3648 pixels. Each model was run with an image scale of one half (default), and points were set to have a minimum of 3 image matches. The contours were generated at 10 cm resolution and 25 cm intervals. To generate the digital surface model (DSM) and the orthophotographs, the pixel size was set to the default setting of 1xGSD, where the GSD is the average ground sampling distance, which was 1.3 cm for these models.

4 CERTIFICATE OF RESTORATION REQUIREMENTS

Version 1.2 of the BCOGC Certificate of Restoration Application Manual (BCOGC 2016) lists the site reclamation objectives for well pads. To assess the capabilities and limitations of the drone for evaluating reclamation, the objectives were summarized into the following five criteria that were used to analyze reclamation at each of the well pads.

4.1 Geotechnical Stability

There should be no visible slope movement, slumping, subsidence, or tension cracks on the well pad surface. On-site cut and fill slopes should be stabilized.

4.2 Land Use and Facilities

Any facilities remaining from oil and gas development activities should not affect the natural drainage or slope of the area. There should be no gravel or rock piles or windrows. There should be no industrial debris or woody debris interfering with reclamation efforts.

4.3 Vegetation

There should be no prohibited nuisance weeds and no infestation of any particular species on the well pad. There should be an even distribution of vegetation density with no bare areas, and a sustainable, intentional plant community growing such that at least 80% of the well pad surface area is covered with vegetation, as outlined in Schedule B, BCOGC Site Reclamation Requirements (BCOGC 2013).

4.4 Slope and Drainage

There should be no areas with water ponding or erosion resulting from the ground topography that would affect land reclamation. The land should be restored to the original slope of the site, or to a condition that is compatible with the surrounding landscape. Ditches should be refilled along the edges of the well pad and access routes.

4.5 Soil Replacement

Any salvaged soil that was piled on the well pad should be spread over the site. Soils that were compacted from oil and gas development activities must de-compacted to assist with plant growth. There should be no evidence of impact from adjacent land use.

5 RESULTS

The well pads considered in this study were each evaluated against the five criteria outlined in Section 4. This section focuses on the use of drone-acquired aerial images and the resulting digital outputs generated with the images to evaluate these criteria.

5.1 Geotechnical Stability

There is no visible slope movement, slumping, subsidence, or tension cracks on or around any of the well pads that were examined, and there are no remaining cut and fill slopes. These features could be easily detected from aerial images taken using the drone if they were present.

5.2 Land Use and Facilities

Approximately 40 m northwest of the well pad 2 there is a large gravel quarry. This quarry is not shown in GoogleEarth where a 2012 image was obtained, but it was detected in the aerial images taken while flying the drone over the well pad. The northwest edge of the well pad and the southeast edge of the gravel quarry are both captured in the aerial image shown in Figure 3. Some fallen trees can also be seen.

Figure 3: Aerial photograph showing edge of a gravel quarry 40 m northwest of well pad 2

Because of the trees between the well pad and the gravel quarry, this quarry is not easily seen at the ground level, and it is not shown in GoogleEarth. The quarry does not seem to affect the drainage of the well pad area. However, it is worth noting the close proximity of the quarry to the well pad. The presence of the open quarry might have contributed to blow-down of trees on the northwest side of the well pad.

Adjacent land use activities can be seen in aerial images taken by the drone over well pad 4. Approximately 25 m north of the well pad, the land is being used for farming as shown in Figure 4. The well pad is separated from the farming field by a strip of trees.

Figure 4: Strip of trees and agricultural land north of well pad 4 captured by the drone

5.3 Vegetation

Well pad 4 had an approximate surface area of 1.8 hectares in July 2017. The area of this well pad was significantly smaller in 2012 as seen in a comparison between images taken 2012 (Google Earth) and 2017 as shown in Figure 5. This well pad was expanded, likely because additional wells were drilled from the same pad.

The contrast in vegetation coverage on this well pad was significant over a two-month period. The aerial images from July and September in Figure 5 show that large areas that were sparsely vegetated in mid-summer had become better vegetated by the fall. Note that 2017 was the first summer after the well pad had been seeded.

Figure 5: Restoration progress of well pad 4: 2012 (Google Earth), July 2017, and September 2017

While plant species growing on well pads are difficult to discern from aerial images taken by a drone, it is easy to spot areas where vegetation is density or height differs. Some vegetation differences in aerial images taken over well pads 3 and 4 are shown in Figure 6. The aerial images taken by the drone over some of the well pads clearly showed evidence of fallen trees and woody debris.

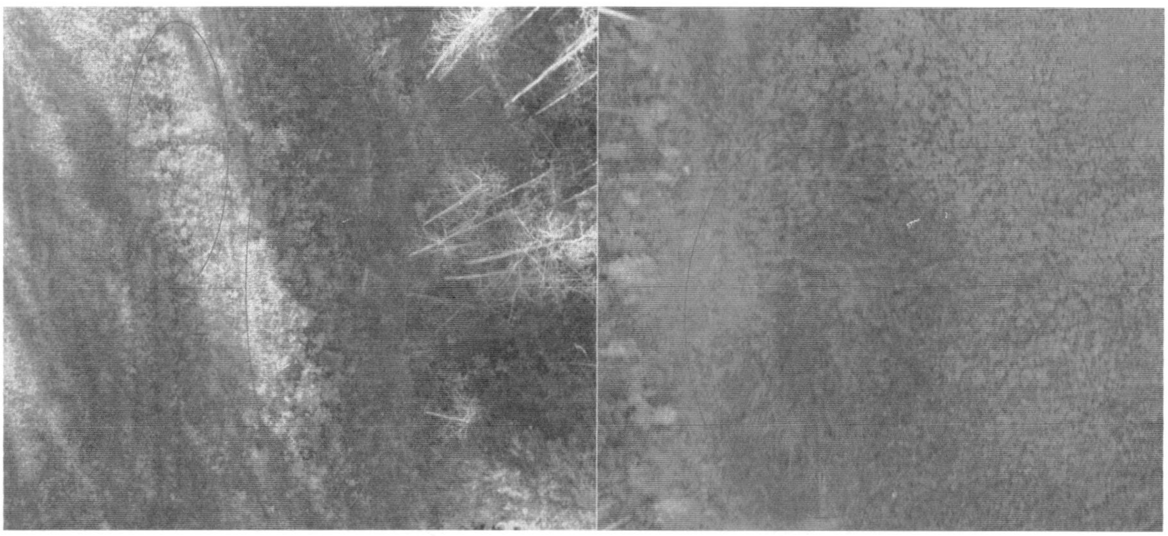

Figure 6: Vegetation differences shown in aerial images over well pads 3 (left) and 4 (right)

5.4 Slope and Drainage

It is easily confirmed with aerial images taken by the drone whether drainage ditches remain at the well pads or whether they have been removed, as required in the Certificate of Restoration criteria. Drainage ditches were not present at any of the well pads examined.

The 2D digital surface model generated using point cloud data from well pad 2 shows that water could pond in a slight depression near the centre of the well pad. The area of potential ponding corresponds to an area with slightly impeded vegetation growth as seen in Figure 7. Lack of drainage from the low area may be affecting the vegetation since there is no apparent link to heavily compacted soil or any other cause. A similar trend can be found in the southwest corner of well pad 4, as shown in Figure 8.

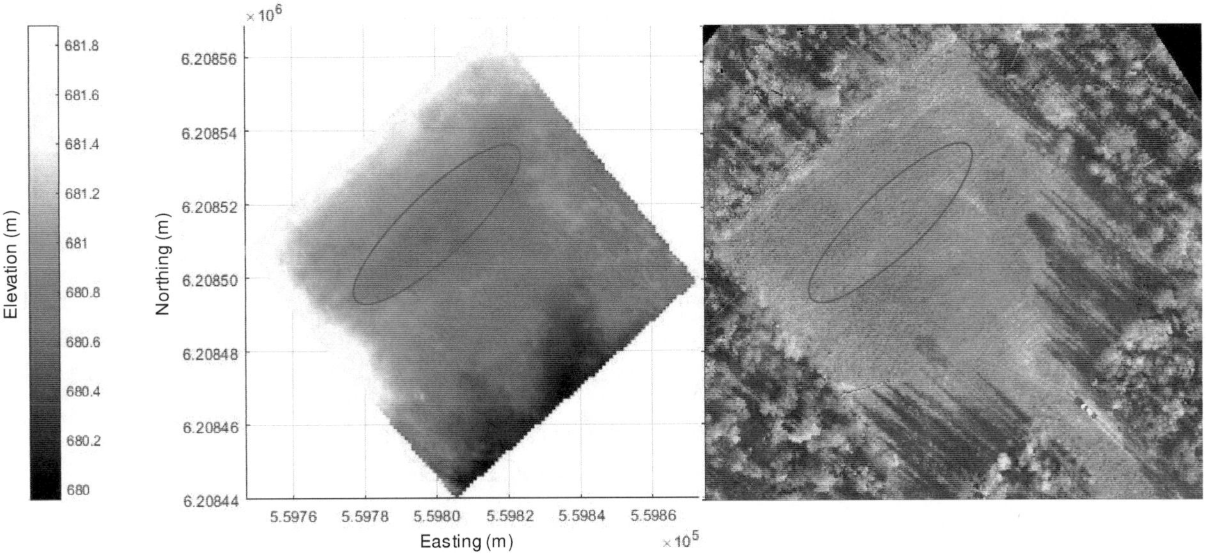

Figure 7: Slight depression and impeded vegetation growth near the centre of well pad 2

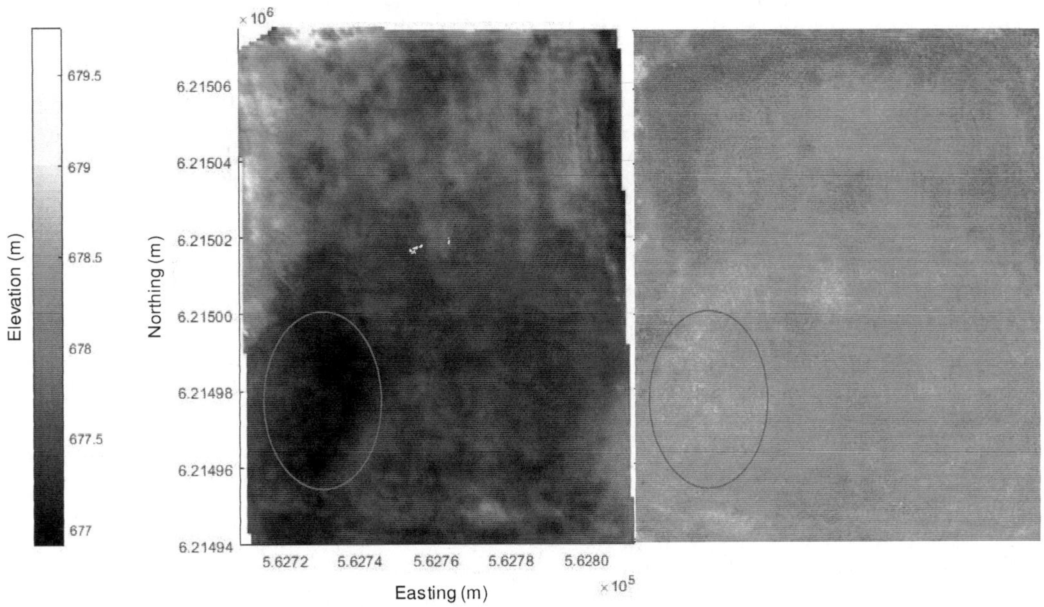

Figure 8: Potential ponding may impede growth on well pad 4

The Certificate of Restoration criteria state that drainage should be restored to the conditions present prior to oil and gas development activities taking place, or to a state that is consistent with the surrounding drainage conditions. The slopes of the examined well pads are relatively flat; they were all less than 5%.

5.5 Soil Replacement

Compaction of the soil beneath a salvaged soil stockpile that once sat on well pad 4 may be a possible explanation for the reduced vegetation along the southeast side of the well pad as seen in Figure 9. The soil pile is no longer present as it was likely redistributed across the well pad as required.

Figure 9: Reduced vegetation density where salvaged soil pile was located on well pad 1

Tire tracks were noted where vehicles have driven over the well pads after the ground surface had been prepared and seeded. There is noticeably less dense vegetation growth in these tire tracks. Soil compaction under the vehicle tires and direct damage to the vegetation is likely reducing the vegetation growth in these tracks. This is shown over well pad 1 in Figure 10. The well pads have nothing around them to restrict vehicle entry, and the access road had not been decommissioned.

Figure 10: Reduced vegetation on soil compacted by vehicle traffic on well pad 1

6 DISCUSSION

Obtaining aerial images of the heavily forested areas surrounding the well pads using the drone enables detection of surrounding land uses that are difficult to see through the trees from ground level. The gravel quarry northwest of the well pad 2 and the agricultural lands north of well pad 4 were noted in the aerial photos. The ability to see adjacent land uses and their location relative to the well pad can give insight as to whether these land uses are impacting reclamation practices.

There does not seem to be a direct correlation between where various equipment or the wells were located on the well pad and vegetation growth. However, vegetation growth does appear to be affected where salvaged soil was piled on the well pad. It can be inferred that the salvaged soil was redistributed over the well pads as soil piles are no longer present. However, the soil compacted from being beneath the piles of salvaged soil may not have been fully de-compacted, and soil compaction does impede vegetation growth.

For the four well pads flown in both July and September of 2017, the contrast in vegetation coverage over this two-month period is significant. In September, the well pads had fewer bare areas and more uniform vegetation as emphasized in the orthophotographs. Photographs taken at ground level can supplement the aerial images regarding what specific plant species are present. Knowing the composition of the seed mixtures planted on these well pads for reclamation purposes would be useful in determining whether the intended plant community is growing and if undesirable plant species are present.

When the structure-from-motion software generates a dense point cloud, the 3D coordinates include all features captured in the aerial images. Therefore the contour plots and surface models may not be representative of bare-earth topography and can be affected by variance in the vegetation height. To create detailed topographic maps, well pads should be flown once reclamation activities have been completed, and before vegetation begins to grow. Later flights could then assess the vegetation growth on the well pad. The point cloud data appropriately captures the surface topography when there is bare earth and in early stages of vegetation growth. However, once shrubs and trees begin to grow on the well pad and the vegetation density and height varies across the well pad it becomes more difficult to generate a DSM that represents the bare earth conditions for analysis of drainage and slope. At these later stages of reclamation the orthomosaics alone may be a more reliable indicator stage of restoration.

7 CONCLUSIONS

For the four well pads included in this study, the drone was flown at a 50 m height above ground using preset autonomous grid patterns. The aerial images captured by the drone were used to assess the requirements for a Certificate of Restoration for the well pads, including geotechnical stability, land use and facilities, vegetation, slope and drainage, and soil replacement. With photogrammetry software and post-processing of the point cloud data, the imagery can be used to identify several quantifiable features, including the health and uniformity of vegetation coverage and occurrence of bare areas, the magnitude and direction of the slope and areas of potential ponding, whether salvaged soil has been redistributed across the well pad, appearance of debris or fallen trees, adjacent or nearby land uses and whether oil and gas facilities have been removed, and other disturbances including tire tracks that compact soil and prevent vegetation growth.

The results of the study suggest that drones can be a valuable tool to evaluate reclamation progress of well pads. In many cases, the data that can be produced from the drone imagery can be used to assess multiple requirements needed for a Certificate of Restoration.

Acknowledgements

The research presented in this report was funded by the BC Oil and Gas Research and Innovation Society and Colleges and Institutes Canada Natural Resources Internship. Kevin Parsonage, Devin Scheck, and others from the BC Oil and Gas Commission assisted in identifying well pads to document and aspects to study. Wenbo Zheng from the UBC School of Engineering assisted during the July fieldwork.

References

BCOGC 2013. Schedule B - Site Reclamation Assessment. British Columbia Oil and Gas Commission. https://www.bcogc.ca/content/schedule-b-site-reclamation-assessment (accessed February 15, 2018)

BCOGC 2016. Certificate of Restoration Application Manual, version 1.2. British Columbia Oil and Gas Commission. http//www.bcogc.ca/node/12445/download (accessed February 15, 2018).

DJI 2017. DJI Phantom 4 Pro. https://www.dji.com/phantom-4-pro (accessed February 15, 2018).

Pix4D 2017. Pix4Dmapper: Professional photogrammetry software. https://pix4d.com/product/pix4dmapper-photogrammetry-software/ (accessed February 15, 2018).

Building Tomorrow's Society
Bâtir la Société de Demain

Fredericton, Canada
June 13 to 16, 2018 / *13 au 16 juin, 2018*

EXPERIMENTAL AND NUMERICAL STUDY OF THERMAL BUOYANT OFFSET JET DISCHARGED INTO STAGNANT WATER

Alfaifi, Hassan[1,2,3] and Mohammadian, Abdolmajid[1]
[1] University of Ottawa, Canada
[2] King Abdulaziz City for Science and Technology (KACST), Saudi Arabia
[3] halfa103@uottawa.ca

Abstract: The aim of this paper is to study the behavior of the offset thermal jet in the mixing zone for the effluent of the industrial outfalls (e.g. desalination plants) on the near-field, experimentally and numerically. Several experiments of the thermal buoyant offset jet discharged horizontally into stagnant ambient water are conducted. The Particle Image Velocimetry (PIV) method is used to measure the time-history of the velocity distribution in the near-field mixing zone. The characteristics of the jet flow with different densimetric Froude numbers (Fd) compared with numerical results are presented in this study. Two different densimetric Froude numbers (Fd) (10 and 20) and difference density (Δρ) (5 and 10) are used. Three Reynolds-averaged Navier-Stokes (RANS) turbulence models; standard k-ε, realizable k-ε and buoyancy-modified k-ε models are applied in this study. The comparison between laboratory and numerical results for jet trajectory as well as with existing experimental and numerical data are presented. In general, a good agreement is observed between all numerical and experimental cases examined here, for both Froude numbers, where the realizable k-ε was shown to be the best among others.

1 INTRODUCTION

Cooling systems such as Multi-Stage Flash distillation system (MSF) of desalination plants produce hot water, known as reject water, which is discharged into the marine environment as a thermal buoyant jet, through a submerged outfall. Some parameters such as densimetric Froude number, pipe discharged diameter, density, and temperature can affect the jet behavior in order to achieve higher dilution, which is the main purpose of outfall facilities. Since most of hydraulic engineering practices happen in sensitive places of aquatic environment, investigation of the jet behavior and its impacts is essential. The outfall system is either made of diffusers with many ports or just one single pipe from which the effluent is discharged. The effluent can be denser or lighter than the receiving water. If the effluent density is less than the ambient water density, the jet behaves as a (positively) buoyant jet. This effluent may affect the surrounding water closer to the exit pipe by increasing the temperature. This might have another adverse impact on the marine ecosystem in the near-field area as well (Mabrook, 1994; Hashim and Hajjaj, 2005; Milione and Zeng, 2008; Lattemann and Höpner, 2008; Roberts et al., 2010). Therefore, it is important to study the characteristics of the near-field mixing zone to prevent any of adverse impacts caused by the outfall systems and to keep the marine environment clean and usable as a sustainable source of drinking water for future generations. A submerged outfall (offset, wall and inclined jets) is typically designed to achieve higher dilution for buoyant jets. An offset jet refers to a flow that is discharged above a horizontal solid wall offset with a certain distance which is parallel to the axis of the jet exit, as shown in Figure 1. Despite the fact that offset jets are currently in use worldwide, little research has been carried out on their properties. Therefore, more laboratory experiments are still required to study the near field flow and mixing characteristics of the thermal jets due to the complex nature of this phenomenon.

Figure 1: An offset buoyant jet discharged

Experimental and numerical studies of negatively-buoyant jets have been conducted extensively (Bosanquet et al., 1961; Hoch and Jiji, 1981; Gu, 1996; Nasr and Lai, 1998; Kanna and Das, 2005; Shao and Law, 2009; Shao and Law, 2011; Agelin-Chaab and Tachie, 2011; Kheirkhah Gildeh et al, 2015a; Assoudi et al., 2015; Kheirkhah Gildeh et al, 2015b; Assoudi et al., 2016), while positive buoyant jets have been studied less and require further investigation. Some of these studies have been performed only using Particle Image Velocimetry system (PIV), or combined with planar Laser-Induced Fluorescence (LIF). Rawn and Palmer (1930) were pioneers in this field, where they made an experimental investigation of the diffusion of horizontal offset jets. Their study included 388 experimental observations of dilution at the surface for the offset jets of fresh water injected into sea water. Abraham (1963) studied jets injected horizontally into homogeneous denser ambient fluid, theoretically. Recently, Cuthbertson and Davies (2008) carried out a set of laboratory experiments to investigate the characteristic features of particle-laden, round, turbulent, buoyant jets discharged horizontally into stationary and coflowing receiving fluids using PIV system. Michas and Papanicolaou (2009) studied horizontal turbulent buoyant jets discharged in a homogeneous calm ambient fluid, experimentally. They evaluated the mixing characteristics such as trajectories, turbulence properties, and dilution factors. However, numerical simulations of buoyant offset jets still require further investigation. A numerical investigation has been carried out by Liu and Lam (2015) for round jet discharge horizontally with sediment particles. They applied Large Eddy Simulation (LES) for fluid phase comparison, while Lagrangian particle tracking was used to calculate the motion of sediment particles.

This paper investigates the behaviors of the buoyant offset jets discharged into stagnant ambient water experimentally, using PIV system technique, and numerically. The numerical results of the centerline trajectory were compared with laboratory experiments

2 EXPERIMENTAL SETUP AND PROCEDURES

In this study, Particle image velocimetry (PIV) system was used to measure the time-history of the velocity distribution of the jet flow. The PIV system is an experimental technique in fluid mechanics which basically involves photographic recording of the motion of microscopic particles that follow the fluid. Image processing methods are then used to determine the particle motion, and hence the flow velocity, from the photographic recordings. PIV system has generally four main components; laser box, laser lens, high-speed CCD camera, and computer for synchronization between camera and lasers. Small tracer particles have to be added to the discharged flow to be able to follow the fluid paths.

All experiments were conducted in a rectangular transparent glass tank with dimensions of 1.2 m length, 0.5 m width, and 0.5 m depth. Both the bottom and side walls of the tank were made of glass panels. Tap water was used for all experiments as an ambient water (tank water) with certain temperature (T_a) and density (ρ_a). A circular jet nozzle with an internal diameter (D) of 5.18 mm was attached horizontally to one end of the tank wall (at the center) with 10 cm distance from the bottom wall.

The jet was parallel to and offset from the bottom of the tank. A buoyant jet with an initial velocity (U_0), temperature (T_0) and density (ρ_0) was generated by discharging heated and fresh water into the tank through this nozzle. The nozzle was connected to a pipe that is equipped with valves to control the discharged water and a flow meter (rotameter) to determine the flow rate of the discharged water. The pipe was connected to a constant head tank with dimensions of 0.75 m length, 0.3 m width, and 0.3 m depth. This tank contained discharged water, which was located about 1.5 m above the discharged nozzle. The head tank had two separate storage tanks, one of which was used as a store for the surplus of the discharged water. The water was then returned from the store tank using a pump to make sure the water level was constant. To ensure that the water temperature in the pipe was equal to that of the heated water in the elevated tank, the pipe was covered and isolated by heat-insulating material. Also, there was another outlet located outside of the experimental tank with a distance of 10 cm upstream of the nozzle to check the temperature of the discharged water before running the experiments. Moreover, the nozzle outlet was equipped with a very thin sensor connected to a thermometer to check the flow temperature during the experiments. For brevity, only the results of three cases are shown in the table and figures.

The same concept is applied in all experiments using hot water (low density) discharged into stationary water as a thermal buoyant offset jet. A summary of the experimental and numerical parameters, same parameter values were used in both, are given in Table 1.

Table 1: Experimental and numerical parameters

Case	D (mm)	U_0 (m/s)	T_0 (°C)	ρ_0 (kg/m^3)	T_a (°C)	ρ_a (kg/m3)	F_d	R_e	$\Delta\rho$
1	5.18	0.16	38.6	992.75	21.5	997.89	10	1222	5.1
2	5.18	0.23	50.5	988.06	21.5	997.89	10.3	2162	9.8
3	5.18	0.45	50.7	987.75	21.5	997.89	19.8	4246	10.1

3 NUMERICAL MODEL

OpenFOAM (OPEN Field Operation and Manipulation) CFD model with structured grid, was used to simulate the buoyant offset jet discharge in this study. OpenFOAM (a free and open-source package) is widely used for modelling and solving scientific problems, for example, positively and negatively buoyant jets (Kheirkhah Gildeh et al., 2015a and 2015b; Zhang et al., 2016). The model solves a set of Partial Differential Equations (PDEs) using the Finite Volume Method (FVM). The pisoFoam solver for incompressible fluid was developed to use in this study. Advection-diffusion equation for temperature was added to handle the transport and dispersion. Three different Reynolds-averaged Navier-Stokes (RANS) turbulence models were chosen to apply in this study due to their accuracy among other RANS turbulence models, as reported on the recent studies of Kheirkhah Gildeh et al. (2014, 2015a and 2015b). These models are; the standard k-ε, realizable k-ε and buoyancy-modified k-ε model.

3.1 Governing Equations

The model uses three-dimensional RANS equations as the governing equations for incompressible fluids:

Continuity Equations:

[1] $$\frac{\partial u_x}{\partial x}+\frac{\partial u_y}{\partial y}+\frac{\partial uz}{\partial z}=0$$

Momentum Equations:

[2] $$\frac{\partial u_x}{\partial t}+\frac{u_x\partial u_x}{\partial x}+\frac{u_y\partial u_x}{\partial y}+\frac{u_z\partial u_x}{\partial z}=\frac{\partial}{\partial x}\left[v_e\left(\frac{\partial u_x}{\partial x}\right)\right]+\frac{\partial}{\partial y}\left[v_e\left(\frac{\partial u_x}{\partial y}\right)\right]+\frac{\partial}{\partial z}\left[v_e\left(\frac{\partial u_x}{\partial z}\right)\right]-\frac{1}{\rho}\frac{\partial P}{\partial x}$$

[3] $$\frac{\partial u_y}{\partial t}+\frac{u_x\partial u_y}{\partial x}+\frac{u_y\partial u_y}{\partial y}+\frac{u_z\partial u_y}{\partial z}=\frac{\partial}{\partial x}\left[v_e\left(\frac{\partial u_y}{\partial x}\right)\right]+\frac{\partial}{\partial y}\left[v_e\left(\frac{\partial u_y}{\partial y}\right)\right]+\frac{\partial}{\partial z}\left[v_e\left(\frac{\partial u_y}{\partial z}\right)\right]-\frac{1}{\rho}\frac{\partial P}{\partial y}-g'$$

[4] $$\frac{\partial u_z}{\partial t}+\frac{u_x\partial u_z}{\partial x}+\frac{u_y\partial u_z}{\partial y}+\frac{u_z\partial u_z}{\partial z}=\frac{\partial}{\partial x}\left[v_e\left(\frac{\partial u_z}{\partial x}\right)\right]+\frac{\partial}{\partial y}\left[v_e\left(\frac{\partial u_z}{\partial y}\right)\right]+\frac{\partial}{\partial z}\left[v_e\left(\frac{\partial u_z}{\partial z}\right)\right]-\frac{1}{\rho}\frac{\partial P}{\partial z}$$

where:

[5] $$g'=g\frac{\rho-\rho_0}{\rho}$$

where u_x, u_y and u_z are the components of the mean velocity in the x, y and z direction, respectively, P is the fluid pressure, t is the time, g' is the modified gravity acceleration, g is the acceleration of gravity, ρ is the fluid density, ρ_0 is the effluent fluid density, v_e is the effective kinematic viscosity of water ($v_e = v_t + v$), and v_t is the turbulent kinematic viscosity. The effect of the variable density (buoyancy) in the vertical direction (y-coordinate) is considered and added in Eq. (3).

Temperature equation:

[6] $$\frac{\partial T}{\partial t}+\frac{u_x\partial T}{\partial x}+\frac{u_y\partial T}{\partial y}+\frac{u_z\partial T}{\partial z}=k_e\frac{\partial}{\partial x}\left[\frac{\partial T}{\partial x}\right]+\frac{\partial}{\partial y}\left[\frac{\partial T}{\partial y}\right]+\frac{\partial}{\partial z}\left[\frac{\partial T}{\partial z}\right]$$

where:

[7] $$k_e=\frac{v_t}{Pr_t}+\frac{v}{Pr}$$

where T is the fluid temperature, k_e is the heat transfer coefficient, Pr is the Prandtl number, and Pr_t is the turbulent Prandtl number. In this study, the Pr_t and Pr values are chosen to be 1.0 based on Kheirkhah Gildeh et al. (2015a). The details of the buoyancy-modified k-ε turbulence model are not described here for brevity, more details can be found in (Yan and Mohammadian, 2017).

3.2 Boundary Conditions

The boundary conditions for the buoyant offset jet discharged in a rectangular tank are discussed in this section. Figures 2 and 3 illustrate the schematic views of the numerical model with the coordinate systems and the domain dimensions, respectively.

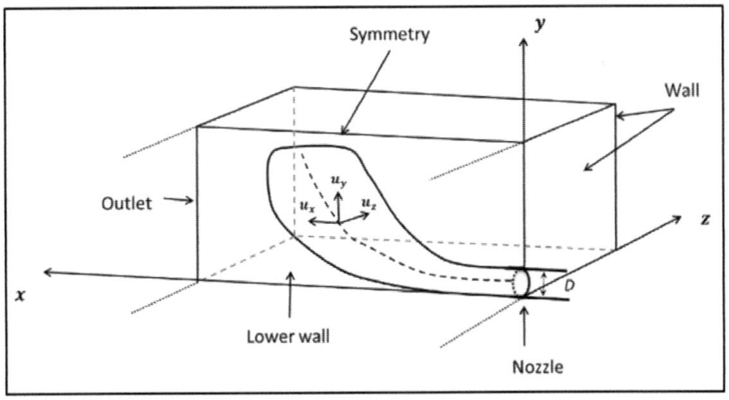

Figure 2: Schematic diagram of the computational domain with the coordinate systems

The parameters used in the numerical model are similar to the ones in the experiments (see Table 1). The nozzle (outfall) boundary conditions are chosen based on Kheirkhah Gildeh et al. (2014), as $u_x = U_0$, $u_y = u_z = 0$, $T = T_0$, $k = 0.06u^2$, $\varepsilon = 0.06u^3 / D$. A zero-gradient boundary condition perpendicular to the outlet plane is defined for u_x, u_y, u_z, k, ε, T and S for the flow at the outlet boundary section. For the walls, boundary conditions defined as $z = 0$, $u_z = 0$ and no-slip condition was considered for this study. Thus, k and ε are assumed to be standard wall function for the thermal buoyant jet. On the other hand, fixed values were assumed to be the initial conditions for all wall boundaries of the non-thermal jet that discharged in dense ambient. Finally, the symmetry boundary was modeled using zero gradient conditions. In the present study, only half of the buoyant offset jet domain is considered because of the symmetry nature of the problem. The dimensions of the computational domain (Figure 4) were chosen to simulate the present experiments. A refined mesh is considered in the near field area of the offset jet to better capture velocity and temperature characteristics for all simulations, as shown in Figure 3.

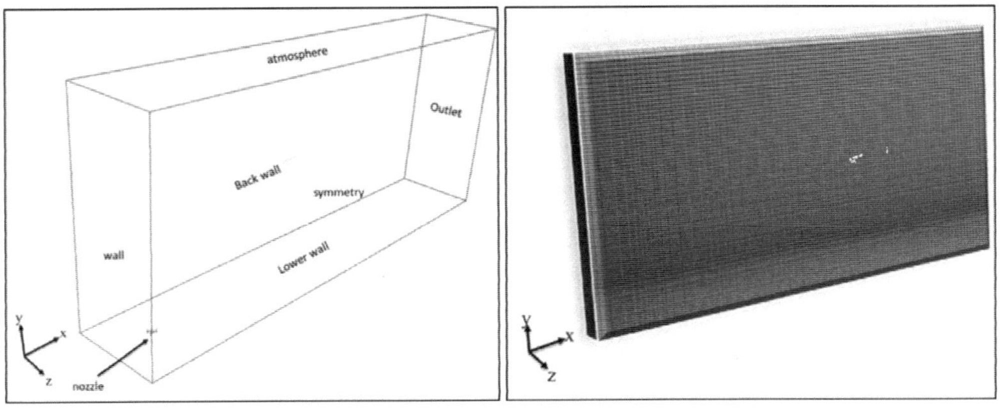

(a) (b)

Figure 3: a) The numerical model domain and boundary definition; b) fine grids mesh near the nozzle at the symmetry plane

4 RESULTS AND DISCUSSIONS

4.1 Experimental Observations

When a horizontal offset buoyant jet discharged into a water body, the entrainment of the water affects the jet and causes a change in jet behavior. Within a short distance away from the nozzle exit, the jet flow is dominated by momentum force. Then, after a certain distance from the nozzle tip, the buoyancy force becomes larger and dominates the jet and enforces it to rise up to the water surface. The experiments in this paper were intended primarily to obtain some quantitative measurements of the trajectory behaviors of buoyant offset jets to be used for validation of the numerical simulations.

In these experiments, development of the jet was recorded using a PIV system. Figure 4 shows three different jet flows of the average velocity superimposed on the velocity filed contours which extracted from PIV data for Fd 10, 10.3 and 19.9. As we can see from the figure, all jet behaves different when the initial velocity increased even if the temperature or Fd are almost the same. Although case 2 and 3 have the same temperature and almost same difference in density, but the Fd in cases 3 is higher, the jet deviates far away from the nozzle as a result of the effect of the Fd. It can be observed from the experiments that when the jet goes farther into the ambient water, the time to reach the surface is increased, thus the dilution may increase (Roberts and Toms (1987). Therefore, studying the parameters that can be more sensitive on the behave of the jet trajectory and then the near-field mixing characteristics are needed. The jet trajectory behaviors are discussed in detail in the following section.

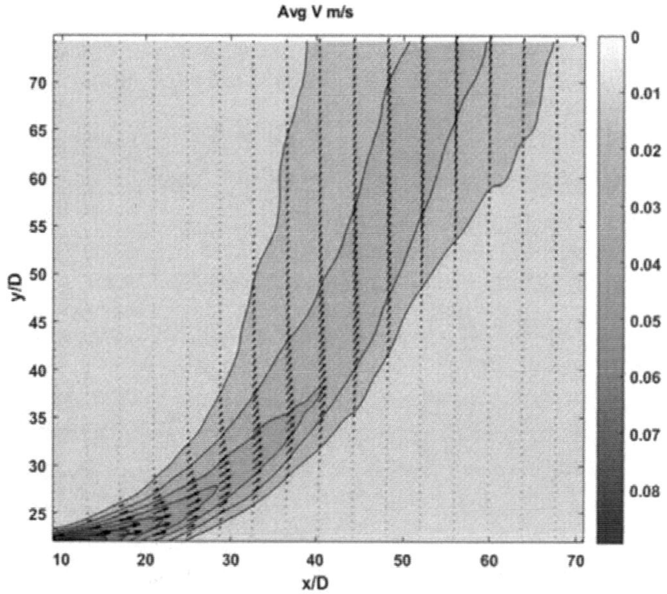

(a)

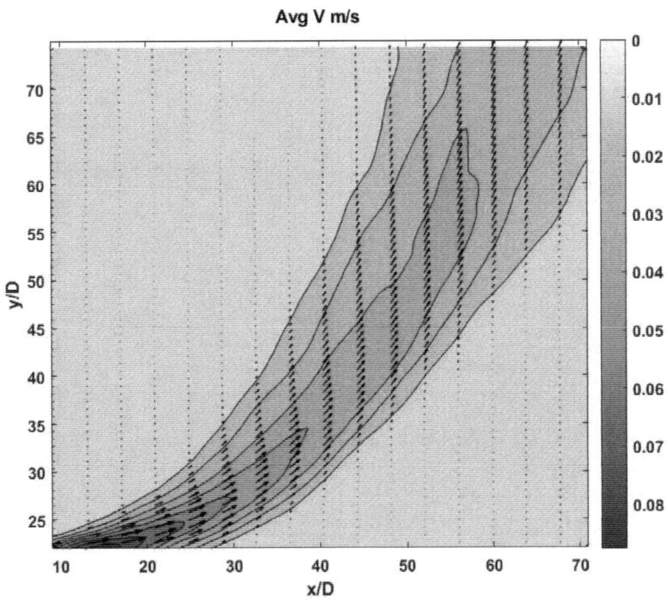

(b)

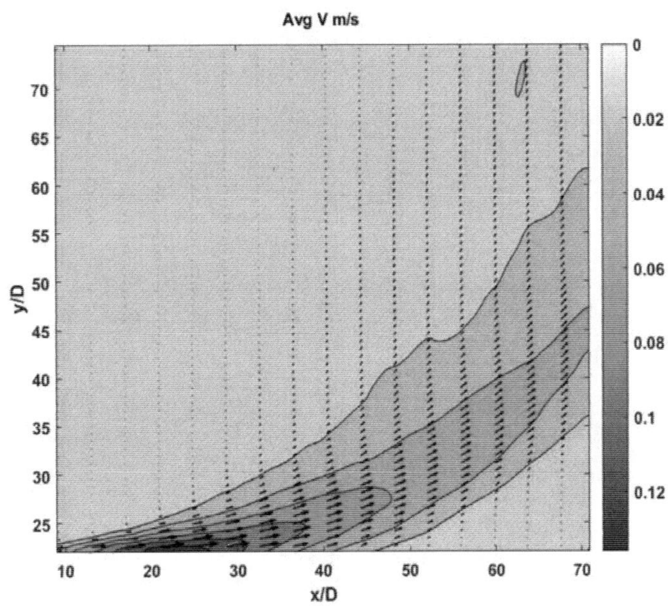

Figure 4: The average velocity overlying on the velocity filed contours extracted from PIV data: (a) u = 0.16 m/s, $F_d = 10$, $\Delta\rho = 5.1$; (b) $u = 0.23$ m/s, $F_d = 10.3$, $\Delta\rho = 9.8$; (c) $u = 0.45$ m/s, $F_d = 19.8$, $\Delta\rho = 10.1$

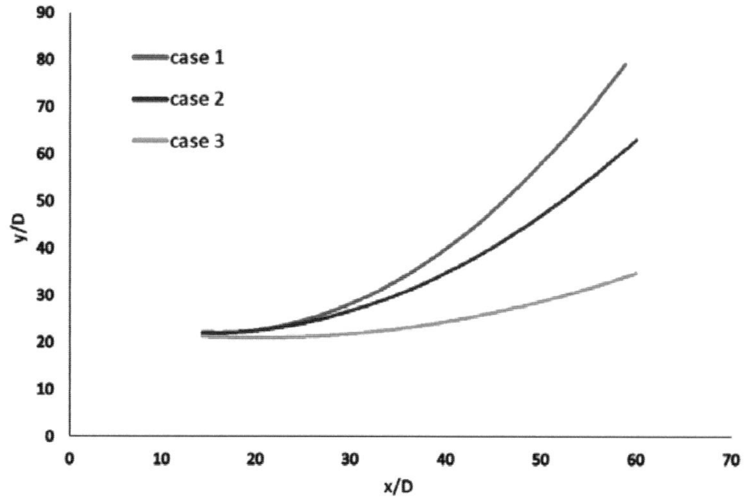

Figure 5: Comparison of the centerline trajectory from PIV data for cases 1, 2 and 3

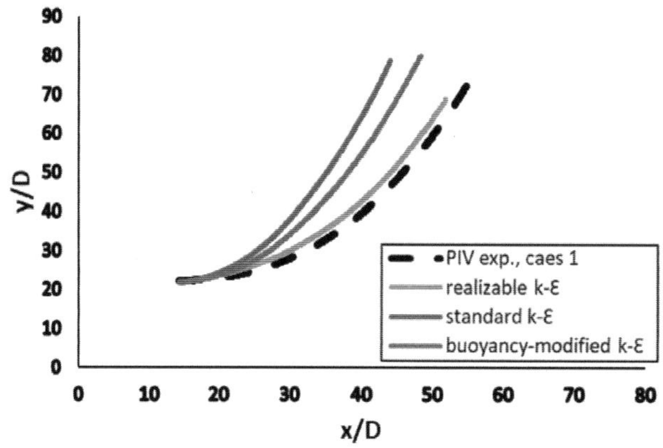

(a)

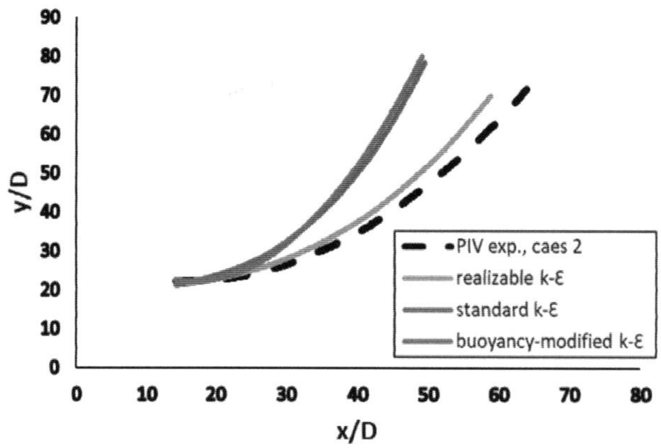

(b)

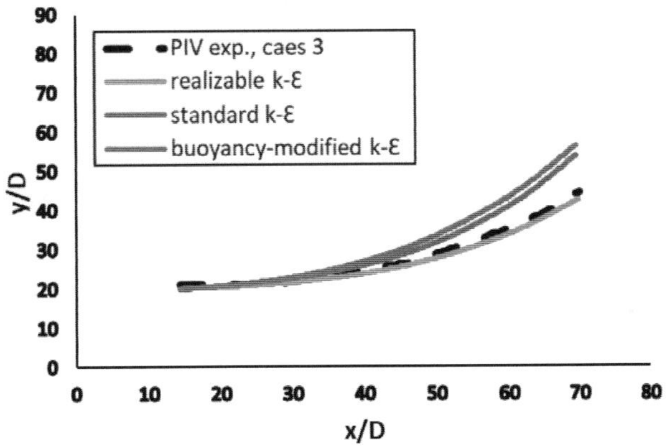

(c)

Figure 6: Numerical results of the centerline trajectory: (a) u = 0.16 m/s, F_d =10, $\Delta\rho$ =5.1; (b) u = 0.23 m/s, F_d =9.8, $\Delta\rho$ =10.3; (c) u = 0.45 m/s, F_d =19.8, $\Delta\rho$ =10.1

4.2 Jet Trajectory

Jet trajectory is a very important parameter in outfall system design as it identifies the path that jet goes through and where it reaches the water surface. The final dilution at the end of near-field is also a function of the jet path and ambient water characteristics (e.g. water depth especially for positively buoyant jets). Jet trajectory can be determined when the jet enters into a region known as the buoyant region or the free jet region (Huai et al., 2010). The PIV measurements for the average velocity vectors of the jet trajectory for the experimental results of for all cases are shown in Figure 5. The figure shows clearly a different behavior in jet trajectory in all experiments where their parameters are different.

4.3 Experimental and Numerical Comparison

The experimental results were used to examine the capability of three different turbulence models, standard k-ε, realizable k-ε and buoyancy-modified k-ε, in the same conditions. The numerical results of the jet centerline trajectory for all models comparing with all experiments are shown in Figure 6. The results show that all numerical results of all models were in a good agreement with the case 3, especially the realizable k-ε model which was the best among others. Also, a good match was observed between the numerical results of realizable k-ε model and other cases (1 and 2). As mentioned, one can clearly see from the figures that a better prediction is obtained using realizable k-ε than standard k-ε and buoyancy-modified k-ε turbulence models. In addition, standard k-ε and buoyancy-modified k-ε models for case 3 show a good prediction than other cases where the Fd and turbulences are increased. Therefore, the numerical results of the realizable k-ε model are in a good agreement with present experimental studies.

5 CONCLUSIONS

This paper presented the results of an experimental and numerical study of thermal buoyant offset jets discharged into stagnant water. The main objective of this study was to obtain experimental data to be used for comparison with the numerical model results. The OpenFOAM software with structured grid was used to simulate thermal buoyant offset jet discharged in stagnant water. Three RANS turbulence models, standard k-ε, realizable k-ε and buoyancy-modified k-ε were applied to predict the behavior of these jets. The numerical results of the realizable k-ε model showed that the jet trajectory was in a good agreement with all experiments. Moreover, a good general agreement of jet trajectories is observed between all numerical models and the experimental case 3 examined here. Finally, the obtained results from standard k-ε in this study, have shown a better prediction for jet trajectory compared to the buoyancy-modified k-ε turbulence model. More investigation for the buoyant offset jet with different higher Froude numbers and various numerical models is currently in progress by the authors.

Acknowledgements

Hassan Alfaifi would like to thank King Abdulaziz City for Science and Technology (KACST) for their financial support.

References

Abraham, G. 1963. Jet Diffusion in Stagnant Ambient Fluid. PhD thesis. Amsterdam University of Applied Sciences.

Agelin-Chaab, M. and Tachie, M.F. 2011. Characteristics of Turbulent Three-Dimensional Offset Jets. Journal of Fluids Engineering, 133: 051203-1– 051203-9.

Assoudi, A., Habli, S., Said, N.M., Bournot, H. and Le Palec, G. 2015. Experimental and Numerical Study of an Offset Jet with Different Velocity and Offset Ratios. Engineering Applications of Computational Fluid Mechanics, 9(1): 490-512.

Assoudi, A., Habli, S., Said, N.M., Bournot, H. and Le Palec, G. 2016. Three-Dimensional Study of Turbulent Flow Characteristics of an Offset Plane Jet with Variable Density. Heat and Mass Transfer, 52(11): 2327-2343.

Bosanquet, C.H., Horn, G. and Thring, M.W. 1961. The Effect of Density Differences on The Path of Jets. Proc. R. Soc. Lond. A, 263(1314): 340-352

Cuthbertson, A.J. and Davies, P.A. 2008. Deposition from Particle-Laden, Round, Turbulent, Horizontal, Buoyant Jets in Stationary and Coflowing Receiving Fluids. Journal of Hydraulic Engineering, 134(4): 390-402.

Gu, R. 1996. Modeling Two-Dimensional Turbulent Offset Jets. Journal of Hydraulic Engineering, 122(11): 617-624.

Hashim, A. and Hajjaj, M. 2005. Impact of Desalination Plants Fluid Effluents on The Integrity of Seawater, with The Arabian Gulf in Perspective. Desalination, 182(1-3): 373–393.

Hoch, J. and Jiji, L.M. 1981. Two-Dimensional Turbulent Offset Jet-Boundary Interaction. Journal of fluids engineering, 103(1): 154-161.

Huai, W., Li, Z., Qian, Z., Zeng, Y. and Han, J. 2010. Numerical Simulation of Horizontal Buoyant Wall Jet. Journal of Hydrodynamics, 22(1): 58-65.

Kanna, P.R. and Das, M.K. 2005. Numerical Simulation of Two-Dimensional Laminar Incompressible Offset Jet Flows. Int. J. Numer. Meth. Fluids, 49(4): 439–464.

Kheirkhah Gildeh, H., Mohammadian, A. Nistor, I. and Qiblawey, H. 2014. Numerical Modeling of Turbulent Buoyant Wall Jets in Stationary Ambient Water. Journal of Hydraulic Engineering, ASCE 140(6): 1-17.

Kheirkhah Gildeh H., Mohammadian A., Nistor I. and Qiblawey H. 2015a. Numerical modeling of 30° and 45° inclined dense turbulent jets in stationary ambient. Environmental Fluid Mechanics, 15(3): 537-562.

Kheirkhah Gildeh H., Mohammadian, A., Nistor, I., Qiblawey, H. and Yan, X. 2015b. CFD modeling and analysis of the behavior of 30 and 45 inclined dense jets—New numerical insights. J. Appl. Water Eng. Res, 4(2): 1-16.

Lattemann, S. and Höpner, T. 2008. Environmental Impact and Impact Assessment of Seawater Desalination. Desalination, 220(1-3): 1–15.

Liu, P. and Lam, K.M. 2015. Large-eddy simulation of horizontally discharging sediment-laden jets. Journal of Hydro-Environment Research, 9(3): 388–403.

Mabrook, B. 1994. Environmental Impact of Waste Brine Disposal of Desalination Plants, Red Sea, Egypt. Desalination, 97(1-3): 453-465.

Michas, S.N. and Papanicolaou, P.N. 2009. Horizontal round heated jets into calm uniform ambient. Desalination 248(1-3): 803–815.

Milione, M. and Zeng, C. 2008. The Effects of Temperature and Salinity on Population Growth and Egg Hatching Success of The Tropical Calanoid Copepod, Acartia Sinjiensis. Aquaculture, 275(1-4): 116–123.

Nasr, A. and Lai, J.C.S. 1998. A turbulent Plane Offset Jet with Small Offset Ratio. Experiments in Fluids, 24(1): 47-57.

OpenFOAM version 5 [Computer software]. OpenCFD Ltd., Bracknell, U.K.

Rawn, A.M., and Palmer, H. K. 1930. Predetermining the Extent of a Sewage Field in Sea Water. Trans. Amer. Soc. Civil Eng., 94(1): 1036-1060.

Roberts, P. J., and Toms, G. 1987. Inclined dense jets in flowing current. Journal of Hydraulic Engineering, 113(3): 323-340.

Roberts, D.A., Johnston, E.L. and Knott, N.A. 2010. Impacts of Desalination Plant Discharges on The Marine Environment: A Critical Review of Published Studies. Water Research, 44(18): 5117–5128.

Shao, D. and Law, A.W.-K. 2009. Turbulent Mass and Momentum Transport of A Circular Offset Dense Jet. Journal of Turbulence, 10(40): 1–24.

Shao, D. and Law, A.W.-K. 2011. Boundary Impingement and Attachment of Horizontal Offset Dense Jets. Journal of Hydro-environment Research, 5(1): 15–24.

Zhang, S., Jiang, B., Law, A.W.K. and Zhao, B. 2016. Large eddy simulations of 45 inclined dense jets. Environmental Fluid Mechanics, 16(1): 101-121.

Yan, X. and Mohammadian, A. 2017. Numerical Modeling of Vertical Buoyant Jets Subjected to Lateral Confinement. Journal of Hydraulic Engineering, 143(7): 04017016.

Building Tomorrow's Society
Bâtir la Société de Demain

Fredericton, Canada
June 13 to 16, 2018 / *13 au 16 juin, 2018*

IMPACT OF AGED CONTAMINATION ON BIOVENTING PERFORMANCE

Zytner, Richard G.[1,3], Xiao, Mei[1] and Mosco, Michael[3]
[1] School of Engineering, University of Guelph, Canada
[2] Brookfield Renewable Energy, Ottawa, Canada
[3] rzytner@uoguelph.ca

Abstract: Bioventing can be an effective in situ remediation method to clean-up a site contaminated with petroleum hydrocarbons. Significant work has been completed on determining the optimum conditions needed to stimulate the native microorganisms so that native bacteria degrade the petroleum hydrocarbons. Understanding on scale-up factors has improved so that smaller laboratory experiments can be run, and then the degradation rate adjusted to represent what would take place at a larger scale. However, much needs to be learned on the impact of aging. Aging happens when petroleum hydrocarbons are released into the soil and stay in contact with the soil for an extended period of time, causing a bond to develop between the contaminant and the soil particles. This bond makes it difficult for the microorganisms to access the petroleum hydrocarbons.

Experiments were completed by spiking loamy sand to about 8000 mg/kg and storing the mixture four to ten months in the refrigerator, followed by testing in an 80 kg bioventing reactor. Two conditions were tested, dry soil storage and wet soil storage. Additional experiments were completed by spiking the soil, aging the wet mixture and then conducting bioventing tests at regular intervals using 150 g respirometers to determine if the time of aging had an impact on degradation. The respirometers showed an increase in degradation rate for the wet loamy sand, starting at 0.079 d^{-1} for freshly contaminated soil to 0.171 d^{-1} for soil aged up to 300 days. In contrast, the 80 kg reactor results showed no change in degradation rate of the aged experiments for the loamy sand soil under similar aging conditions. However, for the dry-aged conditions, the degradation rate was half that of the freshly contaminated reactor test after only four months of aging; 0.053 d^{-1} as compared to 0.118 d^{-1}. In all cases, a sharp reduction in extractable contaminants was observed due to an increase in aging. These results show that the bonding that happens through aging impacts the bioavailability and the rates of biodegradation differently under different aging conditions and soil types, which ultimately increases the time required to attain site closure in large-scale experiments.

1 INTRODUCTION

Sites contaminated with petroleum hydrocarbons are a challenge worldwide. The contamination negatively impacts the soil and water environment, increasing the exposure to all forms of life, with an ultimate impact on humans. In Canada it is estimated that are approximately 22,000 federally contaminated sites, with an estimated liability of $4.9 billion (OAGC, 2012). The Federally Contaminated Site Action Plan (FCSAP) in Canada indicates that there are an additional 2,500 suspected sites that still have to undergo assessment (TBCS, 2015). As a result, site remediation is a big business and governments are always looking for cost effective, yet environmentally responsible methods to clean up these sites.

A popular remediation technology is bioventing. In this process the native microorganisms are stimulated by adding sufficient nutrients, water and oxygen, which allows the microorganisms to environmentally breakdown the contamination into harmless by-products (Zytner et al., 2001). Bioventing is a non-destructive approach that is cost effective and can address the remnants of other technologies like soil vapour extraction that cease to be effective due to mass transfer limitations that develop.

Significant effort has been spent on determining the rate coefficients of the bioventing process as a function of soil types. These rate coefficients help estimate the time of remediation (Frutos et al., 2010; Eyvazi and Zytner, 2009). Sutton et al. (2013), Chien et al. (2009) and Kao et al. (2010) noted that the results and correlations developed in typical laboratory respirometers do not always represent what occurs in the field due to the impact of heterogeneity and aging of contaminants. As a result, Khan et al. (2015) and Mosco and Zytner (2017) determined scale-up factors to assist with the transfer of lab results to the field, using 4 kg and 80 kg reactors respectively with different soils, with the degradation results compared to the results from 150 g respirometers (Eyvazi and Zytner, 2009) for similar soils. The scale-up factors in both sets of experiments were found to be approximately 2 in both cases, where the 4 kg and 80 kg degradation rate coefficients were larger than the 150 g rates.

Determining the scale-up factors showed that the 150 g degradation rates need to be adjusted when estimating bioventing performance. The work by Mosco and Zytner (2017) also showed that experiments with the largest reactor (80 kg) are not needed, as the 4 kg size reactors are easier to use and provide enough degradation information. However, these scale-up experiments did not consider the aging affect. Jain et al (1992) reported that aging causes chemical bonds between soil particles and petroleum hydrocarbons, which becomes stronger with an increase in age. This is supported by the work of Sutton et al. (2013), where aged contamination lead to a reduction in the bioavailability of the hydrocarbons to the microbial population. As a result, remediation efforts would require additional time when using bioventing. Understanding the differences in the degradation rates between freshly and aged contaminated soils is important for evaluating remediation options. Consequently, large (80 kg) and small scale (150 g) experiments were completed to determine the degradation rates of aged contamination and compare it to fresh contaminated soil.

2 METHODOLOGY

The two soils used for the study. One was Delhi loamy sand, collected from the Delhi Research Station, in Delhi, ON, the other Elora silty loam obtained from the Elora Research Station. After collection, both soils were air dried and sieved to pass a 2 mm sieve opening. Khan and Zytner (2015), previously determined that the loamy sand had a porosity of 0.4 and an average bulk density of 1550 kg/m^3, while the silty loam had a porosity of 0.48 and a bulk density of 1350 kg/m^3.

After drying, the soil was spiked to an initial concentration of 8080 mg/kg of synthetic petroleum hydrocarbon in the mixing cans. No water was added to the soil and synthetic gasoline mixture. This higher concentration accounts for some volatilization during the long-term storage of the soil in the refrigerator. After mixing, the soil was placed into aluminum lined containers, sealed and placed in the refrigerator for approximately five months to minimize volatilization during the aging process.

Following five months of storage, the loamy sand soil was placed into mixing cans. Nitrogen in the form of NH4Cl was dissolved into sufficient water to achieve a 15 – 18 % soil water content by weight, which would provide the required 50 % water content of the soils field capacity. This provided the ideal water content and C:N ratio of 10:1 for optimum degradation (Eyvazi and Zytner, 2009). After mixing, the soil was packed in layers in the 80 kg reactor to a bulk density of 1550 kg/m^3. Speed was of the essence to minimize volatilization. The 80 kg reactor was then placed in an experimental chamber within a fumehood. The chamber made it possible to control the air flow rate at 8-10 mL/min. The air was humidified via a bubbler to ensure that the soil did not dry out, which would have stopped the degradation process. The reactor was also wrapped a moist cotton wrap to help minimize the loss of water from the reactor. The reactor also had a vapour sampling port where granular activated carbon sorption could be installed to capture any petroleum hydrocarbons. Analysis showed that volatilization was minimal as expected. Complete details are in Mosco and Zytner (2017).

In order to determine if water content played a role in the aging process, additional loamy sand was taken and spiked to a level of 4000 mg/kg of synthetic petroleum hydrocarbon. In addition to the synthetic gasoline, water was added at a soil water content of 15 % to test aging under wet conditions. The soil was then stored for approximately ten months in the refrigerator. After 10 months, ammonium chloride was added to the soil to stimulate the microbial population and then placed in the 80kg reactor to conditions similar to the dry soil conditions.

For both 80 kg reactor tests, the degradation rate was tracked by taking approximately 5 g of soil sample from the reactor at regular intervals using a specifically design coring device. The soil was placed in 40 mL amber vials, to which 20 mL of methylene chloride was added to extract the petroleum hydrocarbons. The vials were then placed on a shaker for 1 h, after which the vials were placed in a freezer overnight to allow the sediment to settle. The next morning, approximately 2 mL of solvent was removed from the supernatant and placed into gas chromatograph (GC) vials for analysis. Chemical analysis was done on a GG equipped with a Flame Ionization Detector.

Loamy sand soil was also prepared for respirometer testing, where it was wet aged as for the large reactors, but this time a portion of the soil was removed on regular intervals. The respirometers were prepared by placing 150 g of the stimulated soil in 1 L glass bottles that could be sealed. However, before sealing, a test tube of KOH was placed inside the respirometer to carbon dioxide produced by the microbial respiration (Eyvazi and Zytner, 2009). The respirometers were incubated at 25°C. Periodically, the respirometers were vented to ensure the oxygen concentration remained at an adequate level, but air flow through the soil is kept entirely passive. This procedure kept volatilization to a minimum. Every few days, a respirometer was sacrificed to extract soil samples to track degradation rates. The soil samples were analyzed using the same method as described above, for the 80kg reactor.

Extracted soil samples were also tested for the presence of soil microorganisms. The two groups of interest were total heterotrophic bacteria (THB) and petroleum-degrading bacteria (PDB). Gasoline was added to the PDB mixing solution at a rate of 4 g/L to provide the food source. Using a spread plating technique, two different media were selected to grow the two different types of bacteria. Tryptic Soy Agar (TSA) was used for THB and Bushnell-Hass (BH) medium was used to culture the PDB. Serial dilution was used to plate 100 uL of solution on each batch of TSA and BH plates. The TSA plates were counted for colonies after 48 h and then reconfirmed after 72 h in the incubator. The BH plates were counted for colonies after 7 d and then reconfirmed after 14 d in the incubator. The results of the experiment were reported as colony forming units per gram of soil weight (CFU/g).

3 RESULTS

3.1 Large Reactor

The loamy sand soil used in the large reactor aging experiment was spiked to a concentration of about 8000 mg/kg in dry soil. Sampling the soil prior to the degradation experiment, showed that after 4 months of storage, the initial concentration had dropped to 2500 mg/kg. This was deemed acceptable based on all the handling and storage losses, mainly through volatilization, which was reduced due to the refrigerated storage. After sampling the soil, it was immediately stimulated with ammonium chloride and placed in the reactor, where bioventing took place at room temperature.

Figure 1 shows the aging data collected by Mosco and Zytner (2017). Two reactors were run in succession, with the data sets showing good agreement. Figure 1 shows that there is a two stage degradation process; Stage 1 - Day 0 to 11 with solid circles and open triangles; Stage 2 – Day 11 to 34 with open circles and solid triangles. However, the second stage was not as predominant as the non-aged experiments completed by Mosco and Zytner (2017). The difference in second stage degradation was attributed to the fact that the aged soil experiments needed to acclimate themselves first as nutrients were added to the soil just prior to placement into the reactor. For the non-aged soil, the nutrients were added 5 days prior to gasoline spiking. For the aged soil it was assumed that the 4 months of storage in the refrigerator would allow for sufficient acclimation. After the 4 d, the stimulated soil microorganisms dissolved the readily available gasoline in the soil water and the degradation pattern was similar to the non-aged experiments.

The first order degradation rate for the data given in Figure 1 averaged to 0.053 d^{-1} for both experiments. This rate is approximately half the degradation rate of the freshly contaminated Delhi loamy sand, which had a degradation rate of approximately 0.118 d^{-1}. The difference in degradation rates can be attributed to the increased bonds between the gasoline and the soil particles (Jain, 1992). This increased bond strength becomes a limiting factor for biodegradation as microorganisms are unable to easily breakdown the petroleum hydrocarbons. Several studies have been completed indicating that a strong bond exists between soil particles and petroleum hydrocarbons, which can have a limiting factor on biodegradation (Chang et al., 2011 and Sutton et al., 2013). In freshly contaminated soil, the bonds between the soil particles and the petroleum hydrocarbons are weaker and can be dissolved back into the soil water when a concentration gradient exists. Not only does this make the contamination harder to bio-remediate but impacts the solvent extraction efficiency. Studies have shown that extraction efficiency can drop to almost 50% for some types of petroleum hydrocarbons after a period of 28 days (Jain et al., 1992). Aged contaminants require the soil microorganisms to first breakdown the petroleum hydrocarbons from the soil particles by using the microorganisms' bio-surfactants. Once the petroleum hydrocarbons are removed from the soil, it can be continuously biodegraded by the soil microorganisms.

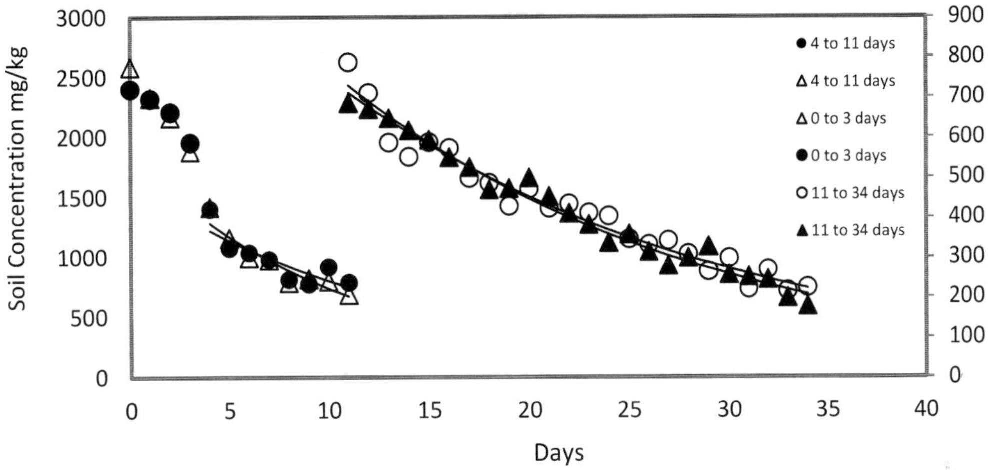

Figure 1: Degradation Trend of 80 kg Reactor – Delhi Loamy Sand

The degradation trend of the wet-aged loamy sand soil that was stored for approximately 10 months was seen to be a single-stage degradation rate. The first order degradation rate was measured at 0.129 d^{-1} in the one reactor that was run. This rate is statistically similar to the degradation rate of the freshly contaminated soil, at 0.118 d^{-1}, and twice that of the 4 month dry aging test. The increase in degradation rate for the wet aging was attributed to the difference in aging process, where wet-aging allowed for the microbial communities to better acclimatize to the contaminants present in the soil, providing for improved biodegradation of the petroleum. The microorganisms were not dormant like in the dry aged soil and were able to start breaking down the contamination once stimulated by the nutrients.

Review of the results for the Elora silty loam soil from the 80 kg reactor showed no measureable signs of degradation. This suggests that the long term aging increased the bond to such a state that the microorganisms were unable to access the synthetic hydrocarbons. Based on the trends observed for both soils, further study into the aging condition is needed. Why does dry aging have such an impact on the degradation rate? What are the governing soil and chemical interactions? The level of spiking may also have an impact. The wet-aging condition was spiked at a lower level and research is needed to determine if this was a governing condition.

As expected, the microbial testing showed that the stimulated heterotrophic and petroleum-degrading bacteria thrived in both soil environments as there was a constant food source such as gasoline. Overall, microbial work was completed to confirm the presence of these microbial species in the soil and that they had increased.

3.2 Respirometers

Respirometer studies were first completed with the loamy sand. Respirometer experiments with 150 g of soil that were exposed to different aging intervals were completed to determine if there was a difference in the degradation trend when compared to the 80 kg reactor. The initial spiking was at 4000 mg/kg, and then over the course of 10 months, bottles were sacrificed to provide the aged soil for the experiment without disturbing the aging process for the rest of the batch. Results showed that the starting concentration (post aging) decreased from around 2000 mg/kg, which account for the losses due to mixing and handling of the soil, to less than 100mg/kg after 9 months.

Figure 2 shows that there a general increase in the measured degradation rates with an increase in contaminant age for the loamy sand soil for wet aging. The degradation rates increase from 0.079 d^{-1} to 0.171 d^{-1} over the course of 10 months. The increase in degradation rate is likely due to the acclimatization of the bacteria during the aging process, creating colonies that thrive better in TPH-rich soil and that are better at the desorption of TPH once stimulated by nutrients. The presence of water also made it easier for the microorganisms to access the contamination. Taking the slope of the line in Figure 2 allows the calculation of the overall degradation for the wet-aged soil. The overall degradation rate was 0.146 d^{-1} at the 150 g scale, which is statistically similar to the 0.129 d^{-1} measured in the 80 kg reactor for the loamy sand. The equal degradation rate, respirometer vs 80 kg reactor for similarity aged soils is different from that found in freshly contaminated soil, where the larger reactors had a degradation rate that is approximately twice that of the respirometer (Mosco and Zytner, 2017).

Respirometer tests were also done with the Elora silty loam soil. As in the 80 kg reactor, no measureable signs of degradation were seen. This suggests that the long term aging increased the bond to such a state that the microorganisms were unable to get any access to the synthetic hydrocarbons.

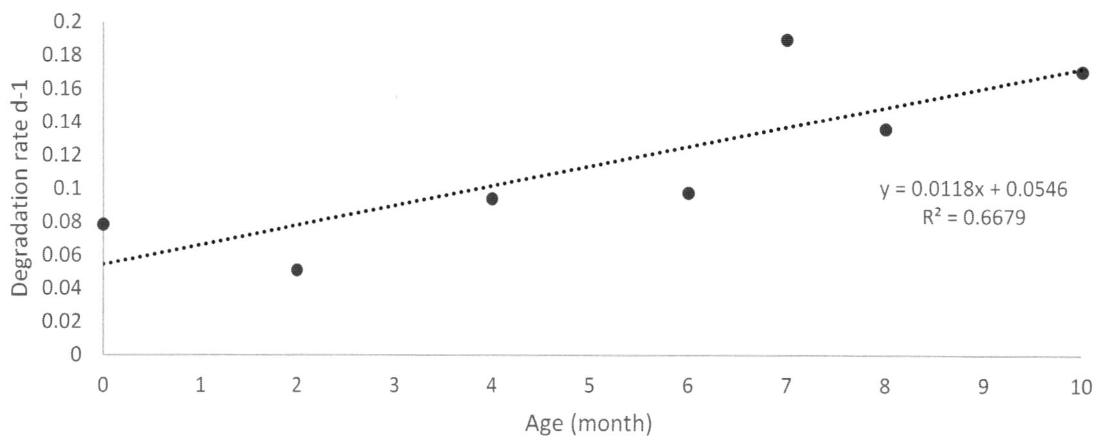

Figure 2: Degradation Rate Increase in 150g Reactor – Delhi Loamy Sand

The change in behaviour in degradation rates between fresh and aged soil highlights the differences between dry and wet-aging. The dry soil was air dried, while the wet-aged soil was at a 15% water content. At the higher water content aged soil there is an increase in degradation rate from both the 150 g experiments and at 80 kg, whereas the dry-aged soil showed a decrease in the degradation rate. It shows that the presence of water allows for better bacterial acclimatization and improved degradation. This behaviour needs to be investigated further for additional soils and conditions.

3.3 Application Implications

The remediation implications of this research suggest that spills in dry soils will take longer to remediate if they are allowed to age for an extended time period. First the microorganisms will have to be reactivated from the dormant state by adding sufficient water, and then stimulated with nutrients to ensure that they degrade the spilled contamination. Conversely, it can be expected that an aged petroleum hydrocarbon spill in a wet unsaturated soil will remediate faster once the appropriate nutrients are added.

The type of soil also plays a role in remediation. It was observed that the loamy sand has a faster degradation rate than the silty loam. This is consistent with the findings of Eyvazi and Zytner (2009) and Khan and Zytner (2015) where the soils with a higher fraction of fine particles are more difficult to remediate. The main cause would be increased sorption, which impacts remediation for both fresh and aged spills. The result is a remediation process that would take longer to reach closure.

4 CONCLUSIONS

Bioventing experiments were conducted with a 150 g respirometer and an 80 kg bioventing reactor by spiking two types of soil four to ten months in advance and storing the mixture in a refrigerated sealed container. Additional experiments were conducted by spiking the soil and letting it age at different intervals to determine the time period where aging had an impact on degradation. The respirometers showed an increase in degradation rate for the loamy sand, from 0.079 d^{-1} for freshly contaminated soil to 0.171 d^{-1} for the wet aged soil that was aged up to 300 days. It was also seen that the degradation rate from the respirometers was statically similar to the degradation rates from the 80 kg reactor, suggesting no scale-up factor under wet aging conditions. The decrease in bioavailability was further seen by the silt loam soil test where no degradation was measured after 250 days of aging in both the respirometers and 80 kg reactors.

The results suggest that water content of the soil during aging is a factor. Dry soil aging minimized microbial acclimatization and the microorganisms most likely went dormant. Thus when remediation began, the microbes first had to be reactivated when stimulated. The result being a degradation rate for the dry aged soil that was half of the freshly contaminated reactor test after only four months of aging; 0.053 d^{-1} as compared to 0.118 d^{-1}. In all cases, a sharp reduction in extractable contaminants was observed due to an increase in aging. The tests and analysis showed that the petroleum contamination was sufficiently tightly bound that microorganisms could not access the petroleum hydrocarbon. These results show that aging impacts the bioavailability and the rates of biodegradation, and the complexities and differences between different aging methods, ultimately increasing the time required to attain site closure in large-scale experiments.

Acknowledgements

The authors wish to acknowledge the support of the NSERC of Canada in the form of a Discovery Grant.

References

Chien, H. Y., Kao, C. M., Liu, J. K., Takagi, K., & Surampalli, R. Y. (2010) Clean up of Petroleum-Hydrocarbon Contaminated Soils Using Enhanced Bioremediation System : Laboratory. Journal of Environmental Engineering, ASCE 136(6):597–606.

Eyvazi, M. J., & Zytner, R. G. (2009) A Correlation to Estimate the Bioventing Degradation Rate Constant. Bioremediation Journal, 13(3):141–153.

Frutos, F.J.G., Escolano, O., Garcia, S., Babin, M. and Fernandez, M.D. (2010) Bioventing remediation and ecotoxicity evaluation of phenanthrene-contaminated soil, Journal of Hazardous Materials, 183:806–813.

Jain, D.K., Lee, H. and Trevors. J.T. (1992) Effect of Addition of Pseudomonas Aeruginosa UG2 Inocula of Biosurfactants on Biodegradation of Selected Hydrocarbons in Soil. Journal of Industrial Microbiology, 10:87-93.

Kao, C. M., Chien, H. Y., Surampalli, R. Y., Chien, C. C., and Chen, C. Y. (2010) Assessing of Natural Attenuation and Intrinsic Bioremediation Rates at a Petroleum-Hydrocarbon Spill Site : Laboratory and Field Studies. Journal of Environmental Engineering. 136(1):54–67.

Khan, A.A., Feng, Z. and Zytner, R.G. (2015) Correlations and Scale-up Factor to Estimate Petroleum Biodegradation Rate in Soil, Bioremediation Journal, 19:32-46.
Mosco, M. & Zytner, R. G. (2017) Large Scale Bioventing Degradation Rates of Petroleum Hydrocarbons and Determination of Scale-up Factors. Bioremediation Journal, 21(3-4):149-162. doi: 10.1080/10889868.2017.1312265.
OAGC (2012) Office of the Auditor General of Canada Spring Report of the Commissioner of the Environment and Sustainable Development. Extracted on September 15, 2014, from: http://www.oag-bvg.gc.ca/internet/English/parl_cesd_201205_03_e_36775.html#ex9.
Sutton, N.B., van Gaans, P., Langenhoff, A.A.M., Maphosa, F., Smidt, H., Grotenhuis, T. and Rijnaarts, H.H.M. (2013) Biodegradation of aged diesel in diverse soil matrixes: impact of environmental conditions and bioavailability on microbial remediation capacity, Biodegradation, 24:487–498.
TBCS (2015) Treasury Board of Canada Secretariat, Find Sites by Classification. Accessed June 12, 2015, from: http://www.tbs-sct.gc.ca/fcsi-rscf/classification-eng.aspx.
Zytner, R.G., Salb, A., Brook, T., Leunissen, M. and Stiver, W.H. (2001) Bioremediation Of Diesel Fuel Contaminated Soil, Canadian Journal of Civil Engineering, 28:131-140.

Building Tomorrow's Society
Bâtir la Société de Demain

Fredericton, Canada
June 13 to 16, 2018 / *13 au 16 juin, 2018*

IMPLICATIONS OF CLIMATE CHANGE TO COASTAL CITIES AND THE NEED FOR ENGINEERS

McBean, Edward[1,2], FCAE, FCSCE, FEIC, D.WRE
[1] University of Guelph, Canada
[2] emcbean@uoguelph.ca

Abstract: Issues influencing the sustainability of coastal cities are characterized, reflecting the combination of impending sea level rise and storm surges, and increasing growth in the world's populations in coastal cities. Geologic-time scales are utilized to draw parallels to characterize relevant historical occurrences that help to understand the context of projections of impending sea level rise issue to year 2100. Given that Antarctica holds sufficient water to raise global sea levels by 58 m if the ice were to melt, this indicates that even a small percentage of melting of the polar ice caps, should this occur, will have enormous implications to the sustainability of coastal cities which are projected to hold 12.4 percent of the world's population by 2060. The result is the combination of predicted sea level rise and associated storm surges indicate that drastic measures need to be promoted to improve the sustainability of coastal cities. The need for engineers, and engineering innovation, is going to greatly increase as the world attempts to address the challenges of sea level rise and storm surges. As a specific example of the storm surge effects, Biloxi, Mississippi during Hurricane Katrina experienced a high water mark of 11.4 m.

1 INTRODUCTION

The impacts of climate change are wide-ranging, including notable increases in the intensity and frequency of storms, floods and forest fires, to name but a few. Evidence of devastation is particularly noteworthy from recent examples such as Hurricanes Katrina (New Orleans in 2005), Harvey (Houston in 2017), and Irma (Florida in 2017), recent flood damages in southern Alberta (2013) and Toronto (2013), and fires in Slave Lake (2011) and Fort McMurray (2016). Widespread evidence points to intensification of issues arising from climate change. The magnitudes of these types of events need to be recognized in engineering education, and the profession needs to be encouraging new thinking processes and approaches in engineering practice, not just rebuilding what has been destroyed in recent events.

2 DIMENSIONS OF CLIMATE CHANGE OF INCREASING DIMENSION

To appreciate the need for increasing focus on the impact of climate change in engineering education and professional practice, it is useful to review the increasingly well-established technical bases which provide important insights to the types of events that are proving especially harmful to coastal cities. The intent in this article is to characterize what is becoming apparent with respect to three dimensions: (i) expected sea level rise over the next century, (ii) the intensification of storm surge impacts, and (iii) land subsidence.

2.1 Sea Level Rise

Sea level rise is one of the lines of evidence that supports that the global climate is warming. For example, there is real concern that the Maldives could be significantly under water by 2100 (see https://blogs.scientificamerican.com/news-blog/maldives-drowning-carbon-neutral-by-2009-03-16/). Evidence is now available which is demonstrating changes in sea level that have occurred over geologic time (see Figure 1, building from base information from Tiner (2016)). While there is strong evidence supporting historical, large increases in sea level (120 m over the time period from 14000 years to 6000 years ago (indicated as Time Span A in Figure 1) as a result of the melting of ice as the last ice age receded, and then were much lower rates of change more recently: 2.0 mm/yr from 6000 years to 2000 years ago. There were then rates of 1.3 mm/yr from 2000 years ago until 1900 (after Horton et al., 2013). These relatively recent rates are indicated within Time Span B in Figure 1.

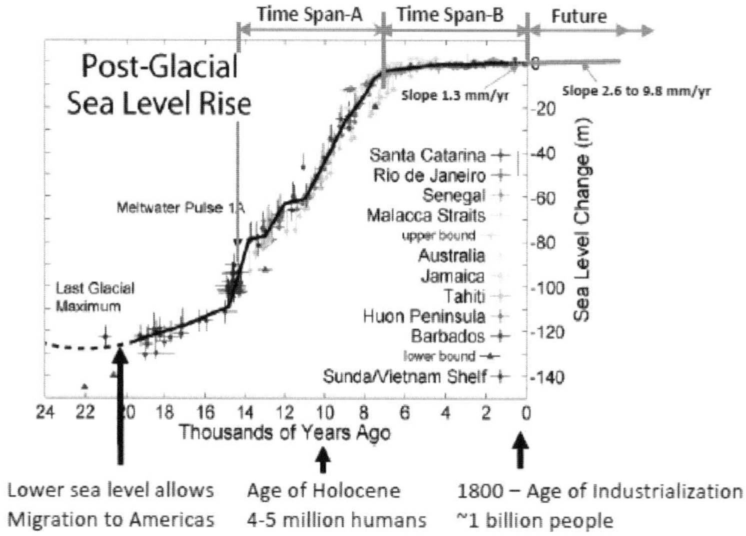

Base Figure Reference: Tiner (2016)
This figure was prepared by Robert A. Rohde from published data, and is incorporated into the Global Warming Art project."
Slopes as added in top right of figure are from Horton et al.,2013 and IPCC, 2013

Figure 1 Post – Glacial Sea Level Rise

Given the expectation that sea level rise over the next century is expected to be between 0.26 m and 0.98 m, there is an interesting comparison between Time Span B and the future. The expected future rates of sea level rise are 2 to 7.5 times the rate of increase per year evidenced over the last century. This prediction into the future includes the effects of both thermal expansion of water as well as the progressive melting of the remaining ice sheets and glaciers (IPCC, 2013).

It is noteworthy that ninety-nine percent of the world's fresh water is in the polar ice caps and the vast majority of this 99 percent is in the Antarctic. The area of Antarctica is 14×10^6 km^2 or approximately twice the size of Australia or the continental US, and is covered on average, by 2750 m (9000 ft) of ice and 150 m (500 ft) of land above current sea level in Antarctica. At the South Pole, ice thickness peaks at 4776 meters (Antarctica Glacier, 2016; http://www.antarcticglaciers.org/antarctica/ten-antarctic-facts/). As a result of these enormous volumes of ice, Antarctica holds sufficient water to raise global sea levels by 58 m if all of the ice melted (Antarctica Glacier, 2016); however, while there is sufficient water in the Antarctic ice cap to raise the sea levels by 58 m, melting of the Antarctic ice cap over the next century is expected to be modest and is incorporated into the above noted increase in sea levels (McBean and Huang, 2017).

2.2 Storm Surges

While the increases in sea level noted above appear initially as relatively modest, one of the most significant dangers to coastal cities are storm surges. To demonstrate the serious magnitude of impacts of storm surges, one only needs to consider evidence from Hurricane Katrina in 2005 which generated an 8.47 m surge (Knabb et al., 2006). When the storm surge magnitude of 8.47 m is combined with the tidal effect plus the storm tide break when the waves reach shallow water, the result in Biloxi, Mississippi was a high water mark of 11.4 m (Fritz et al., 2007). Further, as Tebaldi et al. (2012) indicated, 'Once in a century' storm surges may become 'once in a decade surges'. Clearly, it is not feasible to build walls to protect coastal cities against storm surges.

2.3 Land Subsidence

While tidally-driven coastal flooding is one of the most visible signs of sea level rise, one must also consider the impact of subsidence arising from the extraction of groundwater as an equally challenging problem. The 'pave, pipe and pump' philosophy that has dominated such areas as Houston for over a century has resulted in parts of Houston dropping between 3 and 4 m since the 1920s (Harden, 2016; Galloway et al., 1999).

In Beijing, approximately two-thirds of water supply in recent years for the city of 21 million, has been obtained by overexploiting groundwater at a rate of 10 x 10^8 m^3/yr. The result has been the maximum mean subsidence rate of 0.13 m/yr (Zhou et al., 2016). Further, four hundred of 655 cities in China rely to a large extent on groundwater extraction, with 65 % of municipal water supplies in northern China being derived from groundwater. Examples of subsidence also include Jakarta which has incurred 4 m of subsidence in 35 years (10 to 20 cm/yr). Subsidence now exceeds coastal, absolute sea level rise by a factor of 10 for many coastal cities (Chaussard et al., 2013).

In addition to the subsidence impact, the resulting ingress of salt water into aquifers depleted by overexploitation damages the ability to use groundwater for beneficial purposes.

3 POPULATION GROWTH

Coincident with the challenges that are evolving for coastal regions is the dramatic growth in the world's population now living in coastal low-lying areas. Most of the world's megacities are located in the Low-Elevation Coastal Zone, or LECZ (Brown et al., 2013), and many are situated in large deltas.

Table 1. Populations of LECZ and of the World, with Time (as per Neuman et al., 2015)

Year	Population in the LECZ (in billions)	World Population (in billions)	Percentage of world population living in the LECZ
2000	0.63	6.1	10.3
2030	0.88-0.95	8.7	10.1-10.9
2060	1.4	11.3	12.4

As Neuman et al. (2015) and Merkens et al. (2016) have indicated, the populations residing in the LECZ are predicted to increase by more than half a billion people and reach the proportions listed in Table 1, i.e. 12.4%, of the world's population, is projected to be in the LECZ by 2060.

With increasingly large percentages of the world's population expected to live on, or near, coastlines, and with intensifying impacts arising from climate change (measured by rising ocean levels, stronger hurricanes/typhoons, severe storm surges and land subsidence), there are important implications to coastal cities as a result of their location.

4 COASTAL CITY IMPACTS

The top 20 cities (of 136 port cities around the world that have more than one million inhabitants) most exposed to coastal flooding, are indicated in Figure 2. Logic dictates that the economic threats to these coastal cities are enormous (population levels and level of commerce and trade activity) from rising seas, increased frequency, and subsidence (in cities which are prone to subsidence). It is also noted that Montreal is 101st on the list of cities studied (after Hallegatte et al., 2013). Given both Vancouver's ranking on this list (13th on the list), as well as the substantial international roles of Canadian engineers, the implications of climate change need to be captured in our awareness (McBean, 2017).

Figure 2. The Top 20 Cities Exposed to Coastal Flooding (after Hallegatte et al., 2013)

Sea level rise predictions indicate that a 400 mm (15 in) rise would result in 7 to 10 million refugees in Bangladesh alone, and a rise of 200 mm (8 in) would create 740 thousand climate refugees in Nigeria (IPCC, 2013). Further, the most susceptible countries in terms of populations at risk include China, India, Bangladesh, Indonesia, Viet Nam, Egypt, Nigeria, and The Netherlands. Bangladesh has 40% of its population in the LECZ, and The Netherlands has 73% of its population in the LECZ (Neuman et al. 2015). Clearly, the potentials for climate refugees are not restricted to desertification and economic opportunities.

Responses to these issues are challenging, both in general, and in specifics. Clearly, walls cannot be built high enough to protect coastal cities, and the engineering profession needs to be considering an array of options. A few of the possibilities include:

i. Opportunities to stop or decrease the impacts: better management of subsidence must be addressed (for susceptible cities); re-establish lost wetlands as a friction barrier to protect coastal cities;

ii. Relocating some activities: Land use planning to reduce exposure, including focusing of new development away from the floodplain and preserving space for future infrastructure development; selective relocation needs to be more actively pursued, away from existing city areas to reduce exposure more rapidly than is possible by only focussing on new development;

iii. Making some aspects of coastal cities more robust against failure: since hard infrastructure is only part of the answer, make designs more resilient both with respect to over-topping and when conveyance capacity is insufficient to handle the hydraulic loads; energy grids in North America are fragile and need to become more robust; building regulations need to be upgraded (e.g. flood-proofing) and/or building retrofitting;

iv. Developing more protection and awareness: developing early warning systems and evacuation plans; increasing reliance upon renewable energy; increasing protection through risk-sharing using insurance and re-insurance needs to be expanded.

This list of options needs to continue to be expanded, to encourage innovative thinking.

5 CONCLUSIONS

With the many dimensions as clearly evidenced in the preceding, increased global warming is both obvious and ominous. There is no question that sea levels have risen, and this will continue. The projected increases in sea levels over the next century are expected to be between 0.26 m and 0.98 m from current conditions. However, the implications to coastal cities are much more than sea level rise alone, with storm surges, wave breaking and subsidence, and increased 'populations at risk', all of which are of much greater importance.

Enhanced engineering options need to be considered now, not as just reconstruction of what has been destroyed in the 'last event'. There is no planet 'B', we need to encourage new thinking processes and approaches in engineering practice.

Acknowledgements

Funding from NSERC Discovery and the University of Guelph Leadership Chair are gratefully acknowledged, as contributing to the development of this paper.

References

Brown, S., Nicholls, R., Woodroffe, C., Hanson, S., Hinkel, J., Kebede, A.S., 2013, "Sea-Level Rise Impacts and Responses: A Global Perspective". In: Finkl, C.W., editor. Coastal Hazards. Netherlands: Springer; 2013. pp. 117–149. https://doi.org/10.1007/978-94-007-5234-4_5

Chaussard, E., Amelung, F., Abidin, H., Sang-Hoon Hong, S-H., 2013, "Sinking cities in Indonesia: ALOS PALSAR detects rapid subsidence due to groundwater and gas extraction", Remote Sensing of Environment 128 (2013) 150–161

Fritz, H.M., Blount, C., Sokoloski, R., Singleton, J., Fuggle, A., McAdoo, B.G., Moore, A., Grass, C., and Tate, B., 2007, "Hurricane Katrina Storm Surge Distribution and Field Observations on the Mississippi Barrier Islands", Science Direct, Estuarine, Coastal and Shelf Science 74, pp. 12-20

Galloway, D., Jones, D.R., and Ingebritsen, S.E., 1999, "Land Subsidence in the United States", U.S. Geological Survey Circular 1182, U.S. Geological Survey, 1999, https://pubs.usgs.gov/circ/circ1182/

Hallegatte, S., Green, C., Nicholls, R.J., and Corfee-Morlot, J., 2013, "Future flood losses in major coastal cities", Nature Climate Change 3, 802–806 (2013), doi:10.1038/nclimate1979

Harden, J.D., 2016, For years, the Houston area has been losing ground, May 28, 2016, Houston Chronicle, http://www.houstonchronicle.com/news/houston-texas/houston/article/For-years-the-Houston-area-has-been-losing-ground-7951625.php

Horton, B.P., Engelhart, S.E., Hill, D.F., Kemp, A.C., Nikitina, D., Miller, K.G. and Peltier, W.R., 2013, Influence of tidal-range change and sediment compaction on Holocene relative sea-level change in New Jersey, USA, Journal of Quaternary Science (2013) 28(4) 403–411

IPCC, 2013: "Climate Change 2013: The Physical Science Basis. Contribution of Working Group I to the Fifth Assessment Report of the Intergovernmental Panel on Climate Change", [Stocker, T.F., D. Qin, G.-K. Plattner, M. Tignor, S.K. Allen, J. Boschung, A. Nauels, Y. Xia, V. Bex and P.M. Midgley (eds.)]. Cambridge University Press, Cambridge, United Kingdom and New York, NY, USA, 1535 pp. https://doi.org/10.1017/CBO9781107415324

Knabb, R.D., Rhome, L.R., Brown, D.P. 2006. "Tropical Cyclone Report, Hurricane Katrina", 23–30 August 2005., Published on the Web at http://www.nhc.noaa.gov/pdf/TCRAL122005_Katrina.pdf

McBean, E., and Huang, J., 2017. "Sustainability Risks for Coastal Cities from Climate Change", The Global Environmental Engineers, Volume 4, No. 1, pp. 1-9.

McBean, E., 2017. "Water Security, The Nexus of Water, Food, Population Growth, and Energy", The Global Environmental Engineers Journal, Vol. 3, 33-39 E-ISSN 2410-3624/16.
Neumann, B., Vafeidis, A.T., Zimmermann, J. and Nicholls, R. J., in "Future Coastal Population Growth and Exposure to Sea-Level Rise and Coastal Flooding – a Global Assessment", PLoS One, editor Lalit Kumar, 2015, 10(3), e0118571, Published Online 2015, Mar 11. Doi: 10.137/journal pane 0118571, PMCID:PMC4367969.
Nicholls, R.J., Hanson, S., Herweijer, C., Patmore, N., Hallegatte, S., Corfee-Morlot, N., Château, J., and Muir-Wood, R., 2008. "Ranking Port Cities with High Exposure and Vulnerability to Climate Extremes", *OECD Environment Working Papers*, No. 1, Paris, http://dx.doi.org/10.1787/011766488208
Tebaldi, C., Strauss, B., and Zervas, C., 2012, "Modeling of Sea Level Rise Impacts on Storm Surges Along US Coasts", Environmental Research Letters, 7(1), 014032,https://doi.org/10.1088/1748-9326/7/1/014032
Tiner, R.W., Changing Sea levels During the Past 25,000 Years An Introduction to Changes due to Glaciation and Current Conditions, U.S. Fish & Wildlife Service, https://www.fws.gov/slamm/Changes%20in%20Sea%20Level_expanded%20version_template.pdf (accessed Nov 8, 2016).
Zhou, C., Gong, H., Chen, B., Zhu, F., Duan, G., Gao, M., and Lu, W., 2016. "Land subsidence under different land use in the eastern Beijing plain, China 2005-2013 revealed by InSAR timeseries analysis", GIScience and Remote Sensing, Vol. 53, No. 6, 671-688.

Building Tomorrow's Society
Bâtir la Société de Demain

Fredericton, Canada
June 13 to 16, 2018 / *13 au 16 juin, 2018*

IMPROVING RESILIENCE OF URBAN DRAINAGE SYSTEM IN ADAPTATION TO CLIMATE CHANGE (CASE STUDY: NORTHERN TEHRAN, IRAN)

Binesh, Negin[1], Niksokhan, Mohammad Hossein[2], Sarang, Amin[3,5], and Rauch, Wolfgang[4]
[1] PhD student, Department of Environmental Engineering, University of Tehran, Tehran, Iran
[2] Associate professor, Department of Environmental Engineering, University of Tehran, Tehran, Iran
[3] Assistant professor, Department of Environmental Engineering, University of Tehran, Tehran, Iran
[4] Full professor, Unit of Environmental Engineering, University of Innsbruck, 6020 Innsbruck, Austria
[5] sarang@ut.ac.ir & amin.sarang@gmail.com

Abstract: Given the adverse impacts of climate change on urban water infrastructures' resilience, using adaptive strategies is of great importance. For urban drainage systems (UDSs), despite considerable progress in sustainable management of such infrastructures all over the world, they are still exposed to uncertain drivers of future changes. Quantifying the resilience is an important step for long-term increase in resilience of UDSs. In current paper, the resilience of UDS against flooding is conducted through combining hydraulic performance of storm water drainage network with the system performance function. To do this, baseline and future extreme rainfall time series for different return periods were calculated for West Flood-Diversion (WFD) catchment in Tehran, Iran, and then related IDF curves were obtained. Afterwards, hydraulic and hydrodynamic simulation of storm water drainage system was performed for future time under two scenarios: with and without BMPs. This methodology was applied on one flooded node in the network where the amount of system performance and resilience were estimated under extreme rainfall with different return periods. The results showed that adding a vegetative swale to the drainage system would decrease nodal flooding duration and flood volume up to 30% and 35%, respectively. Moreover, resilience against flooding can increase up to 9.12%. Therefore, the existence of low-impact development measures along with traditional urban drainage can be efficient for increasing system sustainability.

Keywords: Adaptive strategies, climate change, performance function, resilience, urban drainage system (UDS).

1 INTRODUCTION

Recent changes in global climatic conditions, has been followed by increase in extreme rainfall events. This phenomenon along with malfunction of Urban Drainage Systems (UDSs) and lack of capacity due to sedimentation, accumulation of garbage in minor drainage network, entrance of swage in separated drainage networks, etc., is the source of urban flooding and inundation problems (Lee and Kim 2017). Therefore, in recent years, building resilience of UDSs has been taken into consideration as an important approach in reducing floods and their consequences under uncertain future changes (Blockley et al. 2012, Butler and Davies 2011, Djordjević et al. 2011, Gersonius et al. 2013). According to Todini (2000), resilience is referred to the ability of a system to return to the normal level of service, so that it could meet the pre-defined goals of the system after the failure occurs. In other words, resilience is the ability of the system to restore function following the system failure. Therefore, resilience index is determined based on system failure duration (Nazif 2010). In current paper, resilince is defined as: the ability of the UDS for minimising the volume and duration of flooding caused by extreme rainfall events (Mugume et al. 2014). Herein, flooding means the excess runoff volume which overflows from the UDS to the surrounding urban area.

In recent decades, considerable progress have been achieved in context of resilience concept and its quantification in different urban water systems (Jung et al. 2013, Lansey, 2012). However, few studies have been conducted on developing appropriate methods for evaluation and quantification of resilience in storm water drainage infrastructures. Though, the subject is getting more and more interesting for researchers, so that the number of such studies is increasing in recent years. De Bruijn (2004) proposed some indicators aiming at defining resilience of flood risk management systems. The indicators cover three aspects of resilience concept including amplitude, graduality- trend of increasing reaction by increase in severity of flood waves-, and recovery rate. After applying the proposed indicators on simple hypothetical systems, she came to the conclusion that a combination of these indicators can be effective in achieving an overall view of resilience of flood risk management systems. A study performed by Mugume et al. (2014) represents that by increasing rainfall return period, flood duration increases in residential areas of England, and consequently, much reduction occurs in performance of urban stormwater drainage infrastructure. They also showed that urban-drainage performance reduction curve did not change for high return periods (50 and 100 years) and coincided each other. Tahmasebi Birgani (2014) introduced a tool box for quantifying resilience concept in urban-drainage risk management systems (UDRMS). Zhang et al. (2017) compared the vulnerability of tree vs. loop drainage systems under different rainfalls and investigated structural failure due to pipe blockage in such networks. They introduced an index for measuring vulnerability of drainage system before and after applying adaptive measures. The obtained results represented that tree networks possesses higher percent of pipes with critical hydraulic condition compared to loop systems. Therefore, vulnerability of tree systems is significantly greater than loop networks. They also concluded that vulnerability index of tree systems decreases after changing them into loop systems and using adaptive measures.

One of the most solutions that, nowadays, is popular for increasing resilience of UDSs, includes using BMPs for controlling runoff quantity and quality (Jia et al. 2012). Many researchers have considered BMPs in some case studies in order to see the effect on reducing stormwater volume and peak discharge (Åstebøl et al. 2004, Villarreal et al. 2004, Chang and Liou 2010, Stovin 2010, Pyke et al. 2011, Jia et al. 2012). The results are often representative of significant effect of using such management practices.

Despite a large number of studies of which some of them were mentioned, only few studies have examined the performance and resilience of UDS in facing extreme rainfalls. Since urban drainage systems are designed for rainfalls with specific return periods, would be subject to reduced performance in case of extreme rainfall occurrence and in facing phenomenon such as climate change and urban development. Therefore it is necessary that quantifying resilience of such systems in face of extreme events. As mentioned in the literature review above, a study have been conducted by Mugume et al. (2014) in which resilience of an UDS system in England has been examined under higher return periods. In current study it is intended to go further and see the impact of future climatic conditions on a drainage network for a basin in Iran. The effect of using management practices in adapting to climate change will be also investigated. In the proposed methodology, performance of UDS is measured based on hydrologic and hydrodynamic simulations, using flood depth-damage curve and performance function of drainage infrastructure, in two states: with and without using BMPs. At the end, resilience of the system against flooding is quantified.

2 MATERIALS AND METHODS

In figure 1, the methodology and steps for quantifying resilience of UDS is presented. In the following sections, the details of each stage are explained and then, the case study is introduced.

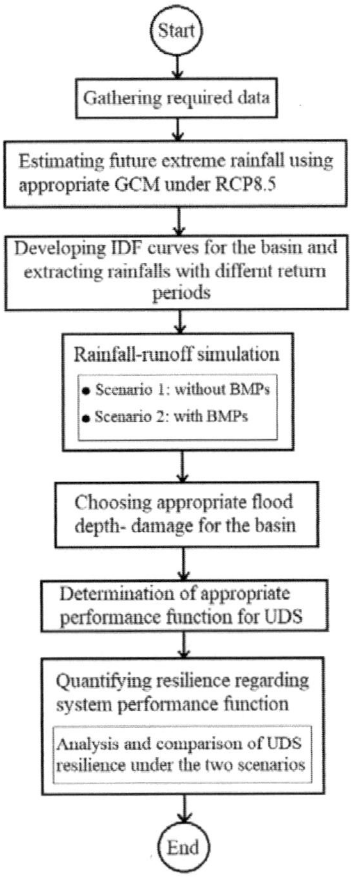

Figure 1: The flowchart of quantifying resilience of urban drainage network

2.1 Input Data to the Model

As mentioned earlier, the goal of current research is to estimate the performance and resilience of UDS under future conditions. According to the literature, climate change impacts mainly emerges as increase in extreme precipitation by increase in rainfall frequency and intensity. Therefore, in current research, in order to consider the impacts of climatic changes on increase of extreme rainfall intensity, future extreme rainfall time series were obtained using the output of General Circulation Models (GCMs), downscaled through Change Factor (CF) method. The reason why downscaling was performed through CF method is its simplicity and ease of use. However, since in current study we deal with extreme precipitation for estimating system flood resilience, an advance version of CF method- in which the changes in extremes, and not just in mean values, are taken into account- was considered (Ruiter, 2012).

After examining different models of 5th Assessment Report (AR5), MRI-CGCM3 was chosen as the best climatic model for simulating future rainfall in the study area (Binesh et al. 2016). Historical data of precipitation as the base of calculations were extracted from two stations: Mehrabad synoptic station in the south and Darakeh hydrometer station in the north, to be able to cover the climatic condition of the whole catchment. The other reason why these two stations were chosen is that they are the only active ones in the basin which included enough data for the purpose of this study. Average precipitation of the area was calculated based on the elevation of each station and daily rainfall recorded in the two mentioned stations, and considering the average elevation of the whole catchment. The characteristics of these two stations has been presented in table 1 according to information obtained from Iranian Water Resources Company and Iranian Meteorological Organization.

Table 1: Characteristics of two stations used in this study

Station Name	Type of the station	Latitude	Longitude	Height (MASL)
Haft-Howz (Darakeh)	Hydrometric	35 °48' N	51°23' E	1700
Mehrabad	Synoptic	35 °41' N	51°19' E	1190.8

Baseline period- according to available data and statistical years the two stations had in common- was considered from 1990 to 2006, and future climate was investigated during the distant future (2084-2100) under RCP8.5 (Representative Concentration Pathway 8.5). In order to have a better view on the system performance under critical future conditions, it is necessary to examine the UDS efficiency in face of more intensified rainfalls with higher return periods. To do this, Intensity-Duration-Frequency (IDF) curves for the catchment were obtained for the distant future (figure 2) using daily extreme rainfall events and relationships proposed by Ghahreman and Sepaskhah (1980) and Ghahreman (1996) who represented equations appropriate for the condition of Iran based on relationships presented by Bell (1969).

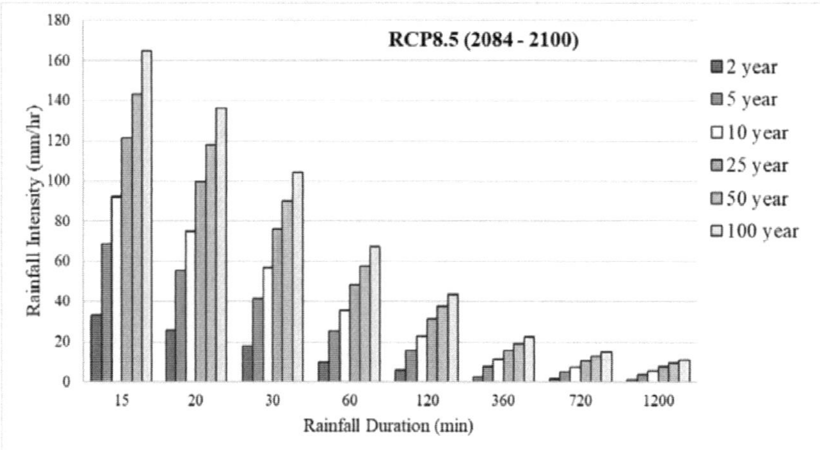

Figure 2: Future IDF curves based on observed data for WFD catchment

2.2 Rainfall-Runoff Simulation

Storm Water Management Model (SWMM) version 5 is used to model UDS. SWMM is a physical model for hydrologic and hydrodynamic simulation, with discrete time step, that can be used for single-event and continuous simulation of storm water quality and quantity in urban areas. In current study, simulating the runoff quantity is considered in a network composed of both open-channels and covered channels, with rectangular and horseshoe-shaped cross-sections, with the width from 1.4 m to 30 m and height of 1.4 meters to 6 meters. System configuration in SWMM software include 73 nodes (apart from the outlet), and 74 links as drainage channels (figure 5). The length of drainage channels and height of the nodes were extracted from DEM of the study area, prepared by Yekom Consulting Engineers (2009). Since curve numbers (CN) of sub-basins were known based on previous studies, infiltration process was simulated through CN method. Flow routing was done through dynamic wave (solving the full Saint-Venant equations), because dynamic wave flow routing is able to consider different types of flow regimes such as backwaters, overflow from the system (i.e. surcharging), reverse flow, and surface ponding (Vanrolleghem et al. 2009). It should be noted that model calibration and sensitivity analysis were done for runoff velocity and depth at the end of the catchment, using data gathered by Mo'afi Rabari (2012).

Since rainfalls in northern Tehran usually last for 6 to 7 hours (Nazif 2010), in current study 6-hour rainfall distribution pattern was used (Mahab-e-Ghods Consulting Engineers 2011) to simulate rainfall-induced runoff in the catchment. In this study, it is assumed that temporal characteristics of extreme rainfall events remains unchanged (Mugume et al. 2013). Rainfall time step was considered 30 min and for output runoff, the time step was considered as 1 hour. Rainfall-runoff modelling was done for a time duration of 3 days.

Simulation was performed under future climatic conditions for two scenarios: 1) for current situation of the system (assuming the system layout and characteristics remain unchanged), and 2) after adding a BMP to the system. In order to investigate the effect of using BMPs on control of nodal flooding, a "Vegetative Swale" of which 50 percent of surface is vegetated was applied as a BMP in all sub-basins. The longitudinal slope is around 1 m, the side slope is 3 m, and the roughness coefficient equals to 0.1. The presence of vegetation in "Vegetated Swale" improves the infiltration of rainwater into the soil. It is also effective in increasing the resilience of the drainage system. The type of herbs used in the swale is mainly adapted to the vegetation of the given area. According to researchers from the University of San Diego, using indigenous plants in this type of BMP rather than a using grass (in Grass Swales) increases the efficiency in terms of reducing runoff rate, as well as removing contaminants and improving runoff quality.

2.3 Flood Depth-Damage and UDS Performance Functions

Due to the lack of recorded flood damage data for different inundation depths in the study area, a depth-damage curve developed by Tajrishi and Malek Mohammadi (2009) for residential areas of Iran (single-storey buildings with and without basement) was used. By linear fitting to the points recorded in this curve (assuming that the variations are linear), the equation for damage versus inundation depth was obtained as follows:

[1] $D_x = 15.369x + 7.1764$

In which x is inundation depth (m) and D_x is the percent of flood damage for the depth equal to x. It should be noted that in Eq. [1], the amount of damage for zero depth is related to the buildings with basement and storeys below the ground floor.

Performance function of the system against flooding is considered as Eq. 2 (Mugume et al. 2014). Performance functions of UDS relate total performance of the system to flood depth. In fact, such functions are mathematical models that relate a given characteristic to a number in range of 0 to 1. The number 1 is representative of perfect performance and zero is related to the least level of system performance (Cardoso et al. 2004, Gharaibeh et al. 2006).

[2] $u(x) = 1 - \frac{D_x}{D_{max}}$

In which is the maximum flood damage that occurs for the maximum inundation depth (considered as 3 m based on previous studies). According to this function, the best system performance happens when inundation depth equals to zero ($x=0$). In such a condition, the value of UDS performance is considered as 1 ($u(x)=1$). The least level of system performance is for the flood depth equals or more than 3 m (Mugume et al. 2014) where UDS performance is considered zero. Figure 3-a shows flood depth-damage curve used for residential areas of Iran, and the figure 3-b is representative of UDS performance function based on the given depth-damage curve and Eq. 2. According to the figure, it is understood that UDS performance has a negative correlation with increase of flood depth. Such that flooding with a depth of more than 60 cm causes the system performance to decrease more than 30 percent.

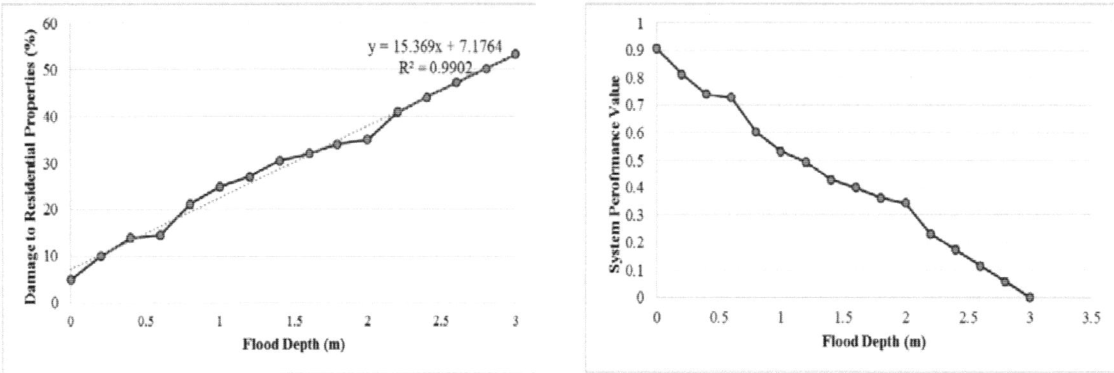

(a) (b)

Figure 3: a) Depth-damage curve for residential areas of Iran and the best-fitting line, b) System performance function values for different depths of urban flooding

2.4 Quantifying the Resilience of UDS

According to figure 4, the concept of resilience could be explained as the following: a system disruption due to an extreme event at the time t_0, causes a reduction in system performance up to ("100" minus "system robustness"). By starting the system recovery, the performance is slowly restored so that returns to the normal level of performance at the time t_r. The upper area of the curve that is formed in such a process for the considered time period, indicates the level of non-resilience of the system (Bocchini et al. 2013). The smaller this area is, the more is the resilience of the system. Vice versa, the area below the performance curve represents the level of resilience of the system. Therefore, after estimating UDS performance function at each time step for different flood depths (x_t), system resilience (Res) can be quantified through equation 3 (Mugume et al. 2014):

[3] $Res = \frac{1}{t_n}\int_{t_0}^{t_n} u(t)dt$

In which t_0 is the start time of simulation, and t_n is the total time past at the end of simulation process. *u(t)* is the value of UDS performance function in time steps with 1-hour intervals.

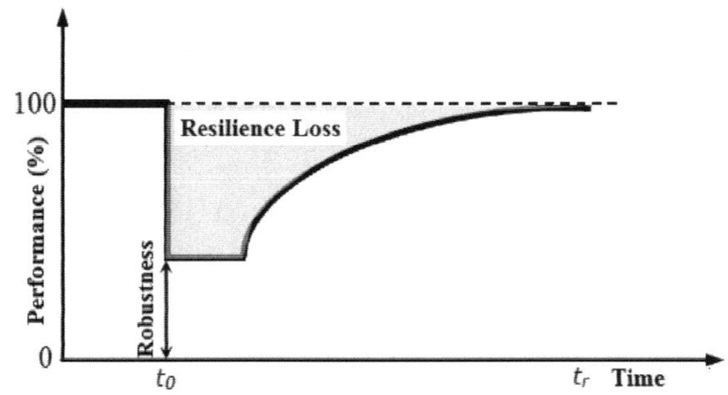

Figure: 4. Schematic of resilience concept (Valizadeh et al. 2016)

3 INTRODUCTION TO THE STUDY AREA

The above methodology was applied on UDS of West Flood-Diversion (WFD) catchment in northern Tehran, Iran, which covers an area of 142.63 km2. This catchment includes three sub-basins namely: Drakeh, Farahzad, and Hesarak (Vesk), and is located in the geographic coordination from 35-48-30 to 35-52-49 N latitude and 51-19-13 to 51-22-29 E longitude. WFD catchment consists of urban and non-urban (mountainous) parts. The storm water runoff from the mountainous areas also incorporates in the urban runoff entering the UDS. Therefore, both urban and sub-urban parts of the catchment were simulated in current study considering the capabilities of SWMM in modelling different types of natural and man-made channels. Figure 5 shows a perspective of urban and mountainous areas as well as the drainage system in SWMM environment.

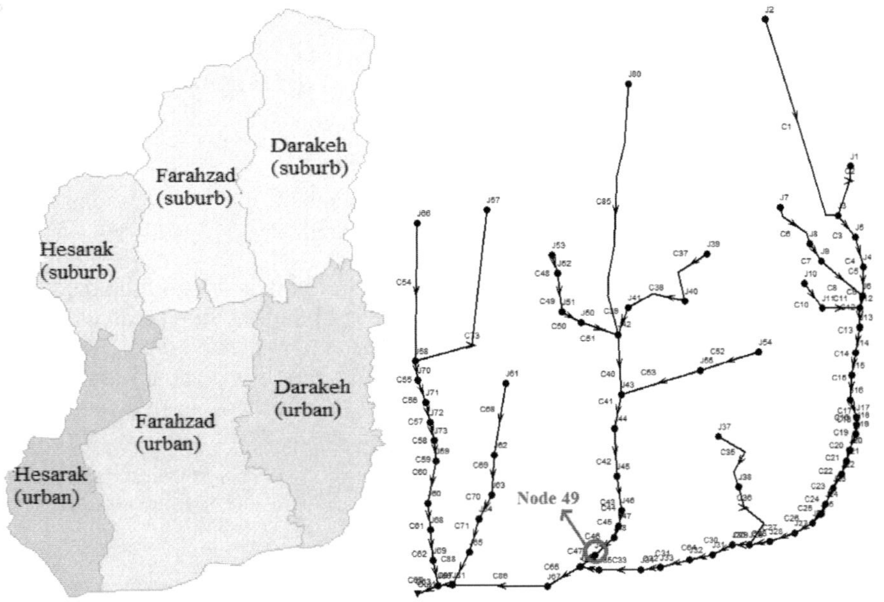

Figure 5: West Flood-Diversion catchment (scale: 1/200000) and modelled nodes and links in SWMM (Source: Authors)

4 RESULTS AND DISCUSSION

4.1 The Results for Hydrologic and Hydrodynamic Simulations

The results of future rainfall-runoff modelling represented that for the first scenario, for the return periods from 2 to 10 years, 11 nodes were inundated. While for 25, 50, and 100-year return periods, the number of flooded nodes were obtained as 16, 19, and 25, respectively. In order to analyse the simulation results in current study, the node number 49 was selected as a sample, whose situation has been shown in figure 5. The reason why we chose this node among the others in the system, is firstly the high volume of flooding, and also being flooded in simulations for all return periods.

Figure 6 shows the system failure time duration and surcharged runoff volume for the two scenarios (with and without BMPs). Based on this figure, it could be said that adding one vegetative swale to each sub-basin for the whole drainage system reduces the average time of nodal flooding between 17-30% and flood volume from 13% to 35%.

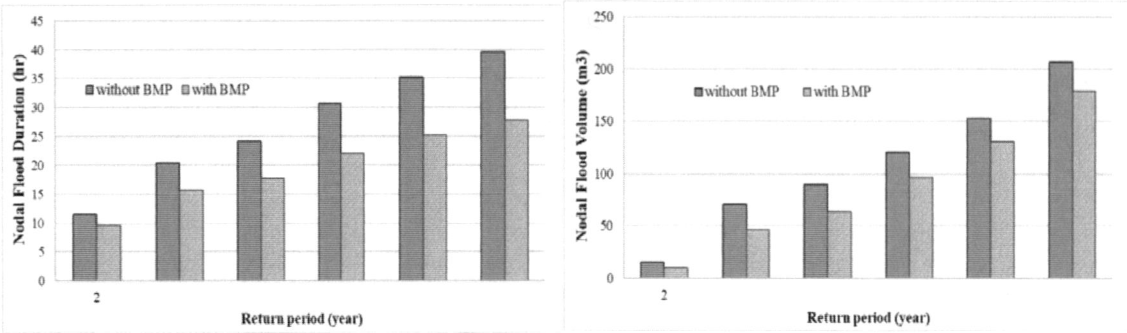

Figure 6: Flood duration and flood total volume for extreme events within different return periods for two conditions: with and without BMPs

Figure 7 represents the curves for the average future flood-depth versus time, for both scenarios: with and without "vegetative swale" alongside the drainage system. It is observed that maximum flood depth for all studied return periods occurs around 5 hours after the beginning of precipitation. The simulation results show that adding one "vegetative swale" to each sub-basin causes a significant reduction of inundation depth and delays the time to peak discharge to some extent.

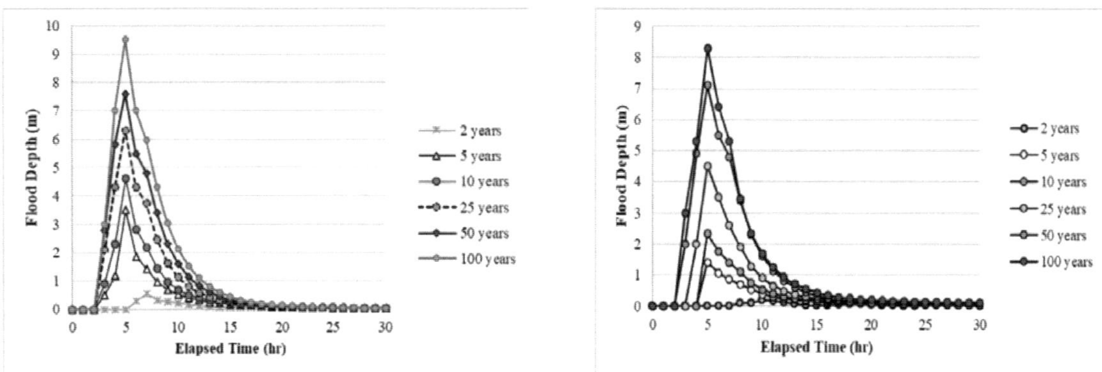

(a) (b)

Figure 7: Depth of nodal flooding for urban drainage system a) without BMP, and b) with BMP

4.2 Estimating the Performance and Resilience of UDS

The value of UDS performance (*u(t)*) in future time was obtained for the two scenarios based on Eq. 2 and inundated depth at each time step. The results are visible in figure 8. It is observed that existence of one BMP at each sub-basin along with the drainage network, has improved the system performance for all studied return periods and reduced the number of failed nodes in the drainage system.

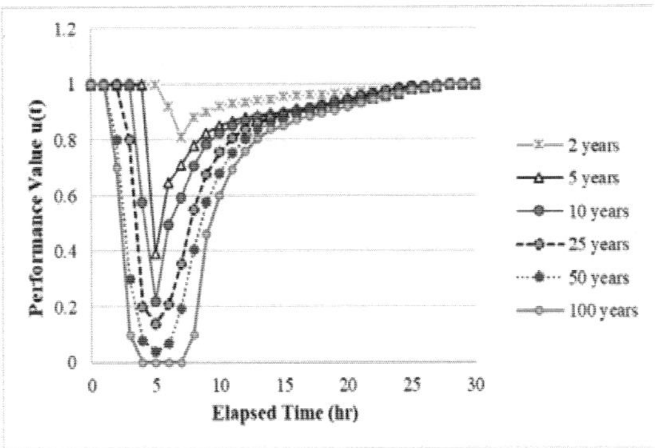

(a)

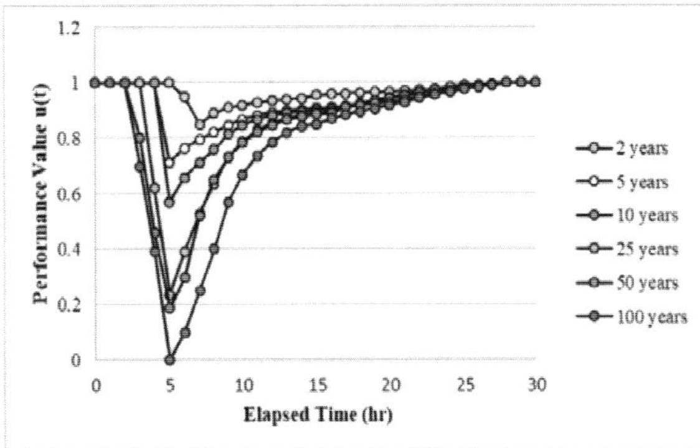

(b)

Figure 8. System performance curves for urban drainage system a) without BMP, and b) with BMP

Based on calculated performance for the drainage network, the amounts of system resilience in future time have been obtained and presented in table 2. According to this table, the range of resilience of UDS, for the without BMP scenario, changes from 96% (2-year return period) to 72.61% (100-year return period). Adding the "vegetated swale" to the system, causes the resilience to fall between 96.37% (2-year return period) and 78.50% (100-year return period). Therefore, the resilience of the system increases between 0.37% to 5.89%, and it could be said that existence of BMPs in sub-basins helps the hydraulic performance of the system return to its original (or acceptable) status in a smaller time duration. Such results agree with the findings by the other researchers (i.e. Mugume et al. 2014) on similar issues.

Table 2: System resilience at given node for different return periods during 2084-2100

Return period (year)	2	5	10	25	50	100
System resilience without BMP (%)	94	87.9	85.32	80.96	75.22	69.38
System resilience with BMP(%)	96.44	92.45	91.24	86.16	83.16	78.50
Improvement of system resilience(%)	2.44	4.55	5.92	5.2	7.94	9.12

5 CONCLUSION

Current research aimed at quantifying resilience of UDS in future time and investigating the impacts of BMP implication as an adaptive strategy on improving system resilience. To this end, flood depth-damage relationship for residential areas and drainage infrastructure performance function were used, and finally, resilience of the UDS was obtained through applying calculated performance values and based on simulation results in SWMM. The results represented that using BMPs along with surface-runoff drainage network can be effective in adapting to future climatic conditions of the basin and in improving the resilience of the system. Such an influence is quite obvious for high rainfall return periods. This means that with increase in extreme rainfall intensity, using management practices is so effective in flood volume reduction in urban areas. However, introducing BMPs to the model may not prevent nodal flooding thoroughly. The degree of flood prevention by such measures is dependent on the number and type of BMPs, their layout in the basin, drainage system configuration (Kim et al. 2013), and type of materials used for BMP construction, and other characteristics.

In order to achieve the best results in reducing runoff peak discharge, it is necessary that BMPs' numbers and layout are determined in an optimized way. As the current study indicated, installing appropriate BMPs in upstream suburban areas can also be effective in improving the system performance and resilience.

In current study, resilience of the system was measured only in one node. However, in order to estimate the resilience of the whole system, it is required to examine and analyze all flooded nodes in the whole system, which is recommended to be considered in future similar studies. In addition, since resilience concept encompasses several indicators such as robustness and recovery rate, it is recommended that ongoing similar researches concentrate on developing new indices for each of these indicators and sub-factors and evaluate the impact of different system failure states on resilience of UDS.

References

Åstebøl, S. O., Hvitved-Jacobsen, T. and Simonsen, Ø. 2004. Sustainable stormwater management at Fornebu—from an airport to an industrial and residential area of the city of Oslo, Norway, Journal of Science of the Total Environment, 334–335: 239-249.

Binesh, N., Niksokhan, M.H. and Sarang, A. 2016. A study of the effect of rainfall variability on Darakeh river flow rate during 1989-2012, Journal of Echohydrology, 3(3): 465-476 (in Persian).

Blockley, D., Agarwal, J. and Godfrey, P. 2012. Infrastructure resilience for high-impact low-chance risks, in: Proceedings of the Institution of Civil Engineers, Civil Engineering Special, 165(6): 13–19.

Butler, D. and Davies, J.W. 2011. Urban Drainage, 3rd edition. Spon Press, Taylor and Francis Group, London and New York.

Bocchini, P., Frangopol, D. M., Ummenhofer, T., and Zinke, T. 2013. Resilience and sustainability of civil infrastructure: Toward a unified approach. Journal of Infrastructure Systems, 20(2), 04014004.

Chang, C.L. and Liou, T.Y. 2010. The placement strategies of structural best management practices for different moving rainstorms, Journal of Environtal Monitoring and Assessment, 166(1-4): 495-502.

Cardoso, M., Coelho, S., Matos, R. and Alegre, H. 2004. Performance assessment of water supply and wastewater systems, Urban Water Journal, 1: 55–67.

De Bruijn, K. M. 2004. Resilience indicators for flood risk management systems of lowland rivers, International Journal of River Basin Management, 2(3): 199–210.

Djordjević, S., Butler, D., Gourbesville, P., Mark, O. and Pasche, E. 2011. New policies to deal with climate change and other drivers impacting on resilience to flooding in urban areas: the CORFU approach, Journal of Environtal Science and Policy, 14: 864–873.

Gersonius, B., Ashley, R., Pathirana, A. and Zevenbergen, C. 2013. Climate change uncertainty: building flexibility into water and flood risk infrastructure, Journal of Climate Change, 116: 411–423.

Gharaibeh, N.G., Asce, M., Chiu, Y. and Gurian, P.L. 2006. Decision methodology for allocating funds across transportation infrastructure assets, Journal of Infrastructure. Systems, 12: 1–9.

Jia, H., Lu, Y., Yu, Sh., and Chen, Y. 2012. Planning of LID–BMPs for urban runoff control: The case of Beijing Olympic Village, Journal of Separation and Purification Technology, 84(9): 112–119.

Jung, D., Kang, D., Kim, J.H. and Lansey, K. 2013. Robustness-based design of water distribution systems, Journal of Water Resources Planning and Management, 140(11): 14pp.

Kim, B., Kim, Y., Sanders, B.F., Famiglietti, J.S. and Han, K. 2013. Calibration of stormwater management model using flood extent data. Proceedings of the Institution of Civil Engineers: Water Management, 167: 17–29.

Lansey, K. 2012. Sustainable, robust, resilient, water distribution systems, 14th Water Distribution Systems Analysis Conference, Engineers Australia, 1–18.

Lee, E.H. and Kim, J.H. 2017. Design and Operation of Decentralized Reservoirs in Urban Drainage Systems, Journal of Water, MDPI, 9(246):14 pp.

Mahab-e-Ghods and Poyry group consulting engineers, 2011, Comprehensive Plan of Tehran's surface water management, Volume III: Basis and regulations of surface water network design and type maps, Technical and Development department of Tehran Municipality and Tehran technical-engineering consulting agency (In Persian).

Moafi Rabari, A. (2012) Optimal design of WFD (West Flood-Diversion) dimensions based on upland catchment' characteristics. Msc. thesis. University of Tehran.
Mugume, S., Gomez, D. and Butler, D. 2013. Statistical methods for climate change impact assessment on urban rainfall extremes for cities in tropical developing countries – A review, in: Butler, D., Chen, A.S., Djordjevic, S. and Hammond, M.J. (Eds.), International Conference on Flood Resilience: Experiences in Asia and Europe. University of Exeter, Exeter, United Kingdom.
Mugume, S., Gomez, D., and Butler, D. 2014. Quantifying the Resilience of Urban Drainage Systems Using a Hydraulic Performance Assessment Approach, 13th International Conference on Urban Drainage, Sarawak, Malaysia.
Nazif, S. 2010. Developing an algorithm for the assessment of climate change impact on urban water cycle. PhD thesis, Faculty of Engineering, University of Tehran, Tehran, Iran (In Persian).
Pyke, C., Warren, M. P., Johnson, T., LaGro Jr, J., Scharfenberg, J., Groth, P., et al. 2011. Assessment of low impact development for managing stormwater with changing precipitation due to climate change, Journal of Landscape Urban Planning, 103(2): 166-173.
Ruiter, A. 2012. Delta-change approach for CMIP5 GCMs, version 3, MSc. Thesis, Department of Physical Geography, UU – Utrecht University, Netherlands.
Stovin, V. 2010. The potential of green roofs to manage Urban Stormwater, Water and Environment Journal, 24(3): 192-199.
Tahmasebi Birgani, Y. 2014. Sustainable management of urban drainage based on resilient risk management, PhD thesis, Faculty of Civil Engineering, Khajeh Nasireddin-e-Tousi University of Technology, Tehran, Iran (In Persian).
Tajrishi, M., and Malekmohammadi, B., 2009, Suitable method to accomplish flood insurance program for crisis management in flood condition of urban areas, Proceedings of the 2nd international conference on integrated national disaster management, Tehran Iran, 1–18.
Todini, E. 2000. Looped water distribution networks design using a resilience index based heuristic approach. Urban Water, 2(2): 115–122.
Valizadeh, N., Zorn, C.R., and Shamseldin, A.Y. 2016. Evaluating the technical resilience of stormwater systems to flooding, Stormwater Conferenc, Australia.
Vanrolleghem, P. A., Kamradt, B., Solvi, A. and Muschalla, D. 2009. Making the best of two Hydrological Flow Routing Models: Nonlinear Outflow-Volume Relationships and Backwater Effects Model, 8th International Conference on Urban Drainage Modelling (8UDM). Tokyo, Japan.
Villarreal, E. L., Semadeni-Davies, A. and Bengtsson, L. 2004. Inner city stormwater control using a combination of best management practices. Journal of Ecological Engineering, 22(4–5): 279-298.
Yekom consulting engineers, 2009, A project on (quantitativa and qualitative) rehabilitation of surface waters in southern Tehran, Volume III (In Persian).
Zhang, C., Wang, Y., Li, Y. and Ding, W. 2017. Vulnerability Analysis of Urban Drainage Systems: Tree vs. Loop Networks, Journal of Sustainability, MDPI, 9(397): 18pp.

Building Tomorrow's Society
Bâtir la Société de Demain

Fredericton, Canada
June 13 to 16, 2018 / *13 au 16 juin, 2018*

IN-SITU REMOVAL OF ALGAE AND SUSPENDED SOLIDS FROM A EUTROPHIC LAKE USING NON-WOVEN GEOTEXTILES

Palakkeel Veetil, Dileep[1], Ghasri, Mina[1], Mulligan, Catherine N.[1,3], and Bhat, Sam[2]
[1] Concordia University, Canada
[2] Titan Environmental Containment Ltd., Canada
[3] mulligan@civil.concordia.ca

Abstract: Lake Caron, a shallow eutrophic lake located in Sainte-Anne-des-Lacs municipality in Quebec, has been the site of algal blooms every summer since 2008. In this trial study, an in-situ floating filtration unit provided with non-woven geotextiles as filter media was tested to treat the water in a small area of the lake enclosed by a floating turbidity curtain. Energy from a solar panel was used to operate the unit. Non-woven geotextiles with different opening sizes were used to capture algae and suspended particles of various sizes. Water quality was monitored inside and outside of the turbidity curtain by examining total suspended solids, turbidity, total phosphorus, COD, total nitrogen and nitrate. Chlorophyll *a* and blue green algae (phycocyanin) concentrations were monitored through out the filtration by deploying a YSI-EXO2 probe. The filters were effective towards algae and suspended particle removals.

Keywords: Eutrophication, Suspended Solids, In-situ filtration, Non-woven geotextiles, Algae

1 INTRODUCTION

As a source of fresh water, lakes are connected to public's health and activities in places where they serve for drinking and recreational purposes (MDDEP 2002). Eutrophication has been reported in many lakes in Canada, including the Great Lakes, because of excess nutrient inputs, especially phosphorus (P), in the lakes through both point and non point sources (De Pinto et al. 1986; Environment Canada 2013). Multiple problems associated with eutrophication include increased primary production, reduced transparency, decreased aquatic biodiversity, low dissolved oxygen (DO) and release of algal toxins (Carpenter et al. 1999; Shinder 2006; Chislock et al. 2013).

The Ministry of Sustainable Development, Environment and Action Against Climate Change of Quebec has taken preventive measurements to address eutrophication in its lakes by monitoring the lake water for total phosphorus (TP), chlorophyll a (Chl *a*), transparency by secchi disk, and dissolved organic carbon through the Quebec Volunteer Lake Monitoring Program (VLMP). Lake Caron is a shallow, artificial, eutrophic lake located in Sainte-Anne-des-Lacs, Quebec. Since 2008, the lake has been in a eutrophic state and listed in the MDDEP's (Ministère du Développement Durable, de l'Environnement et des Parcs (MDDEP) list of eutrophic lakes in Quebec. The lake is currently under a swimming advisory to protect the lake inhabitants from the risks of algal cyanotoxins. The possible sources of P in the lake water include surface run off from the forest area around the lake, excessive fertilizer use and septic tank discharges and internal sediment phosphorus release (Karim et al. 2013, Veetil et al. 2017).

Nutrients entered the lakes through surface run off in to the water column for a long time where they adsorb on to suspended sediment fines (clay particles) and particulate organic matter. The solids settle on the bottom sediment where P can slowly release into the overlaying water (Sondergaard et al. 2003; Mulligan et al. 2009; Inoue et al. 2009). In shallow lakes, like Lake Caron, seasonal recycling and release of sediment P into the overlaying water could be potential reasons for eutrophication (Sondergaard et al. 2003). Thus, removing P-bearing SS, organic matter and algal biomass could reduce the concentrations of TP and organic matter in the water column and sediment that may later reduce the occurrence of eutrophication (Mulligan et al. 2009, Inoue et al. 2009, Sarma et al. 2016).

Filtration is a particle separation technique that uses different materials as filter media in order to achieve specific goals like removing the SS, nutrients and organic matter. Geotextiles are permeable synthetic materials, well known for their use in geotechnical, environmental and hydraulic applications (Mulligan et al. 2009; Inoue et al. 2009). Very recently, geotextiles as filter media have been evaluated for removing the SS and other pollutants including nutrients and heavy metals from storm water and surface water (Alam et al. 2018, Frank et al. 2012, Mulligan et al. 2009, Inoue et al. 2009). Mulligan et al. (2009) reported 93-98% turbidity, 98-99% SS, and 65-75% COD (chemical oxygen demand) removal from river water in a laboratory filtration test with the geotextiles. Similarly, significant removal was reported for TP, COD and SS from a pond water using an in-situ upward geotextile filtration system (Inoue et al. 2009).

Over the past few years, the research team from Concordia University has been monitoring Lake Caron water quality and conducting filtration tests using non-woven geotextiles to improve the water quality. The results from previous on-site studies conducted at the lake were promising for the SS, TP, COD and Chl *a* removals (Veetil et al. 2017, Sarma et al. 2016). In this study, the filtration of the lake water was performed in-situ, using renewable solar energy as source of electricity for the operation. The main objective of this study was to assess the effectiveness of geotextile filtration under real lake conditions for improving the lake water quality by isolating a small portion of the lake by a turbidity curtain.

2 MATERIALS AND METHODS

2.1 Study Area and Experimental Site

The in-situ filtration experiment was conducted in Lake Caron at Ste. Anne-des-Lacs, Quebec. The approximate surface area and water volume of the lake are 35,300 m^2 and 46,400 m^3, respectively. The maximum and average depths are 2.6 and 1.4 m, respectively. For the in-situ filtration test, a small area of the lake was selected and contained in a rectangular shape by deploying a floating turbidity curtain (14.6 m x 1.2 m)(L x W) obtained from Titan Environmental Containment Inc. The turbidity curtain was made to contain and control suspended particles including the algae and fine sediment particles within the enclosure. The volume of water that contained was about 2.88 m^3, which is about 9 times higher than the volume used for the previous on-site test. Figure 1 shows the map of Lake Caron with sampling stations and the enclosure made for the in-situ study.

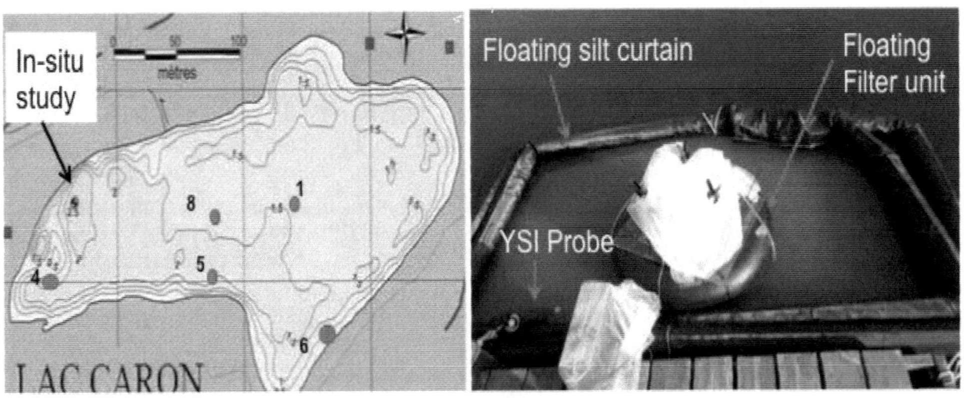

Figure 1: (a) Lake Caron with sampling stations; (b) In-situ filtration set up

2.2 Floating Filtration Unit and Filter Media

The filtration unit is an open squared rectangular vessel, made of plexi glass, with a galvanized copper grid at the base to support the filters (Figure 2). The dimension of the filter unit was 30.5 x 30.5 x 20 cm (W x L x H). A float switch was used to control water level and thus to avoid overflow of water, especially during filter clogging. The whole unit was floated on an inflated synthetic rubber tube (Figure 2). Two submersible water pumps (DC 12V, 4.8 W) having flow rates of 240 L/h were used for pumping the water. The electrical supply to the pumps and float switch was made using a solar panel (AXITEC) (260 W, 30.92 V, 8.43 Ah) and a battery (Crown)(12V, 130 Ah). The dimensions of the solar panel and battery were 164 x 99.2 x 3.5 cm and 33 x 17.1 x 23.8 cm (W x L x H), respectively. The solar panel system was sufficient for supplying electricity to a 10 W system for 24 h for 3 days with low or no light conditions.

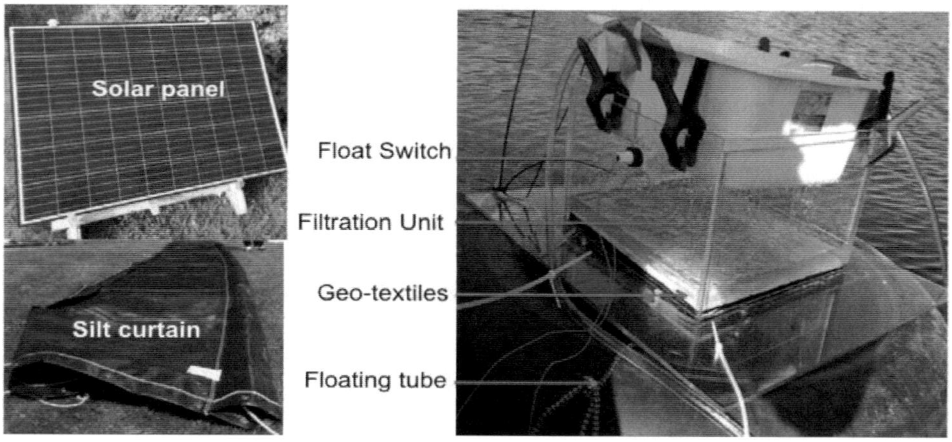

Figure 2. Floating filtration unit

Non-woven geotextiles filters (TE-GTX 300, TE-GTN 300, TE-GTN 350A, TE-GTN 350 B) custom developed and received from Titan Environmental Containment Ltd., MB, were used as filter media to capture the suspended particles, algae and nutrients from the lake water. The filter selection and combination was based on the previous on-site studies during 2015-2016. Table 1 shows the physical characteristics of the geotextiles (based on the data sheets obtained from Titan Environmental Containment Ltd.) used for the study. The filters were cut to area of 30.5 x 30.5 cm (W x L) and arranged in a descending order of their apparent opening size (AOS) for the tests. Different filter combinations were tried during the in-situ test. Upon clogging, used filters were replaced by new ones.

Table 1: Characteristics of the non-woven geotextiles

Filters	Apparent Opening Size (AOS) (μm)	Flow rate (L/m²/min)	Mass/unit area (g/m²)	Material	Permittivity* (sec⁻¹)
TE GTX 300	110	3,900	300	PET	1.62
TE GTN 300	90	3,300	300	PP	0.75
TE GTN 350 A	75	2,700	350	PP	0.56
TE GTN 350B	65	2700	350	PP	0.56

PET: Polyester; PP: Polypropylene

Permittivity*=Volumetric flow rate/unit area of cross-section/unit head

The TE- GTX 300 filter is white in color and made of a continuous filament fibre polyester (PET) with 110 μm AOS. The filters TE-GTN 300 (AOS: 90 μm) and TE-GTN 350-A (AOS: 75 μm) are of black color and made of 100% virgin staple fibre polypropylene.

2.3 Water Sampling and Analysis

Water sampling was conducted at selected stations in Lake Caron from June to September in 2017. During the course of filtration test, water samples from both in and outside of the turbidity curtain were taken and analysed to determine changes in the water quality owing to the filtration. Water samples were collected in 1L brown bottles and stored at 4°C prior to physico-chemical analysis.

The lake water samples were analysed for particle size distribution, total suspended solids (TSS), TP, total nitrogen (TN), nitrate (NO_3^-), COD and turbidity. The dissolved forms of phosphorus (TDP), nitrogen (TDN), COD (D COD), and nitrate (TD NO_3^-) were also measured. For this, water samples were filtered through 0.43 µm pore size filters immediately after sampling and stored under 4°C. Test kits from Hach chemicals were used for analysing TP (TNT 843, Method 10209, ascorbic acid method), TN (TNT 826, Method 10208, persulfate digestion), NO_3^- (TNT 835, Method 10206, dimethyl phenol method) and COD (TNT 820, Method 10221, reactor digestion method) by following Hach protocols (Hach 2018). For the total P, N, COD, and NO_3^- analysis, water samples were acid digested in a Hach DRB 200 acid digestion unit and then measured using a spectrophotometer (Hach DR 2800). Besides, in-situ water quality was monitored throughout the experiment by deploying a YSI EXO2 (E-528-EXO2 YSI, rented from Hoskin Scientific, Canada) water quality probe in the contained water. The YSI probe allowed real-time monitoring of water pH, temperature, turbidity, dissolved oxygen (DO), oxidation-reduction potential (ORP), Chl *a* and blue green algae-phycocyanin (BGA-PC) concentrations. Particle size analysis (PSA) was performed with a laser scattering particle size analyzer (LA-950 V2, Horiba laser particle size analyzer).

3 RESULTS AND DISCUSSION

3.1 Lake Water Quality

The lake water analysis results (Table 2) confirmed the lake's eutrophic state and poor water quality. Except for a couple of instances, TP and Chl *a* concentrations were within in the norm set for eutrophic lakes (TP: 30-100 µg/L and Chl a: 8-25 µg/L) by the MDDELCC, especially during warm weather conditions.

Table 2. Lake Caron Water Quality (Jun.-Sept. 2017)

Parameters	Values
TP (µg/L)	32±5.9
TN (mg/L)	0.84±0.14
COD (mg/L)	26.7±3.5
NO_3^- (mg/L)	0.22±0.06
TSS (mg/L)	3.7±0.91
Temp (°C)	22.7±2.2
Conductivity (µS/cm)	24±1.6
TDS (mg/L)	16.4±16.4
DO (mg/L)	8±0.87
pH	6.8±0.18
ORP (mV)	334±60
Turbidity (NTU)	2±0.7
Chlorophyll (µg/L)	9.2±2.8
BGA-PC (µg/L)	0.5±0.05

The lake water pH was in a slightly acidic to neutral state, within the proposed limit for the Quebec surface water criteria (pH 6.5-9) for protecting aquatic life (CCME, 2009). The DO level was within the lowest acceptable limit proposed to support aquatic life in cold (pH 6.5-9.5) and warm (pH 6-6.5) surface water (CEQG, 1999). The size of the SS in the lake water ranged between 2 and 152 µm. The PSA results showed that 90% of the particle's size (d_{90}) was under the range of 72 to116 µm and 50 % of the particles have (d_{50}) diameters under the 21 to 30 µm range.

3.2 In-situ filtration experiment

The deployment of the turbidity curtain caused high turbidity in the contained water (116 NTU), which was later reduced to 23.1 NTU, after two days of settling. The filtration test was started with a filter combination of two TE-GTX 300 (110 µm), one TE-GTN 300 (90 µm) and one TE-GTN 350 A (75 µm). The experiment was run for 32 days during which filters clogged were replaced by new filters and a total of 34 filters of different filter combinations were used. The lake water quality was found to be much better for the turbidity and algal biomass in 2017 than in 2016. However, the geotextiles, especially the top TE-GTX 300 filter with high permittivity, were found to clog rapidly even at very low turbidity. Figure 3 shows the non-woven geotextiles before and after the filtration. During the test, the algae formed a layer on the filter surface intertwined with the filamentous fibers in the TE-GTX 300 (Figure 3) and this reduced the filter's permittivity significantly to a completely clogged state. While the algae mainly were retained on the top filter, very fine SS were captured on the filters below with small opening sizes.

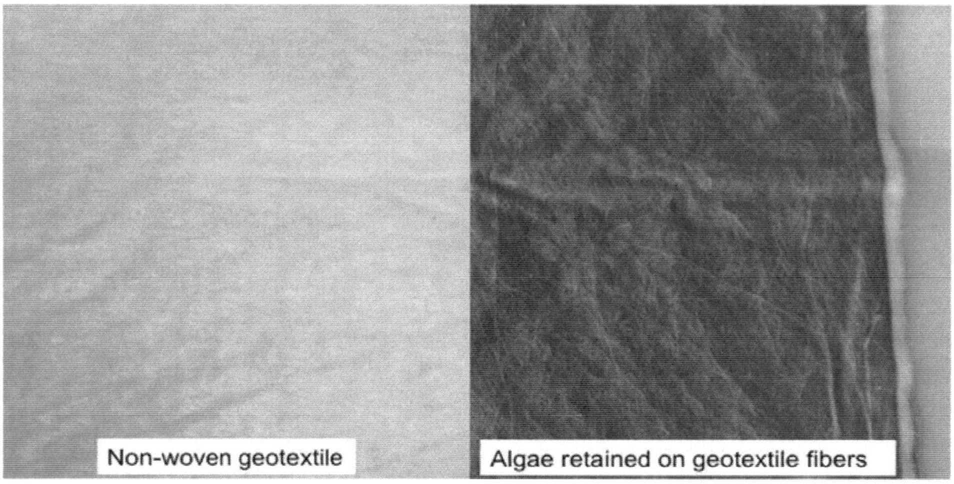

Figure 3. Non-woven geotextile before and after filtration

Figure 4 shows the change in total and dissolved forms of P and COD during the filtration test. As seen in the Figure 4a, the TP concentration in the contained water was initially high due to the high SS and its concentrations inside and outside the turbidity curtain were 75.5 and 28.5 µg/L, respectively. A significant reduction in TP (56%) was observed for the filtered water in the first 7 days, obviously due to the removal of P bearing suspended sediment particles and the algae by both the filtration and settling process. At the end of the test, TP concentration in the treated water was reduced to below the norm set for eutrophic lakes (30 µg/L) in Quebec. The net TP removal was 72%. The concentration of dissolved P (TDP) was similar in both filtered and non-filtered water through out the experiment, indicates that this in-situ test did not cause release of P into the overlaying water. The filtration test was not very effective for COD removal as most of the COD (76-92%) was in the dissolved form (Figure 4b). However, the removal of algal biomass and suspended organic matter resulted in a slight reduction in the COD concentrations in the filtered water.

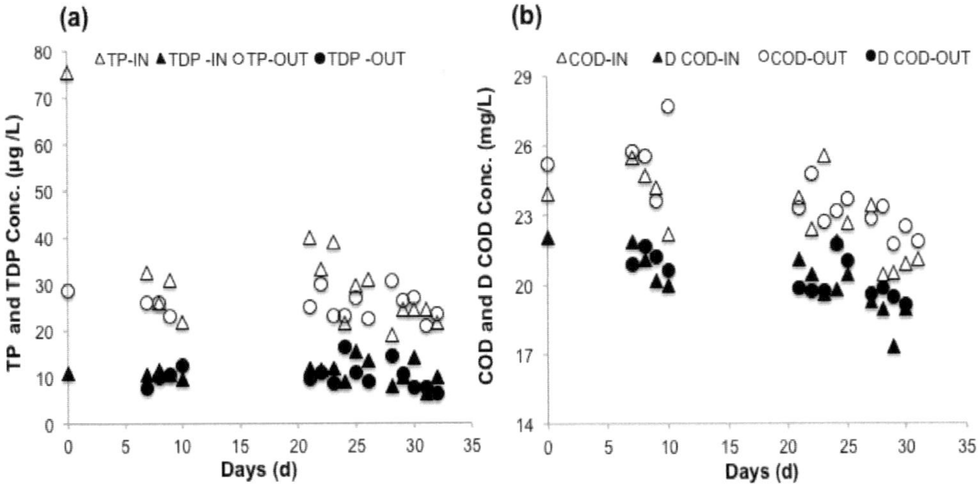

Figure 4. Change in conc. of (a) TP & TDP and (b) COD & D COD during filtration

The change in water turbidity and suspended particles during the filtration test is given in Figure 5. The initial turbidity inside and outside the curtain was 23.1 and 1.48 NTU, respectively. The turbidity removal was significant in the first few days (day 0 to 5) of filtration owing to both the filtration and settling process and thereafter, it was mainly due to the filtration process.

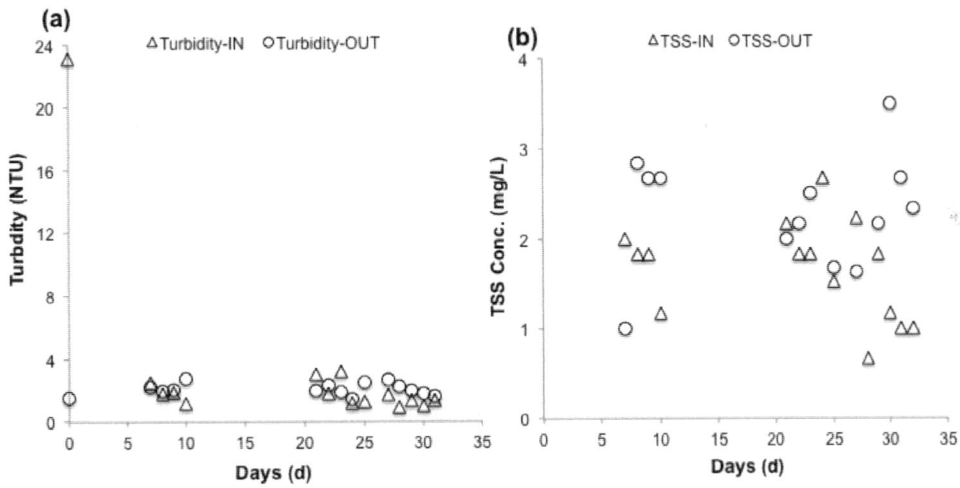

Figure 5. Change in (a) Turbidity and (b) TSS inside and outside the silt curtain during filtration

Considering the initial and final turbidity in the contained water, about 94% turbidity removal was achieved during the test. The final turbidity in the filtered water was 1.3 NTU, which is about 2 times less than those outside the curtain (2.7 NTU). The turbidity of contained water was also monitored hourly during the test by using the YSI probe (Figure 6) and compared the values with those obtained by a hand held turbidity meter (Oakton turbidity meter, T-100). As seen in the Figure 6, the turbidity in the contained water was reduced, from 21.7 to 2.74 NTU, close to the normal lake water turbidity within two days of filtration and settling. The SS data shown in Figure 5b is from day 7 to 32, when about 50% TSS removal was observed. At the end, the TSS concentrations outside and inside the curtain were 2.3 and 1 mg/L, respectively.

Figure 6 shows the changes in temperature, turbidity and Chl *a* concentration in the contained lake water over the filtration test, monitored by the YSI probe. Water temperature is known to have significant role in algal growth and bloom development as the increase in temperature enhances the phytoplankton productivity (Wells et al. 2015). Though the mild weather and low water temperature adversely affected the algal growth throughout the summer in 2017, an increase in the algal growth was experienced later at the end of September due to the warmer weather conditions. As seen in Figure 6, a slight increase in Chl *a* concentration was observed with the increase in water temperature (Day 0-11). While the filtration continues, the chlorophyll *a* concentration in the filtered water was reduced gradually, yielding about 48 % removal. The lake water in the enclosure was in a stagnant condition and this favored the algae in the water to stick on the curtain's inner wall. This reduced the removal of algae to some extent by (i) limiting the algae available in the suspension mode for pumping and filtering and (ii) frequent pump failure due to entrained air bubbles. Therefore, this problem will be addressed in the next phase of this project by changing the pumps in the system.

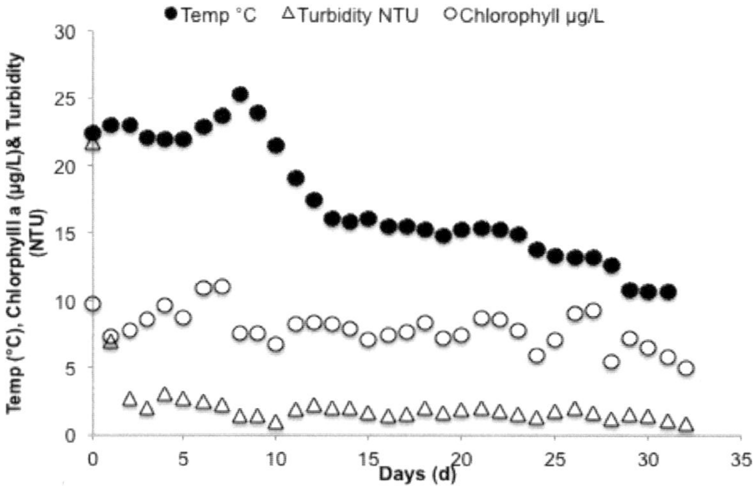

Figure 6. YSI probe data for temperature, chlorophyll a and turbidity of contained water during filtration

The PSA of water samples collected from both in and out of the system during the filtration test is given in Figure 7.

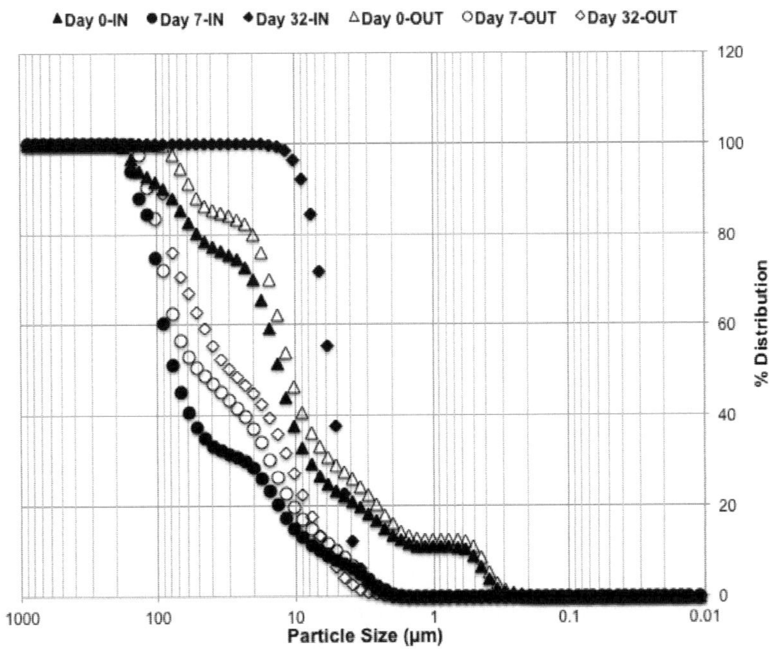

Figure 7. Grain size distribution of SS in Lake Caron water during the filtration test

Figure 7 shows the size distribution of SS in the lake water during the filtration test. Before the filtration, about 90 % (d_{90}) of the particles were below 89 µm and 59 µm for the samples from inside and outside the curtain. Increased algal growth accompanied with high temperature resulted an increase in particle's size in the lake water (Day 7). As can be seen, 90 % (d_{90}) of the particles in the final, filtered water was under 8.5 µm, which shows the efficiency of the geotextiles to capture SS of broader size range.

4 CONCLUSIONS

The results from this preliminary in-situ filtration test have shown the potential use of geotextiles as a filter media to improve the surface water quality under real lake conditions. The in-situ filtration allowed significant removal for TP (72%), chlorophyll a (48%) and turbidity (96%) from the lake water and the results could further be improved in the next phase by solving the problems experienced during the test. The simple design and flexibility make the floating filtration unit ideal for targeting and treating a specific area of the lake with a massive algal bloom and SS, without disturbing the aquatic life, with no chemical addition, and by isolating the treatment area using a turbidity curtain. In the next phase of this project, the filtration test will be evaluated for the flow rate through the filter and solids and P removal rate per filter per area.

Acknowldgements

The authors thank NSERC and Concordia University for financial support for this project. The authors are also grateful to their industrial partner, Titan Environmental Containment Ltd., for supplying geotextiles and providing financial support for this project.

Refrences

Alam, M.Z., Faisal Anwar, A.H.M., Heitz, A. 2018. Stormwater solids removal characteristics of a catch basin insert using geotextile. *Science of the Total Environment*, **618**: 1054-1063.

Carpenter, S.R; Ludwig, D., Brock, A. 1999. Management of Eutrophication for Lakes Subject to Potentially Irreversible Change. 1999. *Ecological Applications*, **9**(3): 751-771.
CCME (Canadian Council of Ministers of the Environment). 2008. Canadian Water Quality Guidelines.1-1484.
CEQG (Canadian Environmental Quality Guidelines). 1999. Canadian water quality guidelines for the protection of aquatic life: Dissolved oxygen (freshwater). Retrieved from ceqg-rcqe.ccme.ca/download/en/17.
Chislock, M.F., Doster, E., Zitomer, R.A., Wilson, A.E. 2013. Eutrophication: Causes, Consequences, and Controls in Aquatic Ecosystems. *Nature Education Knowledge*, **4**(4): 10.
De Pinto, J.V; Young, T.C; Mcllory, L.M. 1986. Great Lakes Water Quality Improvement. *Environmental Science Technology*, **20** (8): 752-759.
Environment Canada. 2013. Phosphrous and Excess Algal Growth. Government of Canada. Retrieved from http://ec.gc.ca/grandslacsgreatlakes/default.asp?lang=En&n=6201fd24-1.
Franks, C.A., Davis, A.P., Aydilek, A.H. 2012. Geosynthetic Filters for Water Quality Improvement of Urban Strom Water Runoff. *Journal of Environmental Engineering*, **138**(10): 1018-28.
HACH. 2018. Water Analysis Hand Book. Retrieved from https://www.hach.com
Inoue, T., Fuke, M., Mulligan, C.N., Uehara, K. 2009. In-situ Removal of Contaminated Suspended Solids from a Pond by Filtration. *Ecological Engineering*, **35**:1249-1254.
Karim, R.A., Mulligan, C.N., Fukue, M. 2012. Preliminary Evaluation of a Sediment Resuspension Technique for Reduction of Phosphorus in Lake Water. *Fifth ASTM Int. Symposium on Contaminated Sediments: Restoration of Aquatic Environment*, Montreal, QC, Canada: 45-65.
MDDEP. 2002. Volunteer Lake Monitoring Program (VLMP). Retrieved from http://www.mddelcc.gouv.qc.ca/eau.rslv.index-en.html.
MDDEP. 2002. Water. Our Life. Our Future. Quebec Water Policy, Bibliotheque national du Quebec, Envirodoq ENV/2002/0310A.
Mulligan, C.N., Davarpanath, N., Fukue, M., Inoue, T. 2009. Filtration of Contaminated Suspended Solids for the Treatment of Surface Water. *Chemosphere*, **74**: 779-786.
Sarma, S., Mulligan, C.N, Kim, K., Veetil, D.P. 2016. Use of Nonwoven Geotextiles for Removing Nutrients and Suspended Solids from a Eutrophic Lake. *Environmental Speciality Conference,* CSCE, London, ON, Canada: 1-10.
Schindler, D.W. 2006. Recent Advances in the Understanding and Management of Eutrophication. *Limnology and Oceanography*, **51**: 353-363.
Sondergaard, M., Jensen, J.P., Jeppesen, Erik. 2003. Role of Sediment and Internal Loading of Phosphorus in Shallow Lakes. *Hydrobiologia*, **506-509**: 135-145.
Veetil, D.P., Mulligan, C.N., Mariana, E., Esteban, C.A., Bhat, S. 2017.Geotextile Filtration for Improving Surface Water Quality of Eutrophic Lakes. *Leadership in Sustainable Infrastructure*, CSCE, Vancouver, BC, Canada: 1-10.
Wells, M.L; Trainer, V.L., Smayda, T.J., Karlson, B.S.O., Trick, C.G., Kudela, R.M., Bernard, S., Wulff, A., Anderson, D.M., Cochlan, W. 2015. Harmful Algal Blooms and Climate Change: Learning From the Past and Present to Forecast the Future. *Harmful Algae*, **49**: 68-93.

Building Tomorrow's Society
Bâtir la Société de Demain

Fredericton, Canada
June 13 to 16, 2018 / *13 au 16 juin, 2018*

INNOVATIVE SLUDGE DISINFECTION APPROACH TO GENERATE CLASS A BIOSOLIDS FOR LAND APPLICATIONS

Luo, Jiaru[1,2], Elektorowicz, Maria[1,3]
[1] Concordia University, Department of Building, Civil and Environmental Engineering, Montreal, Canada
[2] **l_jiaru@encs.concordia.ca**
[3] **mariae@encs.concordia.ca**

Abstract: Wastewater treatment facilities generate a huge amount of sludge from their processes. One of the conventional method for sludge management is the land application as a fertile soil amendment which is an efficient way to recycle nutrients from human waste. However, since sludge also contains pathogens, it is necessary to be disinfected before its usage. Then, a novel system has been demonstrated for sludge disinfection using a treatment supported by electrical field. This paper mainly focuses on optimization of the system to find out the best operating conditions for their further implementations. Investigation of Fe^{2+}/H_2O_2 ratio, pH, current level, and amount of H_2O_2 were conducted to determine effective parameters for sludge disinfection through the Electro-Fenton method. The results showed log 7 reduction of *E. coli* within 30 minutes by applying selected parameters into sludge (WAS) in the electro-reactor.

1 INTRODUCTION

Municipal sludge management might reach up to 60% of the total operating costs in a wastewater treatment plant (Andreoli et al. 2007). Since it should be controlled in a harmless way toward the public, municipal sludge management usually includes dewatering, pathogenic disinfection and nutrient recycling if considered possible. The pathogenic reduction is the most important with respect to public health and environmental protection. Therefore, the regulations regarding the pathogen content in sludge are defined before the sludge land application. In order to satisfy the requirements for Class A biosolids, minimum 5 log reduction of fecal coliform should be achieved (EPA 1993, CCME 2012). The lime treatment, aerobic and anaerobic digestion and composting have been authorized as standard treatment methods. However, these methods require a long treatment time (from 20 hours to months), long retention time of the process, which lead to building facilities with an extensive space demand. Usually, the construction of such expensive facilities for an adequate sludge disinfection is a big financial issue for small wastewater treatment plants (WWTP). . Therefore, electrokinetic technology would be an ideal method to satisfy disinfection requirements because of its relatively low capital and operation costs.

Some previous work have already demonstrated that direct current (DC) applied to sludge generates electrokinetic stressors (Esmaeily et al. 2006), desiccation stressors (Huang et al. 2012), or ohmic heat stressors for disinfection of fecal bacteria (Daneshmand et al. 2012) as well as spores (Safaei et at. 2013). Furthermore, hydroxyl radicals (•OH) showed their high oxidization potential which could rapidly react with organic matrix as an effective disinfectant (Selvakumar et al. 2009). The previous study demonstrated that (•OH) can achieve 2 log *E. coli* reduction with the concentration of 0.8×10^{-5} mg/L.min (Cho et al. 2004). Generation of hydroxyl radicals can be done through Fenton reaction, Eq. [1-2], where ferrous iron, acting as a catalyst, reacts with hydrogen peroxide to generate ferric iron and hydroxyl radicals, while introduced hydrogen peroxide reacts with ferric iron to regenerate ferrous irons:

[1] $Fe^{2+} + H_2O_2 + H^+ \rightarrow Fe^{3+} + H_2O + \bullet OH$

[2] $Fe^{3+} + H_2O_2 \rightarrow Fe^{2+} + HO_2 \bullet + H^+$

Then, the reaction with (•OH) leads to pathogens' inactivation. Fenton reaction has been used in different situations for disinfection, e.g. water. Recent studies on water and wastewater disinfection by using Fenton reaction focus more on Photo-Fenton reaction by applying extra UV light to accelerate the production of ferrous iron to generate more hydroxyl radicals (Polo-López et al. 2012; Ortega-Gómez et al. 2012). Since UV light cannot penetrate through the medium, photo-Fenton is not efficient for treatment of highly turbid water with a high organic load. In this case, Fered–Fenton oxidation process is considered as a substitute technology for the Fenton-based application. Similar to the Photo-Fenton reaction, ferrous iron reacts with hydrogen peroxide to produce hydroxyl radicals. However, reduction of ferric iron is conducted by electro-reduction. Equation [3] shows that ferric iron is reduced at the cathode in the presence of continuously applied hydrogen peroxide and produces hydroxyl radicals (•OH) (Brillas et al. 2009).

[3] $Fe^{3+} + e^- \rightarrow Fe^{2+}$

One of the advantage of Fered-Fenton and Photo-Fenton processes compared to traditional Fenton reaction is the feasibility of using smaller amount of ferrous iron at the beginning as a catalyst, which can be regenerated by the system. For example, Fered-Fenton process were applied for treatment of both leachate and industrial wastewater having obvious turbidity and high concentration of COD (Zhang et al. 2006; Li et al. 2010; Jahromi 2016). However, there is always a concern to find technological conditions which inactivate pathogens during a short retention time at low costs. Therefore, the aim of this study was to find such an electro-Fenton oxidation method which might transform waste activated sludge (WAS) into Class A biosolids during a short exposure period in order to advance the sludge management methods.

2 MATERIALS AND METHODS

Sludge samples were taken from the activated sludge facilities located in wastewater treatment plant in Saint-Hyacinthe (Quebec, Canada) and stored at 4°C. The sampling took place after thickening process where the coagulant (ferrous sulfate) was added before sludge dewatering. The initial characteristic of sludge included: pH, TAN, *E.coli,* Total coliform, Cl^-, conductivity, TS%, while values of these factors were as follow 6.4-6.8, 51mg/l, 4×10^9 cfu/100ml, 4.17×10^9/100ml, 14.5mg/l, 1200µS/cm, and 5.5%, respectively.

A series of batch experiments were conducted in an electro-reactor (12cm × 7.5cm × 10cm) where stainless steel electrodes were placed (Fig. 1). An amount of 1000ml of sludge solution was used for each experimental run. DI water (Milli-Q® Integral, Sigma Inc.) was supplied for sludge dilution to reach a required TS (total solids) content around 0.5-0.8%. The electro-reactor was operated at a constant current density using DC power supply (GENESYS™ 750W/1500W, TDK-Lambda Americas Inc.). To maintain a homogenous sludge solution, a magnetic stirrer (Fisher brand) was used for continuous mixing. Furthermore, a thermocouple was placed in the middle of the reactor for temperature monitoring (Agilent 3450a data logger), and a peristaltic pump (Cole-Parmer Instrument. LLC) was applied for sampling from a middle area of the electro-reactor.

The pH level was adjusted with 1M hydrochloric acid before applying the current as well as during the treatment process. The pH value was monitored by Accumet™ AB2000 pH/conductivity bench meter. Ferrous ammonia sulfate was used as a source of ferrous iron. Total ammonia (TAN) and chloride were measured by selective ammonia membrane probe (IntelliCAL™ Ammonia ISE) and Fisher ISE chloride probe, respectively.

Source of *E.coli,* used as a pathogenic indicator, was the wastewater in a poultry farm (Quebec, Canada). Collected bacteria were subjected to bioaugmentation to reach an adequate density of cells, and then, 1ml of prepare cultured LB was injected into 50ml fresh LB and incubated overnight to maintain the range of bacteria *E.coli* density around 10^{11} cfu /100ml. In order to identify the *E.coli*, MFC agar (BD Difco™) was applied for a standard membrane filtration method with a serial dilution, then, counting on the filters (APHA/AWWA/WEF 2012) was conducted. In this study, for simulating the density of bacteria in the sludge, 10mL of cultured LB broth was injected into the electro-reactor to achieve 10^{-9} cfu/100mL for the initial bacterial concentration.

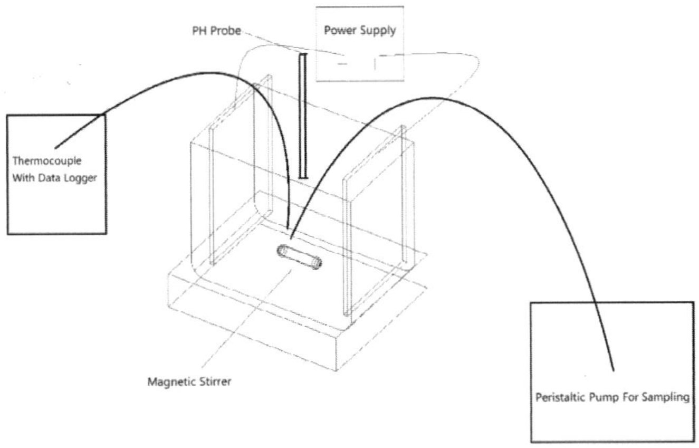

Figure 1 Schema of the experimental setup

Since this study aims to investigate an electro-Fenton like process, an effectiveness of multi-stressors on *E.coli* disinfection should be demonstrated. Subsequently, a series of experiments were conducted in the electro-reactor. Firstly, a disinfection efficiency was assessed in relation to various amounts of H_2O_2 (0ml, 1ml, 2ml), which is shown as cases A, B, C, respectively in Figure 3. Furthermore, Case D (Fig.3) considered the effect of Cl^- ion on *E. coli* inactivation; subsequently, 2.8g/L of KCl was added (at 1.3A) to the electro-reactor. Since hydrochloride was chosen for pH adjustment, Cl^- being oxidized to ClO^-/Cl^- on the anode might generate hypochlorite acid (HClO), which is considered as an effective disinfectant (Huang et al. 2016). Next series of experiments (Fig. 4) considered an effect of various amperage values (0.2A, 0.4A, 0.8A as cases A,B,C, respectively). Comparison of pH values of 4 and 7 (case A and B in Fig. 5) permitted to assess their impact on *E.coli* inactivation. Various ratio of Fe^{2+}/H_2O_2 (no H_2O_2, 1:12, 1:6, 1:3) were applied to demonstrate the Fenton reaction impact on *E.coli* disinfection efficiency (Fig. 6). Such methodological approach permitted to assess impacts of various combinations of stressors on bacteria inactivation and define the best conditions for the proposed disinfection method.

3 RESULTS AND DISCUSSION

3.1 Comparison of Different Applied Conditions on E. Coli Inactivation

The comparison of results from series A (impact of heat as a stressor), B (heat and Fenton reaction), C (stressors from electro-Fenton oxidation), are shown in Figure 2. As can be seen in case A, no significant *E.coli* inactivation during 30 minutes process was observed, which demonstrated that the combination of heat and acid environment was not the main bacteria inactivation factor for this system. However, in case B, an addition of ferrous iron and H_2O_2 permitted to reach 2 log reductions in 15 minutes and 5.2 log reduction in 30 minutes of treatment. This indicates that the combination of ferrous/ H_2O_2 created disinfection conditions, most probably by production of hydroxyl radicals. In such case, the system was exposed to Fenton reaction (Eq. [1]), which contributed to the *E.coli* inactivation. Furthermore, the case C showed that the implementation of electrical field significantly improved *E,coli* reduction by achieving 7.3 log within 30 minutes; while, the required log 5 reduction has been already reached after 15 min. Applied ratio of ferrous iron to hydrogen peroxide was 1:12. Since the ferrous iron concentration is a limiting factor for generation of hydroxyl radicals (•OH) in Fenton reaction, the application of current reduces ferric iron to ferrous iron by electro-reduction at the cathode. Therefore, in case C, ferrous iron was regenerated, additionally reacted with hydrogen peroxide and produced more hydroxyl radicals (•OH).

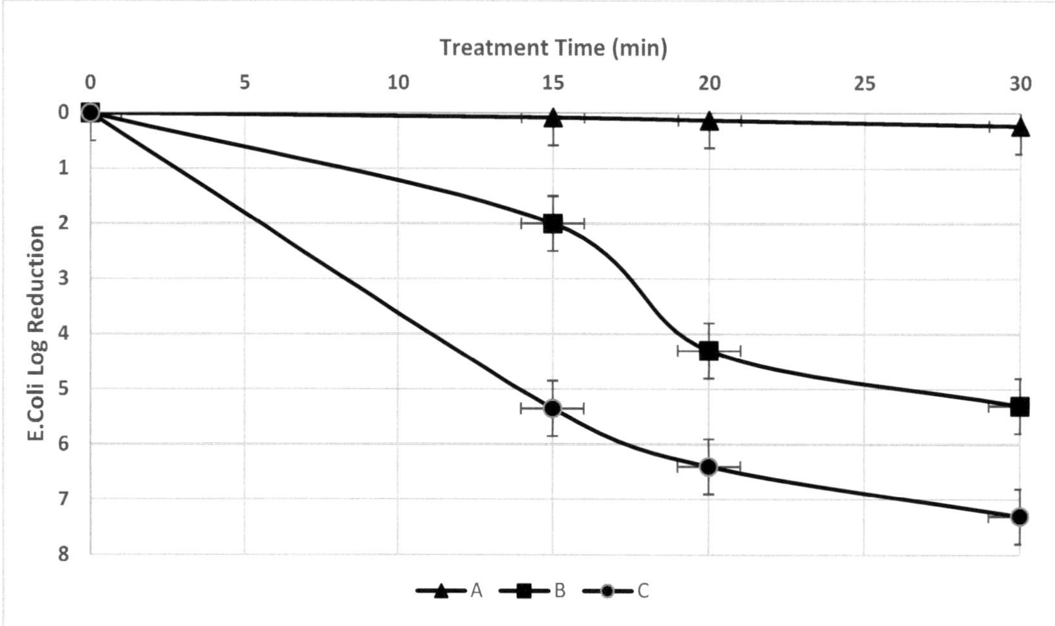

Figure 2 E.coli reduction vs. treatment time and different stressors (A, B, C)

Note: Electro-reactors operated under pH=4, and TS = 0.8%; A: heat as a stressor only; B stressors: heat as well as ferrous iron and H_2O_2 (Fe/ H_2O_2 = 1:12); C stressors: ferrous iron and H_2O_2 as well as a constant current density

The assumption that hydrogen peroxide as a strong oxidant, might also contribute to inactivation of *E. coli* was excluded since other study (Ortega-Gómez et al. 2012) demonstrated that H_2O_2 does not have negative effect on *E.faecails* survival rate. Therefore, it was concluded that hydrogen peroxide acted as a hydroxyl radicals' generator but it was not the main disinfectant.

3.2 Effect of H_2O_2

The experiment B (Fig. 3) shows that the addition of 1ml of H_2O_2 permitted to reach 2 log reduction of *E.coli* in 15 minutes and almost 4 log reduction in 30 minutes compared to 0.2 log reduction without H_2O_2. The application of twice higher amount of H_2O_2 can achieve twice higher inactivation (4 log reduction) in 15 minutes. A higher amount of H_2O_2 generates a greater *E.coli* log reduction due to presence of an organic matter which also reacts with hydroxyl radicals (•OH). The abstraction of protons can be further oxidized to CO_2 and others organic forms (Yoon et al. 2001). Therefore, a high amount of the organic matter might compete with the bacteria inactivation process with respect to (•OH) reactions, and subsequently, might cause a decrease of the disinfection efficiency. Figure 3 does not show an obvious *E. coli* reduction (less than 1 log) in case D for both 15 and 30 minutes. It may be concluded that chlorite is not the main disinfectant in this system. In order to satisfy the conditions of anodic oxidation, the electrode material should have a high oxidation potential. Formation of chlorite on electrode must have higher potential than 1.36V (Martínez-Huitle and Andrade 2011). However, iron electrode achieves 0.77V oxidation potential only, then, no inactivation of *E. coli* can be expected with such low formation of ClO^-/Cl_2. This conclusion was also generated in other tests when additional KCl and seven times higher (1.3A) current than in cases B(0.2 A) and C(0.2 A) were applied. The results still had much lower *E.coli* log reduction (1.3 log) by the end of 30 minutes than in the case C (5.2 log reduction) (Fig. 3). Besides, a poor E.coli reduction might be caused by the presence of ammonia in the medium. Huang et al., 2016 demonstrated that under the presence of ammonia, free chlorine generated in the anode will be converted to chloramine, which has much lower disinfection efficiency for eliminating MS2 (bacteriophage) in wastewater.

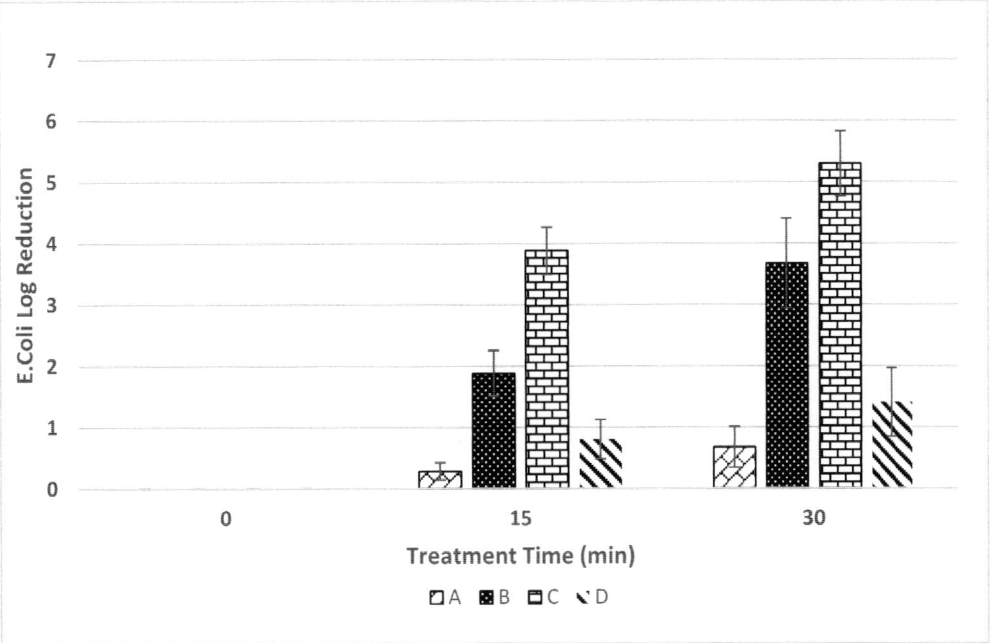

Figure 3 E. coli reduction vs. time of treatment and different dosage of H_2O_2 with additional test applying KCl

Note: All tests operated under 0.2A, TS = 0.8%, pH=4, additional ferrous iron and H_2O_2 in the ratio of Fe/ H_2O_2 = 1:12, while H_2O_2 were 0ml, 1ml, 2ml in case A,B,C, respectively; case D was operated under pH=4 with 1.3A current, 2.8g KCl /L, (pH was adjusted with H_2SO_4)

3.3 Effect of current

Figure 1 has already shown that the highest inactivation of *E. coli* could be achieved when an adequate Fe/ H_2O_2 ratio and DC field were applied. Figure 3 indicates that under constant input of H_2O_2 more ferrous iron is present in the system, thus, better inactivation is achieved. The same observation was presented by others (Brillas et al. 2009). Then, the application of current, which regenerate ferrous, is directly related to disinfection efficiency. Figure 4 shows the relationship between current and *E.coli* log reduction. However, since applied current can also cause ohmic heat effect, higher current comes along with higher temperature in the system; this might be another environmental stressor for bacteria inactivation (Safaei et al. 2013).

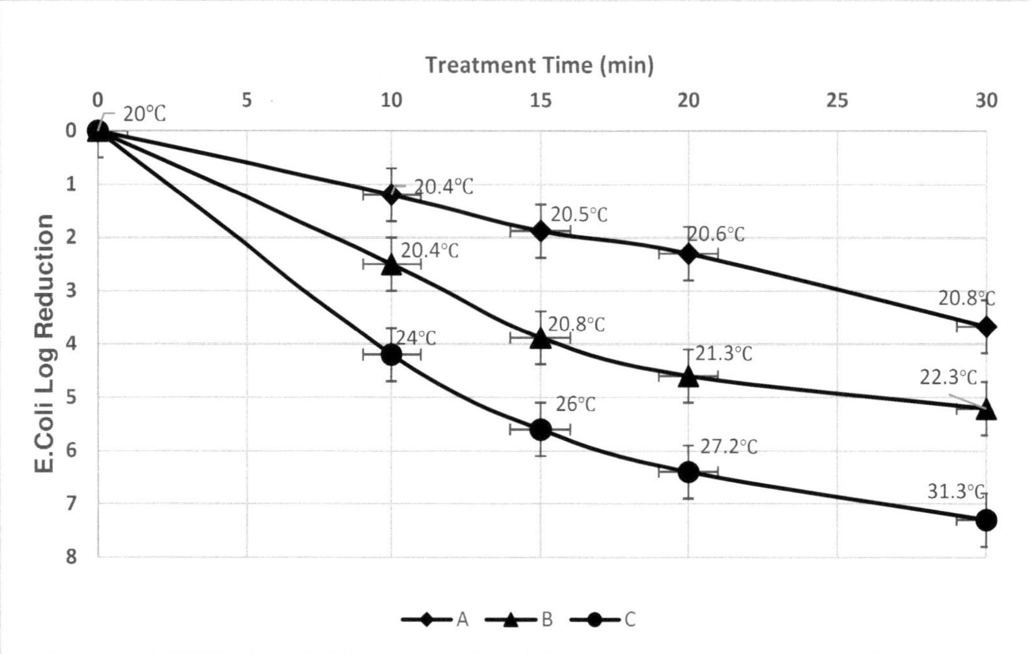

Figure 4 E. coli reduction vs time of treatment for different DC current.

Note: Experiments were conducted under: pH=4, TS: 0.8%; additional ferrous iron and H_2O_2 (1ml) in the ratio of Fe/ H_2O_2 =1:12, while current 0.2A, 0.4A, 0.8A were applied into cases A, B, C, respectively; Temperature was also recorded in different treatment periods

To avoid impact of a heat, the experiments were conducted at a low current level to minimise the temperature effect. Current between 0.2A and 0.4A kept temperature around 20-22 °C. However, the bacteria inactivation was obviously different; the *E.coli* reduction of 3.5 log for 0.2A, and 5.5 log reduction for 0.4A (after 30 minutes) were observed. The current increase to 0.8A did not change significantly the bacteria reduction, although temperature has been simultaneously increased by 6 °C. This proved that ohmic heat was not the main stressor for *E. coli* inactivation. Though, some of the other published articles declared that ohmic heat can also be used for disinfection (Yin et al. 2018). They demonstrated that by using ohmic heat, when temperature went up to 50 °C, the 4.5 log reduction could be achieved, while 0.68 log reduction only for temperature of 28°C. Their results confirmed the outcome of this study. Although raising current density could have a higher reduction rate and regenerate more ferrous iron, however, it would use more energy to produce ohmic heat, thus, it does not seem to be necessary for an adequate inactivation of *E. coli* in thickened WAS. Some investigations on cost-effectiveness should be conducted in the future.

3.4 Effect of pH

As Eq. [1] demonstrated that Fenton reaction should be conducted under acidic environment in order to generate more hydroxyl radicals (•OH). Therefore, pH might directly affect the generation of oxidants. Significant different results of bacteria inactivation were obtained (Fig. 5), under pH=7 and pH=4, where 0.5 log and 5.5 log reduction in 15 minutes were achieved, respectively. No further significant change for pH=7 was observed even when the exposure time was extended to 30 minutes. This demonstrated that contact time in the system was not the main factor affecting disinfection. However, the amount of produced hydroxyl radicals (•OH) defined *E. coli* disinfection. This also was showed by others (Diagne et al. 2007), who removed methyl parathion under pH=3-4 by using hydroxyl radicals (•OH) while a different removal they obtain under pH=7. Furthermore, another explanation (as it has been suggested by Pang et al., 2011) may be attributed to the formation of surface bonding hydroxyl radicals. It was assumed that hydroxyl radicals can be generated even under neutral environment, but they bond with the surface of insoluble metal species (e.g. ferric hydroxides). Subsequently, hydroxyl radicals cannot unselectively oxidize *E.coli*.

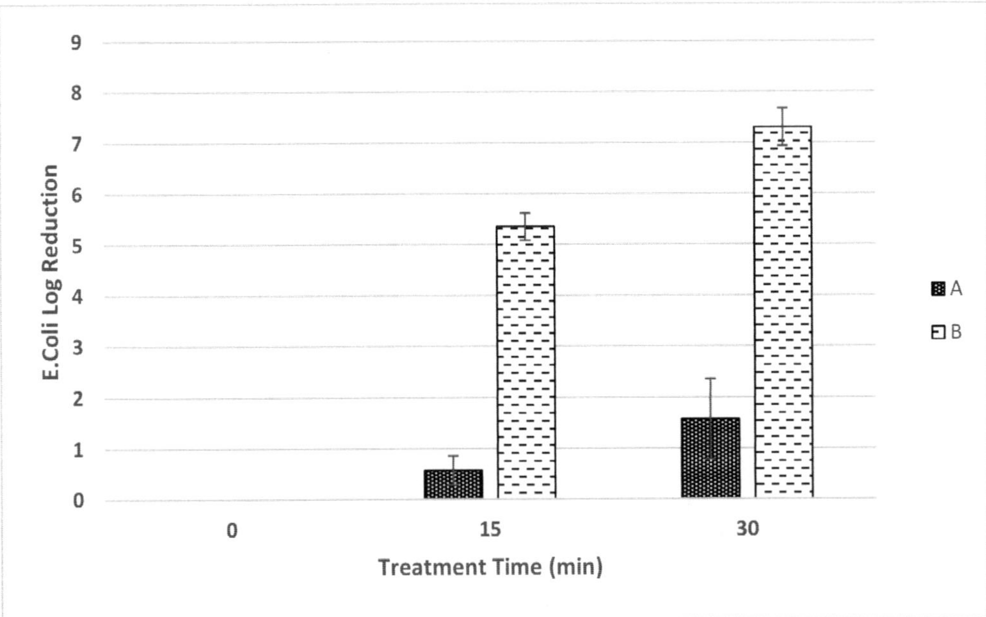

Figure 5 E. coli reduction vs treatment time for different pH

Note: Cases A and B were operated under current 0.8A, ratio of Fe^{2+}/ H_2O_2 =1:12, case A was under pH=7, case B was under pH=4

3.5 Effect of Fe/ H_2O_2 Ratio

Different ratios of Fe/ H_2O_2 were investigated to assess their impact on *E. coli* reduction. No regeneration of ferrous iron was considered in this experiment - no current was applied (Fig. 6). Since the initial thickened sludge has already contained the ferrous sulfate coagulant, the process achieved almost 2 log reduction of *E. coli* without iron additional. The results were improved with the addition of ferrous into the system. This is because (Eq. [1,2]) ferrous was the limitation factor for *E. coli* disinfection when Fe/ H_2O_2 ratio was smaller than 1 (Zhang et al. 2012; Neyens and Baeyens 2003). In such case, an excess of ferrous iron is converted to ferric iron and forms ferric hydroxo-complexes, which aggregate suspend solids and precipitate. The results from other research on Fenton method (Juarez et al. 2010) showed that disinfection efficiency increased as the ratio of Fe/H_2O_2 increases. It was assumed that the inactivation of MS2 was mediated by iron colloids. Therefore, when an amount of ferrous iron increases, and H_2O_2 is kept constant (ratio 1:3), the bacteria reduction reaches 5.5 log within 30 minutes of treatment. It could be expected that when the ratio increases to 1:1.5, a 7.3 log reduction can be achieved. Therefore, without current application, the ratio Fe/ H_2O_2 should be greater than 1:3 to reach 7.3 log reduction. Otherwise, electro-reduction should be used to regenerate more ferrous to compensate insufficient ferrous iron in the system.

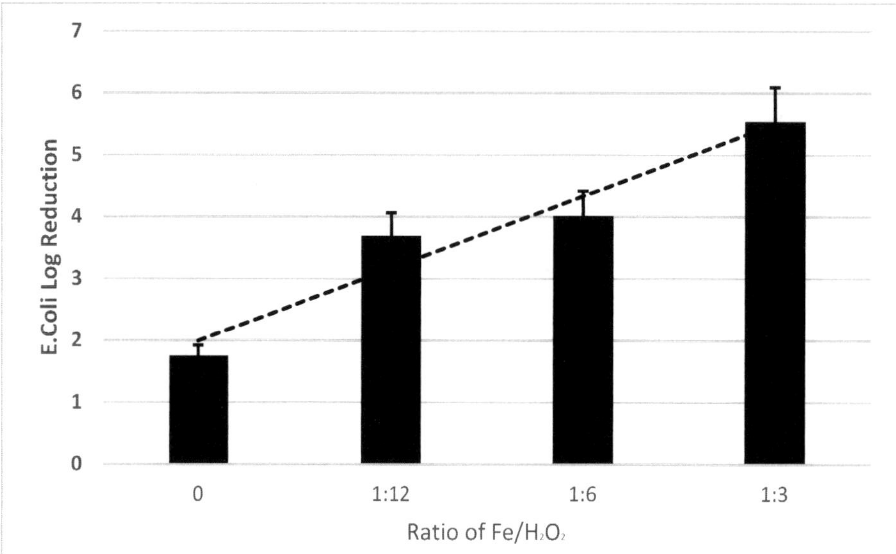

Figure 6 E. coli reduction vs different ratio of Fe^{2+}/H_2O_2 (after 30 min exposure time)

Note: Experiments operated under: pH=4, TS=.8% , addition of ferrous iron and H_2O_2, without current and heat.

4 CONCLUSIONS

- The novel electro-disinfection system was able to achieve log 7 reduction of *E. coli* in thickened sludge (WAS) containing 0.8% TS;
- Under pH=4, current 0.8A, while the ratio Fe^{2+}/ H_2O_2 = 1:6 (contained 1ml H_2O_2), the system could effectively inactivate *E.coli* within 30 minutes;
- Main stressors are in-situ generated hydroxyl radicals (•OH) which interfere with *E. coli* cells; other stressors such as ohmic heat, pH, ClO- ions are minor stressors within this system;
- Higher Fe^{2+}/ H_2O_2 ratio supported by current generate sufficient hydroxyl radicals (•OH) due to regeneration of an adequate amount of ferrous iron in the system to proceed with electro-Fenton like oxidation;
- The proposed method can be applied to enhance sludge management and generate class A biosolids.

Acknowledgements

This research was supported by the Natural Sciences and Engineering Research Council of Canada (NSERC) Discovery Grant awarded to Dr. Maria Elektorowicz. The authors are also thankful for graduate students' financial support (FPRST) provided by the Faculty of Engineering and Computer Science at Concordia University.

Reference

Andreoli C.V., and M. V. Sperling. 2007. *Sludge Treatment and Disposal. Environmental Protection*. Vol. 23. https://doi.org/10.1016/B978-1-85617-705-4.00021-6.

APHA/AWWA/WEF. 2012. "Standard Methods for the Examination of Water and Wastewater." *Standard Methods*, 541. https://doi.org/ISBN 9780875532356.

Brillas E., I. Sire, and M. A. Oturan, 2009. "Electro-Fenton Process and Related Electrochemical Technologies Based on Fenton 's Reaction Chemistry." *Chemical Reviews* 109 (12): 6570–6631. https://doi.org/10.1021/cr900136g.
Cho M., Hyenmi C., Wonyong C., and Jeyong Y. 2004. "Linear Correlation between Inactivation of E. Coli and OH Radical Concentration in TiO2 Photocatalytic Disinfection." *Water Research* 38 (4): 1069–77. https://doi.org/10.1016/j.watres.2003.10.029.
CCME. 2012. "Guildline document for the benificial use of municipal biosolid municipal sludge and treated septage ". In Canada-wide Approach for the Management of Wastewater Biosolids. https://www.ccme.ca/files/Resources/waste/biosolids/pn_1473_biosolids_guidance_eng_1.0.pdf
Diagne M., Nihal O., and Mehmet A. Oturan. 2007. "Removal of Methyl Parathion from Water by Electrochemically Generated Fenton's Reagent." *Chemosphere* 66 (5): 841–48. https://doi.org/10.1016/j.chemosphere.2006.06.033.
EPA. 1993. "EPA Guide to Part 503 Rule." In *EPA Guide to Part 503 Rule*, 23.
Esmaeily A., M. Elektorowicz, S. Habibi, and J. Oleszkiewicz. 2006. "Dewatering and Coliform Inactivation in Biosolids Using Electrokinetic Phenomena." *J of Env Eng Sc* 5 (3). https://doi.org/10.1139/S05-039.
Huang J., M. Elektorowicz, and J. A. Oleszkiewicz. 2008. "Dewatering and Disinfection of Aerobic and Anaerobic Sludge Using an Electrokinetic (EK) System." *Water Sc Tech* 57 (2): 231–36. https://doi.org/10.2166/wst.2008.009.
Huang X, Y Qu, Clément A. Cid, C Finke, M. R. Hoffmann, K Lim, and S. C. Jiang. 2016. "Electrochemical Disinfection of Toilet Wastewater Using Wastewater Electrolysis Cell." *Water Research* 92: 164–72. https://doi.org/10.1016/j.watres.2016.01.040.
Jahromi A. F. 2016. "Enhanced Electro-Oxidation For TKN Removal From Highly Polluted Industrial Wastewater." In. LondonOntario: *14th International Environmental Specialty Conference CSCE Proceeding 2016*.
Li W, Q. Zhou, and T. Hua. 2010. "Removal of Organic Matter from Landfill Leachate by Advanced Oxidation Processes: A Review." *International Journal of Chemical Engineering* 2010: 1–10. https://doi.org/10.1155/2010/270532.
Martínez-Huitle C.A., and L.S. Andrade. 2011. "Electrocatalysis in Wastewater Treatment: Recent Mechanism Advances." *Quimica Nova* 34 (5): 850–58. https://doi.org/10.1590/S0100-40422011000500021.
Navab D. T., R. Beton, R. J. Hill, R. Gehr, and D. Frigon. 2012. "Inactivation Mechanisms of Bacterial Pathogen Indicators during Electro-Dewatering of Activated Sludge Biosolids." *Water Research* 46 (13): 3999–4008. https://doi.org/10.1016/j.watres.2012.05.009.
Neyens E., and J. Baeyens. 2003. "A Review of Classic Fenton's Peroxidation as an Advanced Oxidation Technique." *Journal of Hazardous Materials* 98 (1–3): 33–50. https://doi.org/10.1016/S0304-3894(02)00282-0.
Ortega-Gómez E., P. Fernández-Ibáñez, M. M. Ballesteros Martín, M. I. Polo-López, B. Esteban García, and J. A. Sánchez Pérez. 2012. "Water Disinfection Using Photo-Fenton: Effect of Temperature on Enterococcus Faecalis Survival." *Water Research* 46 (18): 6154–62. https://doi.org/10.1016/j.watres.2012.09.007.
Polo-López M. I., Irene García-Fernández, T. Velegraki, A. Katsoni, I. Oller, D. Mantzavinos, and P. Fernández-Ibáñez. 2012. "Mild Solar Photo-Fenton: An Effective Tool for the Removal of Fusarium from Simulated Municipal Effluents." *Applied Catalysis B: Environmental* 111–112: 545–54. https://doi.org/10.1016/j.apcatb.2011.11.006.
Selvakumar A., M. E. Tuccillo, S. Muthukrishnan, and Asim B. Ray. 2009. "Use of Fenton's Reagent as a Disinfectant." *Remediation* 19 (2): 135–42. https://doi.org/10.1002/rem.20208.
Safaei T. E., Elektorowicz. M, R. Reimers; J. A. Oleszkiewicz, *Water Environmental Federation Proceeding* " Potential Sterilization of Biosolids by the BioElectro™ Process":2771-2777. https://doi.org/10.2175/193864713813674054
Yin Ziqiang, M. Hoffmann, and S. Jiang. 2018. "Sludge Disinfection Using Electrical Thermal Treatment: The Role of Ohmic Heating." *Science of the Total Environment* 615. The Authors: 262–71. https://doi.org/10.1016/j.scitotenv.2017.09.175.
Yoon J., Y. Lee, and S. Kim. 2001. "Investigation of the Reaction Pathway of OH Radicals Produced by Fenton Oxidation in the Conditions of Wastewater Treatment." In *Water Science and Technology*, 44:15–21.

Zhang H, X. Wu, and X. Li. 2012. “Oxidation and Coagulation Removal of COD from Landfill Leachate by Fered-Fenton Process.” *Chemical Engineering Journal* 210. Elsevier B.V.: 188–94. https://doi.org/10.1016/j.cej.2012.08.094.

Zhang H, D. Zhang, and J.y. Zhou. 2006. “Removal of COD from Landfill Leachate by Electro-Fenton Method.” Journal of Hazardous Materials 135 (1–3): 106–11. https://doi.org/10.1016/j.jhazmat.2005.11.025.

Building Tomorrow's Society
Bâtir la Société de Demain

Fredericton, Canada
June 13 to 16, 2018 / *13 au 16 juin, 2018*

LEACHING BEHAVIOR OF NUTRIENTS AND METALS OF GREEN ROOFS: LABORATORY AND FIELD STUDIES

Akther, Musa[1,3], He, Jianxun[1], Chu, Angus[1], van Duin, Bert[2]
[1] University of Calgary, Department of Civil Engineering, 2500 University Drive NW, Calgary, AB, Canada.
[2] The City of Calgary, Calgary, AB, Canada.
[3] msakther@ucalgary.ca

Abstract: Despite the benefits of green roofs in managing urban stormwater, chemical leaching especially at their initial stage can be problematic. Although chemical leaching from green roofs has been acknowledged, to date research in this aspect is still limited. In this paper, both laboratory and field experiments were conducted to investigate and characterize the temporal evolution of nutrient and metal leaching. In the laboratory study, three cells were constructed with three types of growing media and their chemical leaching was monitored under simulated rain events; while the chemical leaching from a full-scale green roof on the Calgary's Municipal Building was monitored in natural events and compared to a nearby conventional roof and the cells. The water samples (collected from the cells, the green roof, and the conventional roof) were analysed for nutrients and metals. Statistical approaches were adopted to investigate their leaching behaviour and to assess the primary influential factor which could be used to capture their temporal evolution. The observations showed that both the laboratory cells and full-scale green roof acted as a source of these pollutants (except two metals in the full-scale green roof). The laboratory observations also showed that nutrient leaching in general decreases with the cumulative inflow and the leaching response to the cumulative inflow appeared to be different for different types of media. Similar results were also observed from the full-scale green roof, especially for PO_4^{3-}, and TP. the results would help in enhancing the design, application and modelling of green roofs.

1 INTRODUCTION

In urban settings, population growth and consequent urbanization have posed significant adverse impacts on urban ecosystem due to the sharpened stormwater runoff hydrograph and degraded stormwater runoff quality. Thus, it is essential to eliminate or attenuate such adverse effects to maintain sustainable urban environment. As the conventional stormwater management technologies have shown their constraints (e.g. space consuming, low efficiency in removing many pollutants, and high cost in construction, maintenance, and rehabilitation) in managing urban stormwater, low impact development technologies (LIDs) as an alternative to the conventional technologies have recently attracted attention. The green roof is one of typical LIDs and has been demonstrated to have dual benefits in managing urban stormwater (in both quantity and quality) (Mentens *et al.*, 2006; Li and Babcock 2014). However in contrast to the concept green roofs could act as a source rather than a sink of various pollutants (especially nutrients) in their early age, as the growing media (GM) of green roofs are often supplemented with compost or fertilizer to sustain vegetation growth (Bliss *et al.*, 2009). In addition, metal leaching is common and its management is challenging (Alsup *et al.*, 2011). The existing body of knowledge demonstrates that the behaviour of a green roof (source or sink) is dependent on GM properties, roof design and age, vegetation type, and the use of fertilizer or compost (Emilsson *et al.*, 2012). However green roofs could be sinks eventually after exchangeable pollutants are exhausted (Alsup *et al.*, 2011).

Although chemical leaching from green roofs has been well acknowledged, to date very limited research has been conducted to link their chemical leaching behavior to potential influential factors including design and hydrological variables qualitatively and quantitatively. Furthermore, to the authors' best knowledge, no research so far has focused on the temporal behaviour of chemical leaching and the investigation of the translatability of knowledge from laboratory to field in this aspect. Therefore, this paper observed chemical leaching of green roofs through both laboratory and field experiments aiming to quantify the effects of the influential factors quantitatively. The results would help in enhancing green roof design, application and modelling to reduce chemical leaching.

2 MATERIAL AND METHODS

2.1 Laboratory Experiments

The laboratory experiments were conducted using simulated events to investigate the effects of design variables (GM type) and hydrological variables (rainfall intensity and antecedent moisture condition (AMC)) on leaching of nutrients and metals. Three laboratory cells were constructed using three types of GM including ZinCoblend-SI (GM1), Eagle Lake rooftop media blend (GM2), and SOPRAFLOR I (GM3)), which are available locally or applied in the City of Calgary. The three cells were exposed to three rainfall magnitudes (2-, 5- and 10-yr storms) and three degrees of AMC (dry (<20%), normal (20-30%), and wet (>30%)). Each cell (Figure 1a) has a surface area of 0.57m x 0.41m, a depth of 150 mm, and a 1.5cm diameter drainage hole at the bottom. The laboratory cells have same structure as the full-scale extensive green roof (GR) investigated in the paper. The cells consist of a layer of 150-mm GM vegetated with sedum, a filter sheet, a drainage layer (FD 25-E), and protection mat (SSM 45). Sedum are locally available and most commonly adopted due to their tolerance to extreme weather condition (e.g., extreme temperature and high wind speed) and the limited water consumption criteria (VanWoert *et al.*, 2005; Berrett *et al.*, 2014). The slope of the cells was fixed at 1% to mimic the slope of the GR. The threshold values of AMC were determined based on average GM moisture (using 2015-2016 data) measured from the GR. The rainfall duration was fixed at 1-hour in all experiment runs as it has been used to examine the operation of source control practices such as green roofs in the City of Calgary (The City of Calgary, 2011). Thus, the rainfall intensities, which is uniform over the duration, corresponding to 2-, 5-, and 10-yr storms are 14.09, 20.27, and 24.31mm/hr, respectively.

Due to varying these design and hydrological variables, a total of 27 experiments were conducted. In each experiment run, deionized water was applied to the cells, thus no additional pollutants were added to the cells. A composite sample from each experiment run was collected and analyzed for nutrients including nitrate (NO_3^-), ammonia (NH_4^+), total nitrogen (TN), orthophosphate (PO_4^{3-}), and total phosphorus (TP), and metals including zinc (Zn), copper (Cu), and lead (Pb). Their Event Mean Concentrations (EMCs) were reported. In addition, the GMs were analyzed for their initial contents of nutrients and metals using the water extraction method (Hurley *et al.*, 2017), which is a modified version of the field leach test of U. S. Geological Survey. All water samples were analyzed for nutrients using Hach methods and for metals using 7900 ICP-MS at Calgary's laboratory in the Bonnybrook Wastewater Treatment Plant.

2.2 Field Experiments

A 335-m^2 extensive GR (Figure 1b) and a 335-m^2 conventional roof as the reference roof (RR), which are situated on the 5th and 4th floor of Calgary' Municipal Building in the downtown core, have been monitored since July 2015. Note that the 50% of GR surface is covered by gravels and hard roof. The GR was constructed using GM1 and its construction was completed in the summer of 2014. The GR and the RR were monitored in natural events using automated flow monitoring (using magmeter) and water sample collection systems. The automatic water samplers collected discrete samples (maximum 24 samples) in each event. A total of nine species of vegetation are randomly distributed in 85 quadrats of the GR. Irrigation was applied to during the vegetation establishment periods. Soil moisture sensors were deployed at five locations and two depths (top and 5 cm deep) at each location to record moisture of GM. The water samples were analyzed for nutrients and metals. During the study period, eight events (five events in 2015 and three events in 2017) were monitored for the GR and the RR and the EMCs of nutrients and metals were calculated and used in the following analysis.

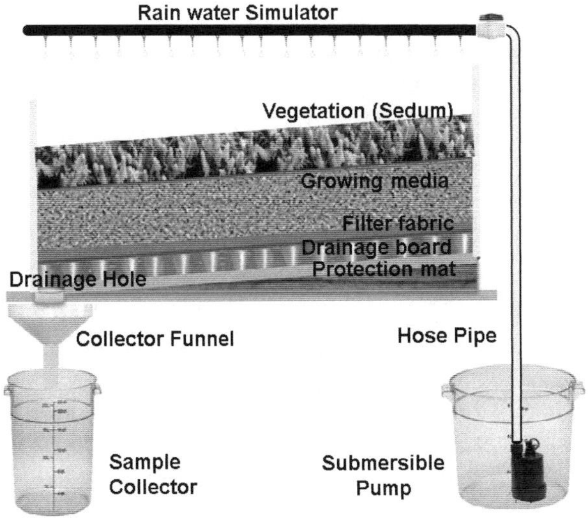

(a)

(b)

Figure 1. (a) The diagram of laboratory cells and (b) the full-scale green roof

2.3 Data Analysis

Considering the small sample size of the datasets, nonparametric Kruskal Wallis test was applied to compare the medians of the datasets. To investigate the sensitivity of the EMCs to the influential factor (cumulative inflow), the linear regression approach was employed. The significance of the regression slopes was tested using the t-test. In addition, the analysis of covariance (ANCOVA) test followed by the multiple comparison was applied to compare the slopes of the regression lines, which were detected to be significant, among or between the GMs for each water quality parameter. In this paper, a significant level of 10%, which is also often used in hydrological studies (e.g., Aziz and Burn, 2006; Hamed, 2008), was used.

3 RESULTS AND DISCUSSION

3.1 Initial Chemical Properties of the Growing Media

GM1 is high quality recycled materials and minerals and enhanced with compost; GM2 is the mixture of peat moss, fir bark fines, compost, sand, pumice and perlite; while GM3 composes of pumice, sand, vegetable compost, perlite, and blond peat mixed with high porous mineral aggregate. The initial chemical contents of the three GMs are presented in Table 1.

As illustrated in Table 1, the chemical properties in terms of the nutrient and metal contents are initially different among the GMs. The initial contents of TN and all metals are highest in the GM2; while the TP content is highest in the GM3. Similarly, a study conducted by the Toronto and Region Conservation Authority (TRCA) (2006) found large variations in both nutrient and metal contents among 11 commercially available GMs, for example TP and TKN (a component of TN) in the ranges of 239 - 1469 mg/kg and 191 - 3324 mg/kg, respectively; Zn, Cu, and Pb in the ranges of 25 - 133 mg/kg, 5 - 49 mg/kg, and 3 - 26 mg/kg, respectively. The initial TP content of the three GMs studied in the paper is below the range reported by TRCA; whereas TN content is within the range reported by TRCA. However, the contents of metals of the GMs (except GM3) are much higher than those reported by TRCA.

Table 1. Initial nutrient and metal contents of the three types of growing media.

Chemical content (mg/kg)	Type of GM		
	GM1	GM2	GM3
NO_3^-	110.00	444.87	147.93
NH_4^+	41.27	39.27	16.53
TN	239.40	819.00	259.80
PO_4^{3-}	109.33	88.87	168.87
TP	130.80	96.80	190.60
Zn	412.00	648.00	14.00
Cu	294.00	488.00	22.00
Pb	318.00	480.00	12.00

3.2 The Degree of Chemical Leaching in the Experimental Study

In a previous study conducted by the authors (Akther *et al.*, 2018), the GM type was identified as the most prominent influential factor of chemical leaching among the investigated design and hydrological variables including the thickness of GM, rainfall magnitude, and antecedent moisture condition. Therefore, in this paper the EMCs of the investigated water quality parameters were pooled and analyzed for individual GM type. Several descriptive statistics of the EMCs for each GM type are presented in Table 2. In addition, the results from the comparison of the medians of the EMCs among the three GMs using the Kruskal Wallis test are also included in Table 2. The presence of all pollutant constituents in outflow demonstrates that all the cells act as the sources of these pollutants. In addition, significant differences in the medians of the EMCs among three or between two GMs were detected in all nutrients and Zn in the Kruskal Wallis test.

As shown in Table 2, the medians/means and standard deviations of the EMCs of TN, NO_3^-, and Zn of GM2 are obviously higher than those of GM1 and GM3; while the highest medians/means and standard deviations of the EMCs of TP and PO_4^{3-} were observed in GM3. In addition, it was found that the medians of the EMCs of TP and PO_4^{3-} are statistically significantly higher in GM3 than GM1 and/or GM2; while the medians of the EMCs of TN and NO_3^- were detected to be statistically significantly higher in GM2 than in GM1 and GM3 in the Kruskal Wallis test. The median of the EMCs of NH_4^+ was found to be significantly lower in GM2 than in GM1 and GM3. Compared to NO_3^-, the degree of the leaching of NH_4^+ is much lower. The results in general support that the GM of higher initial contents of nutrients (TP, PO_4^{3-} and TN) leaches more nutrients and thus the degree of leaching of these constituents is associated with their initial contents in GM. However similar results were not observed for metals. The initial metal contents are highest in GM2. Although the highest medians of the EMCs of the metals were measured from GM2, only the median of the EMCs of Zn was found to be statistically significantly higher than GM1 and GM3. The medians of the EMCs of Cu and Pb are statistically equivalent among the GMs, however their initial contents of GM3 are obviously much lower than those of GM1 and GM2. These results might imply that the leaching of metals (especially Cu and Pb) is less dependent on their initial contents of GM compared to nutrients.

3.3 Characteristics of Chemical Leaching in the Experimental Study

The previous study by Akther *et al.* (2018) concluded that the nutrient leaching is primarily explained by GM type not by the thickness of GM and the hydrological variables including the magnitude of simulated event and antecedent moisture condition of GM. From the model development perspective, it is always desired to know the potential linkage between water quality and water quantity. Thus, the sensitivity/dependence of the EMCs of water quality parameters on the cumulative inflow, which is the sum of water volume applied to the laboratory cells from first experiment run to current experimental run, was investigated herein using linear regression approach for each GM type. The developed regression equations along with their coefficients of determination (R^2) are displayed in Table 3. Sample scatter plots of the EMCs of three selected water quality parameters (including TN, NH_4^+ and Zn) and the cumulative inflow are demonstrated in Figure 2. As illustrated in the results, the EMCs of nutrients (except NH_4^+ of GM1 and GM3, and TP and PO_4^{3-} of GM3) and Zn (except of GM3) decrease with the increase of the cumulative inflow. In contrast, the EMCs of Pb (except in GM1) increase with the increase of the cumulative inflow. No consistent result on the dependence of the EMCs on the cumulative inflow was found for NH_4^+ among the three GMs. Therefore, it can be concluded from the experimental study that the degree of leaching of nutrients (except NH_4^+) in general decreases with time due to the increase of the cumulative inflow. The observed leaching behavior of nutrients and Zn is analog to the pollutant wash-off by stormwater surface runoff, which is related to the amount of pollutants available for wash-off on land surface. Whereas the leaching of Cu and Pb is either not significantly dependent on or significantly positively dependent on the cumulative flow. Thus the "wash-off" concept for the leaching of nutrients and Zn is not applicable for the leaching of Cu and Pb.

Table 2. Descriptive statistics of the EMCs of water quality parameters obtained in laboratory experiments and Kruskal Wallis test results followed by multiple comparison.

Water quality	GM type	Mean	Median	Standard deviation	Min	Max	Sample size	Comparison
NO_3^- (mg/L)	GM1	72.50	42.50	51.55	30.00	170.00	8	GM1<GM2 GM3<GM2
	GM2	408.75	366.25	190.43	227.50	740.00	8	
	GM3	91.88	87.50	34.12	50.00	165.00	8	
NH_4^+ (mg/L)	GM1	11.63	11.50	4.34	7.00	21.00	8	GM2<GM1 GM2<GM3
	GM2	3.72	3.00	1.68	2.25	7.00	8	
	GM3	9.81	9.25	4.14	4.50	16.00	8	
TN (mg/L)	GM1	148.75	129.00	75.16	74.00	248.00	8	GM1<GM2 GM3<GM2
	GM2	682.69	615.00	273.36	325.00	1165.00	8	
	GM3	199.13	207.25	52.95	129.00	285.00	8	
PO_4^{3-} (mg/L)	GM1	28.15	22.50	14.69	19.00	62.00	8	GM2<GM3
	GM2	15.69	13.50	4.19	12.25	23.50	8	
	GM3	66.97	64.58	12.05	52.60	90.00	8	
TP (mg/L)	GM1	28.31	23.83	9.01	22.00	46.00	8	GM1<GM3, GM2<GM3
	GM2	19.82	17.19	6.10	13.50	30.00	8	
	GM3	74.97	71.50	13.45	57.50	91.00	8	
Zn (μg/L)	GM1	131.25	90.00	89.67	50.00	250.00	8	GM1<GM2 GM3<GM2
	GM2	1655.00	1295.00	699.08	1010.00	2520.00	8	
	GM3	188.75	155.00	128.00	60.00	470.00	8	
Cu (μg/L)	GM1	1.63	1.00	2.07	n/a*	6.00	8	--
	GM2	1.88	1.50	1.64	n/a	4.00	8	
	GM3	1.43	1.00	1.27	n/a	3.00	7	
Pb (μg/L)	GM1	146.75	147.00	56.24	79.00	240.00	8	--
	GM2	259.86	291.00	141.56	103.00	404.00	7	
	GM3	228.43	200.00	144.88	63.00	422.00	7	

* n/a: no analysis. The EMC could not be calculated because BDL (below detection limit) was reported in one or more than one samples in some simulated events.

Table 3. Regression equations (whose slopes are statistically significant) of the EMCs of water quality (y) on the cumulative inflow (x in the unit of m^3) for each GM type

Water quality	GM type	Regression equation	R^2
NO_3^- (mg/L)	GM1	$y = -3.77x + 162.93$	0.80
	GM2	$y = -14.53x + 757.10$	0.87

	GM3	y = -2.01x + 139.99	0.52
NH_4^+ (mg/L)	GM1	--	--
	GM2	y = -0.11x + 6.32	0.63
	GM3	y = 0.24x + 4.15	0.49
TN (mg/L)	GM1	y = -4.99x + 268.38	0.66
	GM2	y = -20.66x + 1,178.07	0.86
	GM3	y = -4.13x + 298.23	0.91
PO_4^{3-} (mg/L)	GM1	y = -0.87x + 49.02	0.53
	GM2	y = -0.21x + 20.78	0.38
	GM3	--	--
TP (mg/L)	GM1	y = -0.59x + 42.51	0.65
	GM2	y = -0.36x + 28.48	0.52
	GM3	--	--
Zn (µg/L)	GM1	y = -6.70x + 260.60	0.82
	GM2	y = -49.00x + 2,830.13	0.74
	GM3	--	--
Cu (µg/L)	GM1	--	--
	GM2	--	--
	GM3	--	--
Pb (µg/L)	GM1	--	--
	GM2	y = 9.59x + 26.41	0.80
	GM3	y = 9.98x - 29.13	0.68

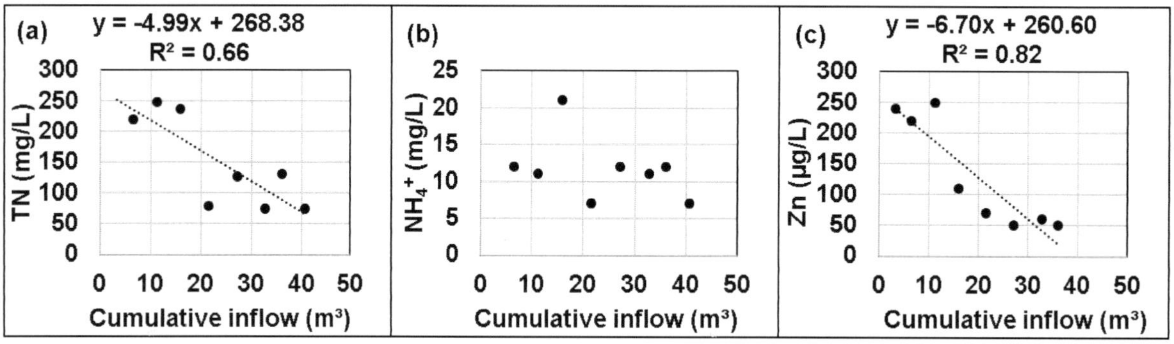

Figure 2. Scatter plots between the EMCs of TN, NH_4^+, and Zn and the cumulative inflow for GM1. The linear regression lines (the dashed lines) whose slopes are statistically significant are included.

To investigate which GM is more sensitive to the cumulative inflow among the three GMs, the regression slopes which are statistically significant were compared for each investigated water quality parameter using the ANCOVA test followed by multiple comparison. The results are summarized in Table 4. No significant differences in the regression slopes were identified for PO_4^{3-} and TP between GM1 and GM2; while these two pollutant constituents were found not to be statistically significantly dependent on the cumulative inflow in GM3 due to the absence of significant regression slopes (Table 3). As discussed previously, highest medians of the EMCs of PO_4^{3-} and TP were observed from GM3. Thus, the contents of PO_4^{3-} and TP in GM3 might be more than enough for wash-off so that their decreases did not show in the experiment study. In contrast, the EMCs of both TN and NO_3^- of GM2 are statistically most sensitive to the cumulative inflow among the GMs; while the degree of sensitivity of their leaching to the cumulative inflow were detected to be similar between GM1 and GM3. The leaching of Zn from GM2 is more sensitive to the cumulative inflow than that from GM1. The results demonstrate that these pollutant constituents leach at different rates among the three GMs, which might be ascribed to the chemical properties of the GMs. For instance, the leach rates of TN and Zn were statistically significantly higher in GM2, which has highest initial contents of TN and Zn. These results argue that the effect of chemical properties of GM on chemical leaching. Furthermore, compared to the leaching of TP, the leaching of TN appears to be faster from all three GMs, as the slopes of regression lines for TN are obviously higher than those for TP. It thus can be concluded that nitrogen is more mobile and thus can leach more easily from the GMs compared to phosphorus.

Table 4. Comparison of the statistically significant regression slopes among three or between two GMs in the laboratory experiments and between the laboratory cell of GM1 and the full-scale green roof (GR).

Comparison	Laboratory (GM1, GM2 &GM3)	Laboratory (GM1) and GR
NO_3^-	GM1<GM2, GM3<GM2, GM1=GM3	n/a*
NH_4^+	GM2<GM3	n/a
TN	GM1<GM2, GM3<GM2, GM1=GM3	n/a
PO_4^{3-}	GM1=GM2	GR <GM1
TP	GM1=GM2	GR <GM1
Zn	GM1<GM2	n/a
Pb	GM2=GM3	n/a

* n/a: no analysis conducted due to the absence of statistically significant regression slopes detected in GR.

3.4 Chemical Leaching from the Full-Scale Green Roof

Several descriptive statistics of the EMCs of water quality parameters for GR and RR as well as the results from the comparison of the medians of the EMCs between GR and RR are presented in Table 5. When using the RR as the reference, the observations show that GR has acted as the source of nutrients since the commence of their operation in May of 2014 as all statistics of GR presented in Table 5 are obviously higher than those of RR. In contrast to nutrients, GR functions as the sink of all three metals as the statistics of metals of GR are lower or slightly lower than those of RR. Furthermore, the medians of the EMCs of all nutrients of GR were detected to be significantly higher than those of RR. The median of Zn EMCs of RR is significantly higher than that of GR; whereas no significant differences in the medians of the EMCs of Cu and Pb were detected between GR and RR. Therefore, it can be concluded that GR functions as the source of nutrients; while GR acts as the sink of Zn during the field study period.

Table 5. Descriptive statistics of the EMCs of water quality parameters obtained in field experiments for both GR and RR and detected significant differences in their medians between GR and RR using the Kruskal Wallis test. Sample size for the analysis for all water quality parameters is 8.

Water quality	Roof type	Mean	Median	Standard deviation	Min	Max	Comparison
NO_3^- (mg/L)	GR	2.26	1.75	1.84	0.81	6.33	GR>RR
	RR	0.82	0.75	0.50	0.19	1.93	
NH_4^+ (mg/L)	GR	0.82	0.49	0.89	0.28	2.94	GR>RR
	RR	0.20	0.13	0.18	0.01	0.49	
TN (mg/L)	GR	4.20	3.62	3.32	1.40	11.21	GR>RR
	RR	1.81	1.63	1.12	0.71	3.91	
PO_4^{3-} (mg/L)	GR	1.32	1.21	0.94	0.43	3.36	GR>RR
	RR	0.27	0.26	0.11	0.12	0.45	

TP (mg/L)	GR	1.45	1.25	1.00	0.45	3.62	GR>RR
	RR	0.31	0.29	0.13	0.15	0.49	
Zn (μg/L)	GR	17.51	15.65	5.13	12.90	28.59	GR<RR
	RR	341.29	363.15	123.83	170.50	488.80	
Cu (μg/L)	GR	9.70	8.28	4.08	5.29	18.04	--
	RR	14.81	13.05	8.67	6.44	33.00	
Pb (μg/L)	GR	1.81	2.00	1.18	0.50	3.82	--
	RR	2.58	2.18	1.61	1.00	5.80	

As the GR and GM1 cell have same growing media and structure, the comparison between GM1 and GR was also conducted. It was found that the medians of the EMCs of all pollutant constituents (except Cu) of GM1 are significantly higher than those of GR. This result is under expectation as the observations from the laboratory experiments represent the initial stage of green roof; while the observations from the GR represents the GR during its age between 1 and 3 years. Therefore, the higher EMCs at the initial stage of green roof operation is speculated and leaching is reduced with time as illustrated in the experimental study, in which the EMCs of many constituents decrease with the increase of the cumulative inflow. Similar to the experimental study, it was observed that the EMCs of all nutrients and metals (except Cu) decrease with time, namely with the increase of the cumulative inflow. The decline (from 2003 to 2004) in nutrient leaching was also observed (TRCA, 2006; Harper *et al.*, 2015). The decline of chemical leaching of GR with time was also identified through comparing the EMCs of water quality parameters between 2015 and 2017, when field data are available. The medians of the EMCs of NH_4^+, PO_4^{3-} and TP were found to be significantly higher in 2015 than those in 2017. On the other hand, opposite results were obtained for RR. Significant increases with time were observed in the EMCs of PO_4^{3-} and TP for RR. In addition, it was also observed that the EMCs of other water quality parameters appear to increase from 2015 to 2017 for RR. The observed increases of the EMCs of some pollutants for RR could be due to the change in atmospheric deposition which is the primary source of pollutants. These results demonstrate that the chemical leaching from GR in general decline with time, which is similar to the observations obtained from the experimental study. However, in the experimental study, Pb was shown to increase with the cumulative inflow from laboratory cells GM2 and GM3, which should be further investigated.

To investigate whether the progression of chemical leaching of GR can also be explained by the cumulative inflow, which was identified in the experimental study, the linear regression equations of the EMCs of water quality parameters on the cumulative inflow were also developed and are presented in Table 6. The cumulative inflow is the accumulation of precipitation (rainfall and snowfall) and irrigation water for GR and the accumulation of precipitation for RR since the start of their operation. As shown in Table 6, the significant regression slope was identified for TP, PO_4^{3-}, and Pb for GR; while it was identified for TP and PO_4^{3-} for RR. The results demonstrate that the leaching of these pollutant constituents can be largely explained by the cumulative inflow. However significant slopes were not detected in the regression equations for TN and NO_3^- from GR, which is different from the laboratory results (GM1). The removal efficiency of nitrogen of various LID (such as bioretention systems and green roofs) have been reported to be also associated with temperature (Buffam *et al.*, 2016; Van Seters *et al.*, 2009). The laboratory experiments were conducted at controlled and approximately constant temperature. Therefore, the role of temperature might be significant in the field. In addition, comparing the regression slopes, which indicate the leaching rates, between GM1 and GR (Table 4), both TP and PO_4^{3-} leach faster from GM1 than from GR, This, result implies that the decline of nutrient leaching could be slower in the late stage compared to the leaching at the beginning. Furthermore, the EMCs of TN and NO_3^- appear to be similar in magnitude in the events collected in 2017 (results not shown), which might explain the absence of the significant slopes for nitrogen for GR. All these results support that nitrogen leaches faster than phosphorous. Phosphate, which often is the major component of TP (approximately 90% in average in this study), can be retained in the ground by biological activities, absorption, and mineralization (Lehmann and Schroth, 2003).

Table 6. Regression equations (whose slopes are statistically significant) for the EMCs of water quality (y) on the cumulative inflow (x) for GR and RR, respectively. Note that the regression equations whose slopes are significant are presented here.

Water quality	Roof type	Regression equation	R^2
PO_4^{3-}	GR	y = -0.004x + 3.73	0.53
	RR	y = 0.001x + 0.037	0.70
TP	GR	y = -0.005x + 3.99	0.52
	RR	y = 0.001x + 0.033	0.78
Pb	GR	y = -0.007x + 5.45	0.77
	RR	--	--

Although the observations from the GR could not be quantitatively explained by the observations from GM1, the qualitative results are similar in general between GR and GM1. In the field, the temporal behavior of nitrogen leaching might not be simply described by the cumulative inflow. However similar to pollutant wash-off by stormwater surface runoff, the chemical leaching especially nutrient leaching could be modeled based on the concept of pollutant wash-off combining with the other processes (e.g., biological and chemical processes).

4 CONCLUSIONS

The laboratory experiments showed that all green roof cells (GM1, GM2, and GM3) are the source of pollution of both nutrients and metals in their early stage. The laboratory study results also demonstrated that the leaching of nutrients (except NH_4^+) in general decrease with time as the cumulative inflow increases; however, their leaching responds to the cumulative inflow at different rates (the regression slopes) in different GMs. Furthermore, the degree of nutrient leaching is qualitatively associated with GM chemical properties. The leaching degree of metals appeared to not be related to their initial contents in GMs, and their leaching (except Zn) is either not significantly or significantly positively dependent on the cumulative inflow. The results supported that different physical processes govern the metal leaching and nutrient leaching, which could be modeled by pollutant "wash-off" process. From the field experiments, GR was identified to be the source of nutrients; whereas it acted as the sink of metals when using RR as the reference. The leaching of TP and PO_4^{3-} from GR responded to the cumulative inflow similarly as observed in the laboratory experiments. However significant dependence between the leaching of TN and NO_3^- and the cumulative inflow was absent, which might be explained by the important role of environmental variables (e.g., temperature). In addition, the comparisons between GR and GM1 and between GR in 2015 and GR in 2017 supported that the nutrient leaching is prominent in the very early stage of green roof establishment. Similar progression of the nutrient leaching was observed from both laboratory and field experiments. The results clearly showed that the progression of nutrient leaching would be primarily captured with the concept of wash-off combining with the other processes (e.g., biological and chemical processes).

Acknowledgements

The authors would like to acknowledge Natural Sciences and Engineering Research Council of Canada (NSERC) and the City of Calgary for the financial support, and the valuable contributions of personnel of the City of Calgary, who have involved with the project.

References

Akther, M., He, J., Chu, A., Valeo, C., Khan, U.T. and van Duin, B., 2018. Response of green roof performance to multiple hydrologic and design variables: a laboratory investigation. *Water Science and Technology* (Submitted).

Alsup S. E., Ebbs S. D., Battaglia L. L. and Retzlaff W. A. 2011. Heavy metals in leachate from simulated green roof systems. *Ecological Engineering*, **37**(11), 1709-1717.

Emilsson T., Berretta C., Walker R., Stovin, V. and Dunnett N. 2012. Water in greenroof substrates – Linking physical measurements to plant performance. *Proceedings of the World Green Roof Congress, September 18–21, 2012, Copenhagen, Denmark.*

The City of Calgary. 2011. *Stormwater management & design manual.* Water Resources, the City of Calgary.

VanWoert, N. D., Rowe, D. B., Andresen, J. A., Rugh, C. L., Fernandez, R. T. and Xiao, L. 2005. Green roof stormwater retention. *Journal of Environmental Quality*, **34**(3), 1036-1044.

Hurley, S., Shrestha, P. and Cording, A. 2017. Nutrient leaching from compost: Implications for bioretention and other green stormwater infrastructure. *Journal of Sustainable Water in the Built Environment*, **3**(3), p. 04017006.

Berretta, C., Poë, S. and Stovin, V. 2014. Moisture content behaviour in extensive green roofs during dry periods: The influence of vegetation and substrate characteristics. *Journal of Hydrology,* **511**, 374-386.

Lehmann, J. and Schroth, G., 2003. Nutrient leaching. *Trees, crops, and soil fertility: concepts and research methods. Wallingford, UK: CAB International*, pp.151-166.

Mentens, J., Raes, D. and Hermy, M., 2006. Green roofs as a tool for solving the rainwater runoff problem in the urbanized 21st century?. *Landscape and urban planning*, ***77***(3), 217-226.

Li, Y. and Babcock, R.W., 2014. Green roofs against pollution and climate change. A review. *Agronomy for Sustainable Development*, **34**(4), pp.695-705.

Aziz, O. A. and Burn, D., 2006. Trends and variability in the hydrological regime of the mackenzie river basin. *Journal of Hydrology,* **319**, 282–294.

Hamed, K.H., 2008. Trend detection in hydrologic data: the Mann–Kendall trend test under the scaling hypothesis. *Journal of hydrology*, **349**(3), pp.350-363.

Bliss, D.J., Neufeld, R.D., and Ries, R.J., 2009. Storm water runoff using a green roof. *Environmental Engineering Science* **26** (2), 407-417.

TRCA (The Toronto and Region Conservation Authority), 2006. *"Evaluation of an Extensive Green roof, York University, Toronto, Ontario."* Available at: https://sustainabletechnologies.ca/app/uploads/2013/03/GR_york_fullreport.pdf, last visited: 26 January 2018.

Buffam, I., Mitchell, M.E. and Durtsche, R.D., 2016. Environmental drivers of seasonal variation in green roof runoff water quality. *Ecological Engineering*, ***91***, pp.506-514.

Van Seters, T., Rocha, L., Smith, D. and MacMillan, G., 2009. Evaluation of green roofs for runoff retention, runoff quality, and leachability. Water Quality *Research Journal of Canada*, **44**(1), p.33.

Harper, G.E., Limmer, M.A., Showalter, W.E. and Burken, J.G., 2015. Nine-month evaluation of runoff quality and quantity from an experiential green roof in Missouri, USA. *Ecological Engineering*, **78**, 127-133.

Building Tomorrow's Society
Bâtir la Société de Demain

Fredericton, Canada
June 13 to 16, 2018 / *13 au 16 juin, 2018*

MAJOR CIVIL ENGINEERING PROJECTS BOOST LOCAL WORKFORCE COMPETENCIES

Diane Ingraham[1], Ph.D., PMP, Wayne Tucker[2], M.E.Des (ES) and Willie McNeil[3], B.Tech. (Env.), CET
[1] Stantec Consulting Ltd, Dartmouth Nova Scotia, Canada
[2] Stantec Consulting Ltd, Corner Brook, Newfoundland and Labrador, Canada
[3] Stantec Consulting Ltd, Sydney Nova Scotia, Canada
[4] Diane.Ingraham@Stantec.com

Abstract: Major civil engineering projects provide opportunities to develop new competencies through strategic government policies and corporate mindfulness. The experiences gained on two such recent projects in Atlantic Canada are examined. The projects are: the "Tar Ponds Project", a major environmental remediation and cleanup of the Sydney Tar Ponds and Coke Ovens site after 100 years of steel-making in Sydney, Nova Scotia, and the "Muskrat Falls Regulatory Compliance Project" for the 824 MW Hydroelectric Development on the Lower Churchill River in Labrador. Both multi-year heavy civil engineering projects shared challenges of large footprint on the land, multiple regulators and government levels, and very engaged stakeholders, including Aboriginal communities. This paper compares these projects to extract and consolidate practical project management insights gained to meet benefits agreement requirements through incorporating and coordinating local community workforces with specialist consulting engineers and scientists. These practical lessons learned inform civil engineering managers responsible for managing similar heavy civil and environmental engineering projects.

1 INTRODUCTION

It is good business for large complex engineering projects (major projects) to build in and incorporate mechanisms for sharing needs, understandings and boosting competencies: from job entry to highly specialized and skilled workers, to the professional development and evolution of engineers, scientists, and managers. Major projects can be a factor in community sustainability through improved "job-worthiness".

Local workforce competencies are enhanced on two levels on major projects—Leadership and Community Workforce. Leadership: engineers and engineering project managers develop leadership capabilities on major projects where there is uncertainty, ambiguity, and high-risk potential (financial and/or reputation). Many levels of leadership must coalesce individuals, teams, organizations, and environment. Leaders in these projects need to be different, flexible, responsive, adaptive, inclusive, and communicative. (Remington 2011) Sense and Kiridena discuss current theories of project management competencies for managing complex projects noting the "greater recognition of the significance of human, behavioural and social dimensions of contemporary project-based undertakings" (Sense and Kiridena 2014). From a practitioner's perspective, both Remington and, Sense and Kiridena provided useful insights during the lifetime of their major projects. Community Workforce: Benefits to local communities has been seen by governments as key to the long-term sustainability of the communities affected by major projects. Legislation at provincial/state and federal government levels provides a catalyst for change by project owners and their contractors. The strategy and legislation are evolving. The United Nations 2030 Sustainable Development Goal #8: "Promote sustained, inclusive and sustainable economic growth, full and productive employment and decent work for all" is a lofty one (United Nations 2018) and one response invites governments to rethink local economic development for major projects (Ovenden 2018). From labourers to highly skilled personnel, major projects offer employment and financial stability. They are also rich training environments presenting individuals with new ways of knowing and behaving in the workforce. The insights the authors' project teams gained on two major, transformative projects in Atlantic Canada, were to get involved early before the major project starts, to partner with training and community employment interest organizations, to provide internal training, to embrace diversity, to build individual competencies, resilience, and retention, and to build job awareness with opportunities offered to expand career capabilities for local workforces. It starts with people first. Providing local workforces with opportunities and experience in large transformational projects appears to have benefits that last beyond the project. Governments can provide leadership and incentives through enabling legislation and training programs available to project owners and developers to encourage engagement of local workforces, provide appropriate training, and operate in a manner that transfers and retains expertise to local workers.

2 TWO MAJOR ENVIRONMENTAL / HEAVY CIVIL ENGINEERING PROJECTS

The two projects discussed in this paper are both transformational projects in building communities and infrastructure that are liveable and sustainable. The projects technically transformed areas, through the engineering and construction that occurred there, and they impacted workforces in communities with sensitive culture and traditions by implementing programs to attract, train and hire local workers at all levels. This paper considers two major engineering projects with similarities and differences that supported competency augmentation in the surrounding communities.

2.1 Tar Ponds Project

The Sydney Tar Ponds and Coke Ovens Remediation Project (Tar Ponds Project) was an environmental clean up of a tidal estuary and lands impacted by 100 years of steel-making and coke production in Sydney, Cape Breton, Nova Scotia. The funding for the cleanup was split between the Canadian federal government (70%) and the Nova Scotia provincial government (30%) under a cost-sharing agreement. Together these two levels of government formed the "Owners" of the project. In 2004 at $400M, it was the largest remediation project active in Canada. The project affected local workforces in the First Nations communities of Cape Breton including Membertou and Escasoni as well as non-aboriginal communities of Sydney centre, the Pier, and Ashby. (Ingraham, McNeil and Burke 2014)

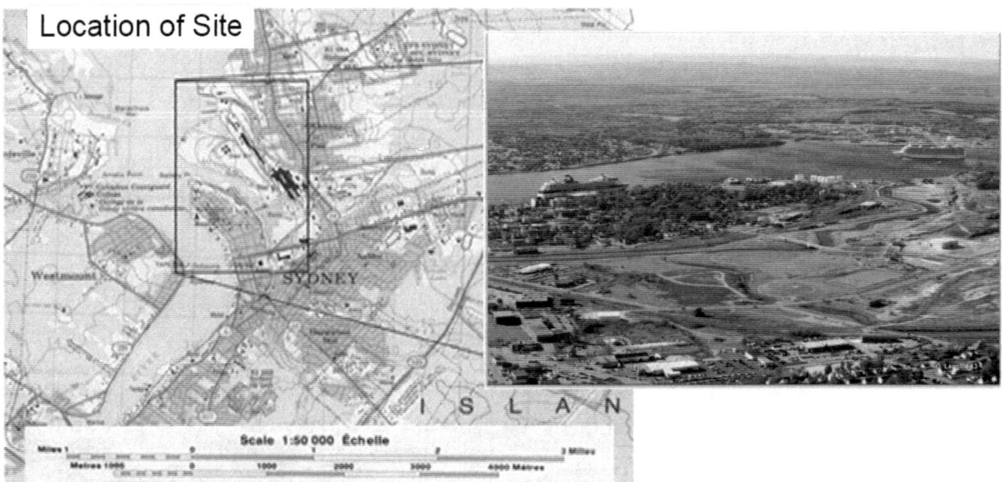

Figure 1 Sydney Tar Ponds Location (Ingraham, McNeil and Burke 2014)

The governments of Canada and Nova Scotia used the Sydney Tar Ponds Agency (STPA), a Special Operating Agency of the government of Nova Scotia, to manage and implement the remediation. The remediation plan, which underwent an environmental assessment (EA) required by the Canadian Environmental Assessment Act and the Nova Scotia Environment Act, described the project and served to determine the need for an environmental assessment and to promote efficient coordination of the environmental assessment. (AMEC 2004) The authors Ingraham (as STPA's Quality Contracts Manager) and McNeil (as Independent Quality Assurance Consultant - Manager) were responsible for quality compliance of work performed by the remediation contractors as mandated by the Environmental Assessment commitments for the project.

Composition of Local Workforce and Initiatives: The area boasts a diverse heritage of English, Scottish, Acadian, African Canadian, Mi'kmaq and Eastern European people. It has been economically depressed since the decline and eventual closure of the coal mining industry, steel making industry and the commercial fishery. While the workforce was well trained in specific aspects of commercial and industrial roles, initiatives relating to engaging and developing workforce competencies were a requirement to train and engage the appropriate employees to work on the remediation project. Some of the initiatives undertaken to engage and prepare the work force and the outcomes are listed in Table 1.

2.2 Muskrat Falls Regulatory Compliance Project

The Muskrat Falls Regulatory Compliance and Environmental Effects Monitoring Project (Muskrat Falls Regulatory Compliance Project) is a part of the larger Nalcor Energy Lower Churchill Project (LCP) Hydroelectric Power Development in Newfoundland and Labrador that was sanctioned in December 2012 (Government of NL 2012) by the provincial government (NL). The LCP is located within the traditional lands of Innu Nation. Locally, it has direct impacts on the communities of Happy Valley – Goose Bay, Sheshatshiu, Mushuah, North West River, Mud Lake, and influences Inuit, Métis, Settler and other surrounding communities and peoples. The area around the project is remote and sparsely settled. (Ingraham and Tucker 2017)

The LCP obtained agreement for Government of Canada guaranteed debt (through long-term (>35 years) bonds) and various other debt vehicles, and equity agreements (Nalcor, Government of Newfoundland and Labrador, and Emera) (Nalcor Energy 2018a) (Nalcor Energy 2018b). The LCP has a Regulatory Compliance Plan for Generation that includes Environmental Compliance (Nalcor Energy 2014) which outlines the framework covering environmental regulatory requirements and permitting required during construction as well as Environmental Assessment (EA) and Joint Panel (CEAA 2018) commitments. As part of the Consultant's team, the authors Ingraham (Senior Project Manager) and Tucker (Project Manager) were responsible for the overall planning and design leadership for the Environmental Compliance of three Muskrat Falls Regulatory Compliance Project programs: Avifauna Management, Historic Resources Management, and Environmental Effects Monitoring.

The Muskrat Falls Regulatory Compliance Project inherited the various requirements of the larger LCP, including compliance with the commitments made regarding employment, procurement, gender equity and diversity and monitoring and reporting. These are consistent with the Impact and Benefit Agreement (IBA) between Nalcor, NL government and Innu Nation ratified June 2011 and LCP's collective agreements set up as a part of the Construction Benefits Strategy (Nalcor Energy 2015). The Muskrat Falls Regulatory Compliance Project also inherited commitments made under the *Tshash Petapen* Agreement ("New Dawn" Agreement) with Innu Nation (ECNL 2008).

Table 1: Some Initiatives on Tar Ponds Project

Initiative	Outcomes
Local Economic Benefits (LEB) requirement for local employment	Tenders specified targets with defined priorities: Cape Bretoners first, Nova Scotians second, and others third); requirement for employment of African Nova Scotians; a built-in 15% weighting in the evaluation scores; LEB statistics were reported monthly to the STPA by every company working on the project.
Nova Scotia's first **Aboriginal Set-Aside** Procurement Strategy (from October 2005 Protocol Agreement with First Nations) (Collier MacDonald 2005)	Allocated $20M for Mi'kmaq First Nations owned companies for capacity and skills development. First Nations businesses received a total of $71.1M out of $323.5M as the experience gained on smaller works generated opportunities for partnering with other firms on larger works.
$1M **Funding** for Unama'ki Economic Benefits Office	Funding provided to upgrade skills of First Nations individuals in Cape Breton and identifying business opportunities for First Nations firms now has expanded operations to all of Nova Scotia and changed its name to Mi'kmaq Economic Benefits Office of Nova Scotia (MEBO)
Training programs – Diversity and Inclusion: African Nova Scotians; Older Workers program and skills gap program; women in non-traditional careers directly or through the Anne Terry Project, a local women's employment and development support centre	<u>Organization</u>: Involvement with Nova Scotia Community College (NSCC), Cape Breton University (CBU), Building Environmental Aboriginal Human Resources (BEAHR), Aboriginal Skills and Employment Partnership (ASEP); portfolio development program (24 participants); 14-week career decision-making program in trades and technology; leadership and sexual harassment awareness workshops (78 participants); 6 student scholarships ($18K); Sydney African Nova Scotian Employment Centre (ANSEC); *Women in Trades*; *Techsploration* for high school women; *Women Unlimited* training
Training Programs – Engineering Consultant Initiative: Work in Partnership with NSCC to develop a customized project training program: Construction-Enviro Worker Training Program	<u>Course Purpose</u>: A 1-year fast-tracked accredited program to train up to seventeen students from the five First Nation communities within Cape Breton in specifically identified areas of heavy civil construction and site remediation. This met the objective to provide inclusion opportunities and the necessary training to assist learners in obtaining supervisory or oversite positions within the project.
OVERALL	<u>First Nations:</u> 85 persons employed (70 Full Time Equivalents - FTEs) (5.4% of total workforce); 2.6M person hours (1301 FTEs) of which 89% were in Cape Breton. <u>Contractors:</u> 28 out of 32 contractors were from Cape Breton. Local Project Expenditures (as at March 31, 2013) 52% Cape Breton, 11% other Nova Scotia, 93% to Canadian firms or individuals. <u>Women</u>: $409K – women in trades and technology; 1-day annual Career Course for Young Women (38 participants) (offered 3 years). <u>African Nova Scotians</u>: $241K – ANSEC; 10 student jobs; tours and presentations to students; $25K for tuition of a NS engineering student/intern of African descent. The Economic Development Officer obtained outside funding for training, including an Older Worker program (36 participants) and skills gap training (16 participants); 28 bursaries
While these initiatives were successful in terms of participation, overall, African Nova Scotians did not maximize opportunities, due to capacity issues and conflict in the community. The African Nova Scotian Employment Centre struggled with communication, building awareness, recruitment, and responsiveness to businesses. A few interviewees felt more initiatives/opportunities should have been provided. It was noted that for women and African Nova Scotians, transforming acquired skills into employment was a challenge. (PWGSC 2014)	

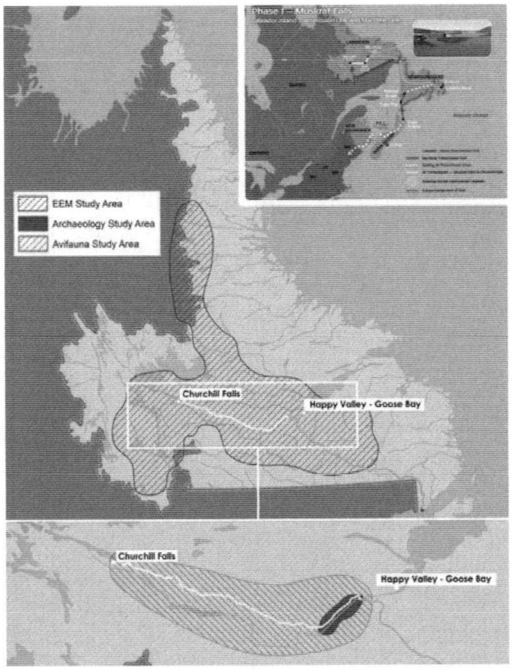

Figure 2: Muskrat Falls - Project Site Location (Ingraham and Tucker 2017)

<u>Composition of Local Workforce and Initiatives:</u> The local area had limited availability to meet the demands of the project. Using local people wherever possible, supported by training programs to prepare them was a requirement. Human resources were drawn locally, regionally, and beyond with professionals, technical staff, and skilled labourers employed. The make-up of the workforce was self-identified Aboriginal (Innu, Inuit, Mi'kmaq), Métis, and forces drawn from corporate resources mostly from Atlantic Canada. The gender ratio was an approximately even split. (Ingraham and Tucker 2017)

Table 2: Muskrat Falls Regulatory Compliance Project: Composition of Human Resources

Year	Male	Female	Aboriginal	Non-Aboriginal	Labradorean	Non-Labradorean
			Employee Numbers			
2013	21	22	26	17	28	15
2014	36	26	25	37	34	28
2015	23	19	24	18	31	11
2016	29	20	31	18	36	13
2017	20	11	20	11	20	11
Total	129	98	125	101	149	78
% Ratio*	57%	43%	56%	44%	66%	34%
			Employee Total Field Hours			
2013	6894.0	6,929.0	8,196.0	5,627.0	9,057.0	4,766.0
2014	13,122.5	11,153.5	12,134.0	12,142.0	15,437.5	8,838.5
2015	9,295.0	6,424.5	8,633.0	7,036.5	10,471.0	5,248.5
2016	13,160.5	7,629.0	13,314.0	7,475.5	15,156.0	5,633.5
2017	9,328.5	5,056.5	10,431.5	3,953.5	9,896.0	4,489.0
Total	51,800.5	37,192.5	52,708.5	36,234.5	60,017.5	28,975.5
% Ratio*	58%	42%	59%	41%	67%	33%

*Ratios are determined for pairs - Male:Female; Aboriginal:Non-Aboriginal; and, Labradorean:Non-Labradorean

Table 3 Muskrat Falls: Field Team and Field vs. Project Hours Breakdowns for all Components

	Field Team Breakdown				Field vs. Project Hours Breakdown			
Year	Tech Hours	Team Lead Hours	Total Field Hours	Lead vs. Total	Field Hours	Non-Field Hours	Total Project Hours	Field vs. Total
2013	9,668.5	4,145.5	13,823.0	30%	13,823.0	10,408.2	24,231.2	57%
2014	17,731.0	6,545.0	24,276.0	27%	24,276.0	18,560.5	42,836.5	57%
2015	9,874.0	5,845.0	15,719.5	37%	15,719.5	13,764.3	29,483.8	53%
2016	16,552.5	4,237.0	20,789.5	20%	20,789.5	11,834.5	32,640.0	64%
2017	11,203.5	3,181.5	14,385.0	22%	14,385.0	8,076.3	22,461.3	64%
Total	65,029.5	23,954.5	88,993.0	27%	88,993.0	62,643.8	151,652.8	59%

This resulted in several initiatives relating to developing workforce competencies and their outcomes summarized in Table 4.

3 PRACTICAL CIVIL ENGINEERING PROJECT MANAGEMENT INSIGHTS

Major projects offer engineers, scientists, and managers the chance to develop practical skills and behaviours that enhance their abilities as leaders. Globally, the technical professions are moving to be relevant and to be forces for change. Publicly visible major projects cannot be planned, designed, constructed nor maintained without considering the communities they affect. Major projects need the "social license" to proceed, which can be earned through generating new learning opportunities and increasing the public understanding of overall long-term community benefits and the environmental safeguards. Engineering companies should set up projects from the start to improve competencies for Aboriginal or First Nations persons, women, visible minorities, and youth. These projects provide opportunities for mentoring, role modeling, direct training to improve skills and fill gaps, to build individual and community resilience, and to have a diverse workforce free from paternalistic and out-moded policies and practices. (For more about individual resilience in the construction industry, see (Chen, McCabe and Hyatt 2017).) Mandatory local benefits programs and agreements may still be needed in the short term with Special Projects Legislation (Oakley 2012), Local Economic Benefits, (e.g. (STPA 2018)), Impact and Benefit Agreements (e.g. (Nalcor Energy 2015)), and Set-Asides (e.g. (Nova Scotia 2018)). Major projects help develop a culture of inclusion, respect and understanding. The engineering profession and practice can enhance these opportunities and local benefits to garner further support for projects, sustain their local workforce, and potentially deliver projects at lower costs through a locally trained and available workforce.

Figure 3 The "Pyramid Model" of Social License (adapted from (Thomson and Boutilier 2011))

<u>Government and corporate mandates for local benefits</u>: The Social License to Operate (Boutilier 2011) (Colton, et al. 2016) (Gehman, Lefsrud and Fast 2017) (Thomson and Boutilier 2011) has gained some momentum as communities, governments and consultants evolve relationships that are sustainable and where major projects, such as the Tar Ponds Project and the Muskrat Falls Regulatory Compliance Project have legitimacy, credibility and trust (Figure 3). The Government of Canada has been referring to Corporate

Social Responsibility (Government of Canada 2018) and the Social Licence for energy and extractive projects as well as projects determined to be of National interest. These relatively new components to decision-making (loosely interpreted as policy changes) necessitate the need to approach projects differently than in the past. In addition, many rural and First Nation communities recognize that they have not always maximized local benefit from projects, and are requiring a more meaningful means of engagement. Corporations also recognize the need to maximize economic efficiencies. Pulling together these three aspects (policy, community benefits and sustainability, as well as project economics) makes good sense for major projects.

Good sense for management: Projects are successful for numerous reasons, but in the end, capable, well-trained, affordable workforces are critical. Through investments and partnerships with local training institutes, hiring agencies, and meaningful trustworthy community relationships, the two projects in our study have shown both local and project benefits in longer term employment, higher than normal retention and a trained and experienced workforce to support the project. A trained workforce that evolves with the project reduces new annual investments for seasonal staff by companies, increasing productivity and reducing annual start-up training costs. Partnering experienced staff with new staff has increased ownership and on-site leadership skills to empower further growth and led to greater staff retention. In the longer term, higher staff retention not only adds to project stability and efficacy, but also to social stability from predictable and meaningful work. These staff become local advocates for the project and the company.

Table 4: Some Initiatives and outcomes specific to the Muskrat Falls Regulatory Compliance Project

Initiative	Outcomes
Innu-owned business	Stassinu Stantec Limited Partnership holds the contract.
Culture and Heritage – strong historical, cultural and spiritual significance	In support of the NL *Historic Resources Act* – Innu Elders are included in reviews of historic and archaeologic findings; school children are exposed in special presentations and activities.
Training Programs:	Site Orientation and Cultural Awareness Training: Innu cultural awareness for all employees – 389 Innu Awareness Orientations; on-the-job training for Avifauna Management and Historic Resources Management for both the project team as well as for other contractors of the LCP, cross-training (field work: Avifauna Management, Historic Resources and Environmental Effects Monitoring for economic and responsiveness gains), field supervisory training, mentoring and coaching that enhanced retention and development of field team leads from year-over-year returning field staff
Gender Equity and Inclusion	Hired 98 women, 125 Aboriginal people, 227 total hired to date
Community networking	Through the Consultant's longstanding office presence in the community of Happy Valley - Goose Bay and the involvement of their partner – Stassinu Services Inc. (an Innu-owned and operated local company) they were able to access the local community through various means and established relationships. These longstanding relationships allowed them to leverage local suppliers and human resources (staff) in effective ways.
OVERALL	Training and employment are welcomed opportunities in a region with some of the highest unemployment and poverty rates in Canada. The legacy is almost 100 local staff trained in new techniques related to biological field studies, archaeological excavations, and historic resource conservation techniques. The Innu-owned business strengthened ties in the entire community; increased understandings of the biophysical environment, as well as expanded the cultural and heritage knowledge the lower Churchill River over the past 3,500 years.

Collaboration Benefits—one example (from the Tar Ponds Project): The composition of the material which required remediation was a very unique product which consisted primarily of the combined by-products of steel making, coking and sewer outfall discharge into the same receiving environment. Given this product's unique nature, it created workability challenges for the proposed Remediation Action Plan (RAP). To address these challenges, extensive bench scale testing for solidification/stabilization was completed to optimize the test results which included specifications associated with compressive strength of the material, permeability, and leachability. One discovery of the bench scale testing and subsequent Phase I field testing was that sample preparation processes and conditions had a large impact on test results. The initial quality control and quality assurance test results from Phase I revealed significant variability in test results. Workability and preparation of the material and test sample preparation caused costly variations in results. The Consultant collaborated with the STPA, the design engineer, and the solidification/stabilization contractor and their quality control consultant to refine the material preparation and sampling process by **creating Standard Operating Procedures (SOPs)** for preparing the solidified/stabilized material and its subsequent test sample collection, handling, preparation, and storage. Once the SOPs were refined and tested, the process (considered unique to the characteristics of the material being solidified and stabilized) was shared by the Consultant with all stakeholders through several half-day **"Sample School"** training sessions. Sample Schools were provided to all those involved in the project including Consultant personnel, STPA personnel, the Independent Engineer (IE), the Design Engineer, and the prime Contractor and their Quality Control Consultant. After completion of the SOP collaboration and the Sample School training, significant improvements in the homogeneity of the processed material and subsequent test samples resulted in testing consistencies that were achieved immediately. The **benefit of collaboration** in this project was to build competence and improve results, which led to improvements in schedule and decreases in costs due to reduced re-working and re-testing of material (which saved many thousands of dollars to rework a cell realizing there were potentially hundreds of cells that might have required rework without this collaborative effort).

3.1 Key Lessons Learned and Opportunities for Improvement

Long-term Investment in Community: Both projects were successful because the Consultant was invested in the community and had its physical office embedded there for decades prior to start of the major project. Thus, the Consultant was able to leverage long-established relationships with local businesses, educational institutions, service providers and community workforces. **Long-term investment in Client:** Consultant and client were known to each other for decades prior to start of the major project and had an established understanding of each other. **Stable, predictable work**: Consultant invested in individuals to build skills and understandings about the nature of major projects and the availability of secure work over the project lifespan. In both projects, there was good staff retention and for the Muskrat Falls Regulatory Compliance Project there was no field staff turnover in the final year—a significant cost savings. **Open communications**: Transparent, honest, and frequent communications with staff, contractors, suppliers, clients/owners are a necessity. Both formal (reports, change orders, field instructions) and informal (conversations, observations, information sharing) communication pathways kept the projects current for those who needed to know information for decision-making. At times, shedding the hierarchical approach expedited solutions on the ground by making use of knowledge held by individuals and teams in the field. Highlighting issues immediately was followed up through more formal mechanisms later. **Adaptable and fluid processes**: Processes that are prescribed on paper at the start of the project must be adaptable to, and realistic for, evolving conditions and situations in the field in real time with direct and immediate communications. More efficient processes and collaborations arose from knowledge exchanged between management and field staff in real time. **Get early buy-in**: Understanding project goal(s) as early as possible benefited outcomes and deliverables of the project and the teams evolved with collaborative and cooperative goal setting and problem-solving. Trust developed early reduced costs of mindless and repetitive rework of tasks. Open communications between ground crews/teams reduced or eliminate suspicions and resistance to work in the field amongst different contractors/consultants. **Interpersonal relationships and experience**: mis-alignment of teams within a company or between a company and the client can negatively impact the progress of the project through withholding of vital information, delayed decision-making, higher costs, slipped schedules and pressures on staff. The project management team must constantly adjust to effect alignment and to communicate assertively the impacts that can result from failure to do so.

4 CONCLUSIONS

The insights that the authors' project teams gained were to get involved early, to partner with training and community employment interest organizations, to provide internal training, to embrace diversity, to build competencies and retention, and to build job awareness with opportunities offered to expand career capabilities for local workforces. It starts with people first. Providing local workforces with opportunities and experience in large transformational projects appears to have benefits that last beyond the project. Governments can provide leadership and incentives through enabling legislation and training programs available to project owners and developers to encourage engagement of local workforces, provide appropriate training, and operate in a manner that transfers and retains expertise to local workers. Long before either of the projects mentioned were awarded, strategic local relationships were well established. Working in the local communities on smaller scale projects built sustainable offices and networks which were heavily leveraged through trusted relationships as major projects in both areas were released. The success was realized not from 'fly by night' organizations looking just to maximize profits, but through working with neighbours as existing community members. Applying this local first approach proved to be both economically more sustainable through high retention and lowering travel associated costs, and potentially increasing community sustainability and resiliency by having a trained and experienced local workforce. Both project areas have historically had seasonally depressed workforces, and although these projects were also seasonal, they allowed for a level of security for workforces in anticipation of the next season of work. This allowed families to invest and settle more fully into their local communities. By having professionals identify the types of skills required to successfully deliver projects, Consultant worked with human resource staff, training agencies, and community resources to find and train local staff who were subsequently fairly compensated and valued throughout the project. Both the company and the individuals mutually benefited from this relationship, further embedding the company into the community and ensuring it is poised to work together building on these meaningful relationships on other projects in the future. As described in (Sense and Kiridena 2014), the projects highlighted in this paper benefited and they support the finding of the value of the human, behavioural and social dimensions of contemporary project-based undertakings.

Acknowledgements

The authors would like to thank the support of their teams and colleagues as well as their colleagues in the former Sydney Tar Ponds Agency (now NS Lands Inc.) and on the Nalcor Energy Lower Churchill Project.

References

AMEC. 2004. Remediation of the Sydney Tar Ponds and Coke Ovens Sites Project Description Executive Summary. Sydney, NS: AMEC Earth and Environmental, 6. Accessed February 2018.

Boutilier, Robert. 2011. "A Measure of the Social License to Operate for Infrastructure and Extractive Projects." SociaLicense.com. Accessed February 15, 2018. https://socialicense.com/publications.html.

CEAA. 2018. http://www.ceaa.gc.ca/050/documents/51706/51706E.pdf. February 14. http://www.ceaa.gc.ca/050/documents/51706/51706E.pdf.

Chen, Yuting, Brenda McCabe, and Douglas Hyatt. 2017. "Impact of individual resilience and safety climate on safety performance and psychological stress of constructions workers: A case study of the Ontario construction industry." Journal of Safety Research 61 (2017): 167-176.

Collier MacDonald, Tanya. 2005. Mi'kmaq sign tar ponds agreement. October 29. Accessed February 15, 2018. http://www.safecleanup.com/jag/ps102905.htm.

Colton, John, Kenneth Corscadden, Stewart Fast, Monica Gattinger, Joel Gehman, Martha Hall Findlay, Dylan Morgan, Judith Sayers, Jennifer Winter, and Adonis Yatchew. 2016. "Energy Projects, Social Licensem Public Acceptance and Regulatory Systesms in Canada: A White Paper." The School of Public Policy Research Papers 9 (20). Accessed February 15, 2018. https://papers.ssrn.com/sol3/papers.cfm?abstract_id=2788022.

ECNL. 2008. Tshash Petapen Agreement - Government of Newfoundland & Labrador and Innu Nation. Accessed February 15, 2018. http://www.releases.gov.nl.ca/releases/2008/exec/0926n07agreement.pdf.

Gehman, Joel, Lianne M. Lefsrud, and Stewart Fast. 2017. "Social license to operate: Legitimacy by another name?" Canadian Public Administration 60 (2). Accessed February 15, 2018. doi:10.1111/capa.12218.
Government of Canada. 2018. Corporate Social Responsibility - Supportive Information and Links. Accessed February 15, 2018. https://www.ic.gc.ca/eic/site/csr-rse.nsf/eng/h_rs00018.html.
Government of NL. 2012. Government of Newfoundland and Labrador Announces Sanction of the Muskrat Falls Development. December 17. Accessed February 15, 2018. http://www.releases.gov.nl.ca/releases/2012/exec/1217n11.htm.
Ingraham, Diane V., and Wayne Tucker. 2017. "Well-Managed Environmental Compliance Program Supports Delivering Sustainable Energy from a Large Hydroelectric Development in Labrador." Proceedings CSCE 2017 - Leadership in Sustainable Infrastructure GEN_207-1-GEN_207-10.
Ingraham, Diane, Willie McNeil, and Donnie Burke. 2014. "New Heart for Sydney in Wake of Tar Ponds and Coke Ovens Cleanup." CSCE 2014 General Conference May 28 to 31, 2014. Halifax, NS: CSCE. GEN_212-1-GEN_212-1-.
Nalcor Energy. 2014. LCP Regulatory Compliance Plan - Generation (LCP-PT-MD-0000-EV-PL-0024-01). Lower Churchill Project, St. John's Newfoundland & Labrador: Nalcor Energy. Accessed February 11, 2018. http://www.mae.gov.nl.ca/env_assessment/projects/Y2010/1305/1305_regulatory_compliance2014.pdf.
—. 2015. "Lower Churchill Project Construction Benefits Strategy." St. John's, NL, October.
—. 2018a. "Boundless-Opportunities-Brochure.pdf." Accessed February 14, 2018a. http://muskratfalls.nalcorenergy.com/wp-content/uploads/2013/04/Boundless-Opportunities-Brochure.pdf.
—. 2018b. Muskrat Falls Project Sources & Uses Overview April 2014_fiinal. Accessed February 14, 2018b. https://muskratfalls.nalcorenergy.com/wp-content/uploads/2013/03/Muskrat-Falls-Project-Sources-Uses-Overview-April-2014_fiinal.pdf.
Nova Scotia. 2018. Aboriginal Set-aside Procurement Strategy Signed. February 14. https://novascotia.ca/news/release/?id=20080925004.
Oakley, James C. 2012. "Review of Special Project Order Legislation in Newfoundland and Labrador." Accessed 2014. http://www.gov.nl.ca/lra, pdf,Review_SPOL-Oakley.pdf.
Ovenden, Mat. 2018. Rethinking Construction Workforce Development, and Local Economic Development. February 14. http://blog.bechtel.com/build-100/november-2017/sustainable-economic-growth/.
PWGSC. 2014. Final Report 2013-602 Final Evaluation of the Sydney Tar Ponds and Coke Ovens Remediation Project. Ottawa: Public Works and Government Services Canada - Office of Audit and Evaluation.
Remington, Kaye. 2011. Leading Complex Projects. Farnham, Surrey: Gower Publishing Limited.
Sense, Andrew, and Senevi Kiridena. 2014. "Building Worforce Competencies through Complex Projects." University of Wollongong - Research Online 17.
STPA. 2018. Sydney Tar Ponds Agency - Procurement Rules & Economic Benefits (2008). February 14. http://www.tarpondscleanup.ca/upload/reports/Procurement_LEB_Vendor_Feb_08.pdf.
Thomson, Ian, and Robert G. Boutilier. 2011. "Modelling and Measuring the Social License to Operate: Fruits of a Dialogue between Theory and Practice." SociaLicense.com. Accessed February 15, 2018. https://socialicense.com/publications.html.
United Nations. 2018. http://www.un.org/sustainabledevelopment/economic-growth/. February 14. http://www.un.org/sustainabledevelopment/sustainable-development-goals/.

Building Tomorrow's Society
Bâtir la Société de Demain

Fredericton, Canada
June 13 to 16, 2018 / *13 au 16 juin, 2018*

MORPHODYNAMIC RESPONSE OF LOWER SPENCER CREEK TO LARGE PRECIPITATION EVENTS AND CHANNEL MODIFICATIONS

Martin, Emily Mae[1], Binns, Andrew[2] and Newson, Tim[1]
[1] Western University, Civil and Environmental Engineering, Canada
[2] University of Guelph, Water Resources Engineering, Canada

Abstract: Severe weather events are occurring more frequently with significant repercussions as our climate changes. Between 2004 and 2012 the city of Hamilton, Ontario experienced 17 storm events that caused flooding. Of these 17 events, five were classified as either 50- or 100-year storm events, with one classified as a storm with a return period greater than 100 years. Events of this magnitude can lead to rapid morphological adjustments in rivers, including instability of river banks, bank collapse, and alterations in stream channel alignment. It is important to understand the hydraulic and geomorphic response of rivers and streams to future storm events to predict long term morphological change, protect hydraulic structures, and manage aquatic ecosystems. The goal of this research is to examine the influence of extreme precipitation events on the morphodynamic response of Lower Spencer Creek under current channel morphological conditions and simulated modifications. Two events are simulated using the one-dimensional computational model, HEC-RAS: an averaged base flow event and a 100-year return period event. Additionally, channel modification simulations were made to four cross sections at the downstream end of Lower Spencer Creek. The first modification widened the channel at these sections, while the second modification lengthened the channel at these sections to be representative of increased meandering of the stream. Based on the flow events, the first modification to the channel showed the largest reduction in velocity, sediment concentration, and shear stress at all cross sections. The results from this study will contribute to building a framework for analyzing the morphodynamic response of urban streams to storm events of varying return periods and durations, and will assist river engineers and hydrologists in managing and restoring urban creeks to mitigate flooding while balancing erosion and ecological processes.

1 INTRODUCTION

Due to numerous natural and anthropogenic effects, urban rivers and streams are often challenged to satisfy hydraulic, geomorphic, and ecological functions. The risk of flooding in urban rivers and streams is increased as a result of climate change and land use changes, such as urbanization and river channel modifications (Khattak et al., 2016; Ashmore and Church, 2001). Heavy rainfall and flooding affect the channel morphology, sediment characteristics, soil erosion, and aquatic habitat (Rae, 1987; Blom and Voesenek, 1996; Ashmore and Church, 2001).

Channelization is used as an approach in river engineering to control floods, stabilize banks, and modify navigation routes. Historically, channelization was utilized to manage rivers in urban environments. However, such modifications to channels often have an unfavourable impact on sedimentologic, hydrologic, and biologic channel properties (Kroes and Hupp, 2010; Bukaceckas, 2007; Soar and Thorne, 2001). Impacts include, but are not limited to, loss of aquatic habitat, increase in water temperature and sediment concentration, and changes in water levels and flow conditions (Shields and Palermo, 1982).

Mathematical models are often used as cost-effective tools for understanding river channel dynamics, such as channel morphology, flow, and sediment transport. Mathematical modeling tools can be used to evaluate stream restoration options to help determine an optimal stream morphological design to balance hydraulic, geomorphic, and ecological functions. Additionally, mathematical modeling tools can be used in the investigation of stream hydraulic and geomorphic response to changes in flow regime. These tools have been widely used in theoretical and practical applications of flood prediction and floodwave routing in urban rivers and streams (Horrit and Bates, 2002).

The Hydraulic Engineering Centre River Analysis System (HEC-RAS), created by the U.S. Department of Defense, Army Corps of Engineers, is widely used by river engineers in stream flow and floodplain analyses. This model is capable of simulating steady flow, unsteady flow, and sediment transport with one-dimensional and two-dimensional approximations. HEC-RAS has been used in a variety of applications relating to flooding in channels. For example, Hicks and Peacock (2005) used this program to evaluate flood routing and water level forecasting, Sholtes and Doyle (2011) applied the program to model synthetic and field-based stream reaches for impaired and restored reaches, and Khattak et al. (2016) combined this program with ArcGIS to develop floodplain maps for the Kubal River.

The city of Hamilton in Southwestern Ontario, Canada, is susceptible to flooding due to rainfall events and snowmelt, with 17 storm events resulting in flooding between 2004 and 2012 (Citizens of City Hall, 2012). A well-known example of flooding in Southwestern Ontario is Hurricane Hazel, which occurred on October 15 and 16, 1954. This event was found to have a 100-year return period, with 200 mm of rain in 24 hours and wind speeds as high as 124 km/h. Events of this magnitude have since occurred in in Southwestern Ontario and are likely to occur in the future. Therefore, it is important to understand and evaluate the effects of large precipitation events on the morphology and ecology in urban streams (Marsh, 2012).

This research provides part of a framework to evaluate hydraulic, geomorphic, and ecological performance of rivers under varying conditions. With the use of HEC-RAS, two channel modifications are made to the downstream cross sections of Lower Spencer Creek. These channel modifications aim to reduce velocity and sediment transport within the channel, thereby reducing the erosion and adverse ecological impact. Due to the expected increase in frequency of heavy storm events, simulations are conducted for each channel modification of large flow events to determine which modification could be a sustainable solution for Lower Spencer Creek.

2 STUDY AREA

Spencer Creek, located in the Lower Spencer subwatershed in Hamilton, Ontario, Canada, is presently under investigation by the Hamilton Conservation Authority (HCA) to return the channel to a more natural channel pattern. The plan proposes a relocation of the existing Spencer Creek Trail near Cootes Drive, replacement of dead ash trees, and removal of invasive grasses. These transformations will attempt to balance sediment transport, flood prevention, and ecology. By creating nesting areas, the plan has potential to protect the existing turtle population in the Spencer Creek area. Lower Spencer Creek, shown in Figure 1, is 7 km in length; however, the present research only examines the lower 4.25 km of the creek between King Street (i.e., XS 2) and Cootes Drive (i.e., XS 18). The widths of cross sections within this reach range from 4 m to 15 m. The initial phase of alterations performed by the HCA will focus on 300 m of Lower Spencer Creek closest to Cootes Drive, with future restoration moving upstream towards Thorpe Street (Leitner, 2014).

3 METHODS

The present research utilizes the one-dimensional computational model, HEC-RAS 4.1, to investigate the morphological performance of Lower Spencer Creek to large simulated precipitation events and multiple channel modifications. A complete description of the methodology followed in this work is provided in Martin (2017).

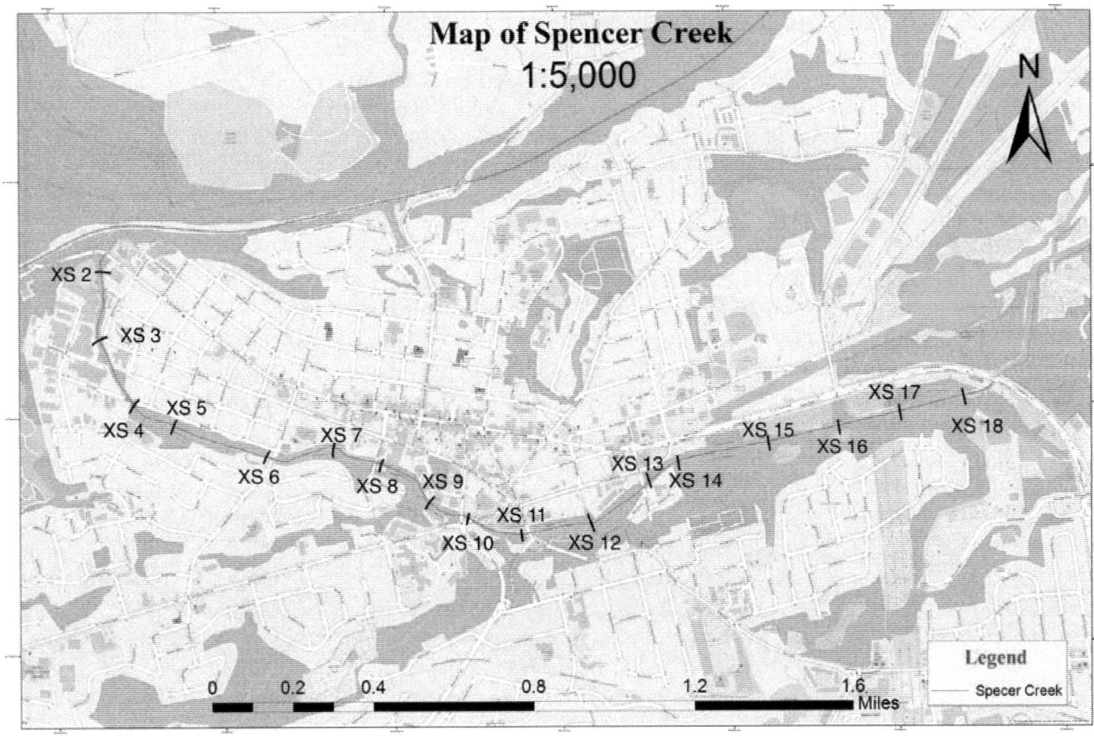

Figure 1: Map of Lower Spencer Subwatershed and Cross Sections in Lower Spencer Creek

3.1 Description of data

Field measurements were conducted to collect cross sectional geometry and substrate data required for modeling. Field measurements, including bankfull widths, cross-sectional depths, and sediment samples were collected at cross sections 15, 16, 17, and 18 of Lower Spencer Creek. A bank-to-bank width was measured at each section, and depth measurements were taken at five locations across each section. Sediment samples at each location of depth measurement were collected to characterize the substrate conditions in the reach. Five (5) samples were collected at each cross section for use in analysis.

Hydrometer tests were performed according to ASTM D422 on the collected sediment samples to determine the grain size distribution for the sediment passing through a No. 200 sieve. This data was used to provide more accurate sediment data at cross sections 15, 16, 17, and 18.

3.2 Channel Modifications

With the use of the channel modification extension in HEC-RAS, simulated modifications were made to the cross sections between Thorpe Street and Cootes Drive, namely, cross sections 15, 16, 17, and 18. The goal of the modifications was to evaluate the degree of improvement in flood control, sediment transport, and channel stability in these sections. These sections were selected for proposed modifications due to the high risk of flooding and stream instability in this reach.

Modification 1 examines widening downstream sections in Lower Spencer Creek (cross sections 15, 16, 17 and 18) by 15%, in efforts to reduce velocity and sediment transport in the channel. Modification 2, which lengthened the distance between cross sections 15, 16, 17, and 18 by 10%, thereby reducing the slope and simulating the creation of a more sinuous channel, aimed to return Lower Spencer Creek a more natural, pre-channelization condition. These modifications both aimed to reduce velocity within the creek, thereby decreasing its sediment transport capacity; this will result in less erosion and channel instability

(Watson et al., 1999). Sediment conditions for the modified simulations remained consistent with the existing channel simulations.

3.3 Precipitation Events

Rainfall events with long durations and high intensities are a common cause of flooding. A well-known example of this is Hurricane Hazel, which was classified as a 100-year event (Marsh, 2012). Storm events of this magnitude are expected to occur more frequently in the future (Takemura and Fukoma, 2014). For this reason, simulations with a 100-year return period were performed on Lower Spencer Creek.

Hydrographs were created for two events using the Soil Conservation Service (SCS) method (see Figure 2). The first event, henceforth referred to as Event 1, had a constant discharge of 1.25 m^3/s over a 24-hour period; this flow was calculated by assessing the average of the peak flows over a 5-year period between 2011 and 2015 in this reach of Lower Spencer Creek. The second event, henceforth referred to as Event 2, was a 100-year event over 24-hours with a peak flow of 12.23 m^3/s. This event was chosen to simulate the Hurricane Hazel event.

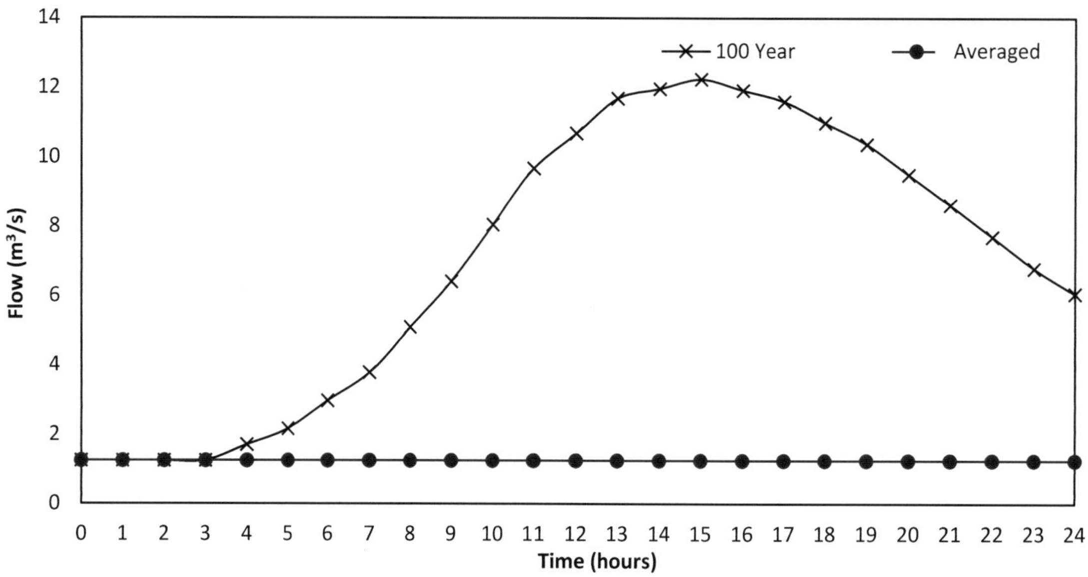

Figure 2: Storm Hydrographs

4 RESULTS AND DISCUSSION

4.1 Results

Channel modifications were performed on the existing channel with the intention of decreasing velocity and sediment concentration within the channel. Applying Modification 1 to the channel confirmed a decrease in velocity at cross section 15, 16, 17, and 18 for Event 1 compared to the existing channel configuration; the largest decrease occurred at cross section 15, with a decrease in velocity by 51% as shown in Figure 3. Similarly, applying Modification 1 to Event 2 produced a decrease in velocity at all cross sections, with the largest change occurring at cross section 18 (40%) as shown in Figure 4. Modification 2 resulted in a minor change in velocity at cross section 15 for Event 1, and no change at subsequent cross sections. Applying Modification 2 to Event 2 resulted minor changes at cross sections 16, 17, and 18, and resulted in an increase in velocity at cross section 15 by 4%.

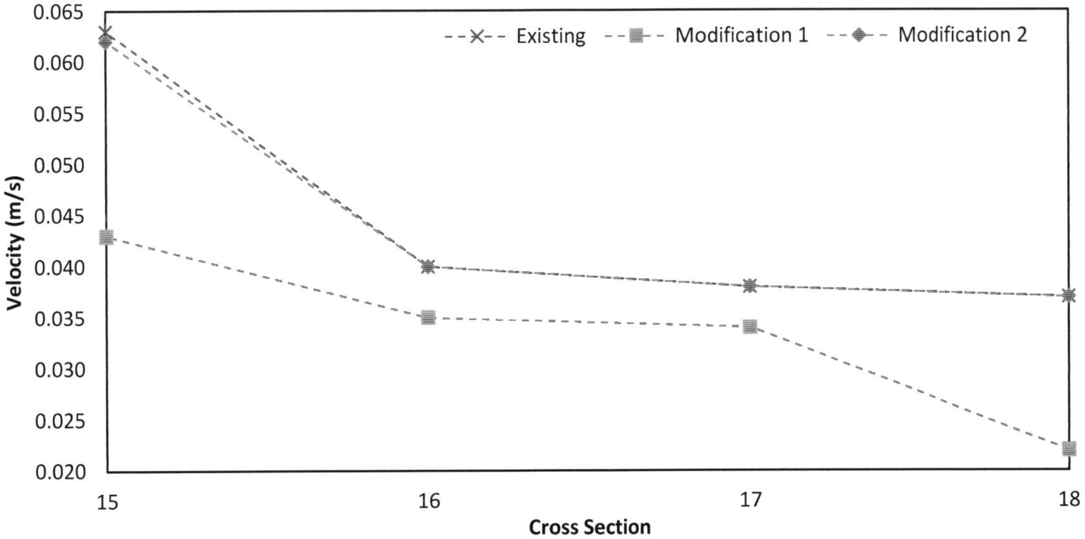

Figure 3: Comparison of Maximum Velocity at XS 15-18 for All Channel Configurations for Event 1

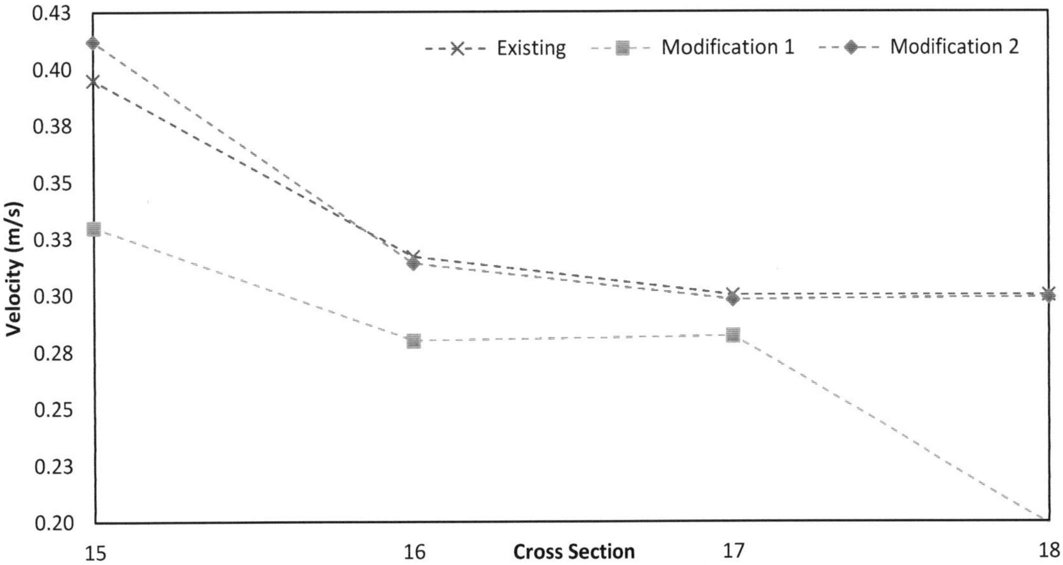

Figure 4: Comparison of Maximum Velocity at XS 15-18 for All Channel Configurations for Event 2

In addition to decreasing the velocity within the channel, channel modifications also reduced sediment concentration within the channel. Modification 1 and Modification 2 resulted in a decrease in sediment concentration at cross sections 15, 16, 17, and 18 for Event 1, with the largest change occurring at cross section 15 for both modifications. For Event 2, however, the sediment concentration was reduced at cross section 15 for both events, and only minor changes (i.e. sediment concentration changes of less that 1%) were observed at cross sections 16, 17, and 18.

All channel configurations resulted in aggradation within the channel at cross sections 15, 16, 17, and 18. As predicted, Modification 1 increased the amount of aggradation occurring at these cross sections; this was predicted due to the decrease in velocity. Modification 2 produced minimal change in aggradation patterns. The resulting aggradation implies these sections may be at risk of harming the fish habitat in the channel, the vegetation on the bed and banks of the channel, and have an adverse impact on stream stability, increasing the potential for erosion.

Erosion in a channel can occur due to numerous factors, such as stream velocity, sediment characteristics, climate, and vegetation. Riverbank erosion is likely to occur in Lower Spencer Creek during rainfall events with a long duration and high intensity. Due to the range of sediment sizes present on the stream bed and banks, hydraulic erosion is likely to occur within the channel. Lower Spencer Creek is largely composed of gravel, sand, silt, and clay. Cross sections 15 and 16 are composed of gravel (76%), sand (17%), and silt and clay (7%). Due to the silt-clay content at these sections, erosion occurring will likely be due to hydraulic processes. Cross section 17 is made up of sand (68%), silt (31%), and clay (1%), while cross section 18 is made up of sand (76%), silt (23%), and clay (1%). These sections have higher silt-clay contents, meaning occurrence of erosion will likely be caused by hydraulic and subaerial processes (Julian and Torres, 2006).

Erosion rates were calculated at cross sections 15, 16,17, and 18 using Equation 1, below, where E is the lateral erosion rate (m/yr), k is the erodibility coefficient, τ is the shear stress by flow (N/m^2), and τ_0 is the critical shear stress (N/m^2) (Julian and Torres, 2006). Negative values resulting from this equation indicate deposition is occurring within the channel, while a positive value indicates erosion is occurring in the channel.

$$[1]\ E = k(\tau - \tau_C)$$

In the existing channel, Event 1 resulted in negative erosion rates, indicating deposition occurring at cross section 15, 16, 17, and 18. Event 2, however, resulted in high erosion rates, indicating there may be instability in the channel during high flow events. Applying Modification 1 resulted in a decrease in erosion rates for cross section 15, 17, and 18 for Event 1 and Event 2. Conversely, an increased erosion rate during Event 1 and Event 2 was observed at cross section 16. Modification 2 produced in a decrease in erosion rates at all cross sections for Event 2, and no change in erosion rates in the channel during Event 1. A comparison of the velocity in the channel and the erosion rates is shown in Figure 5. As anticipated, these results demonstrate an increase in erosion rate as the velocity increases. Conversely, a reduction in velocity in the channel suggests greater potential for deposition in the channel.

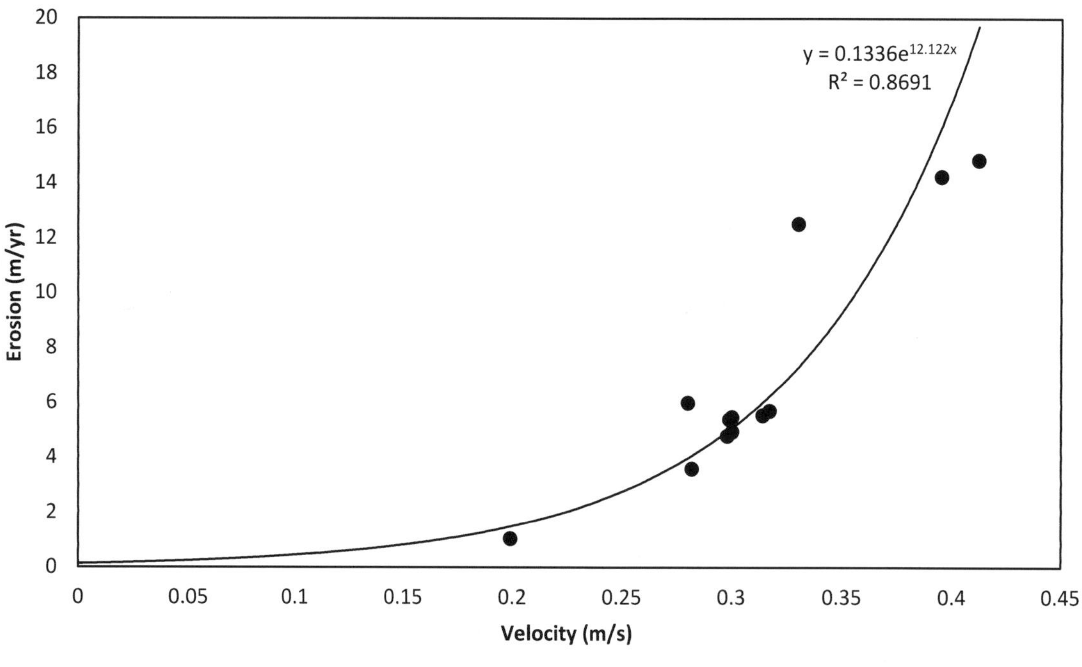

Figure 5: Velocity vs Erosion Rate for Lower Spencer Creek

Erosion rates and sediment concentrations are plotted in Figure 6. Lower erosion rates resulted in lower sediment concentrations. However, when erosion rates are negative and deposition was occurring in the channel, the sediment concentrations began to rise. The lowest sediment concentration of 420.6 mg/L occurred when the erosion rate was 1.05 m/yr. As the erosion rate began to rise, indicating increased erosion was occurring, the sediment concentration began to rise as well. The highest sediment concentration of 516 mg/L occurred at 14.25 m/yr.

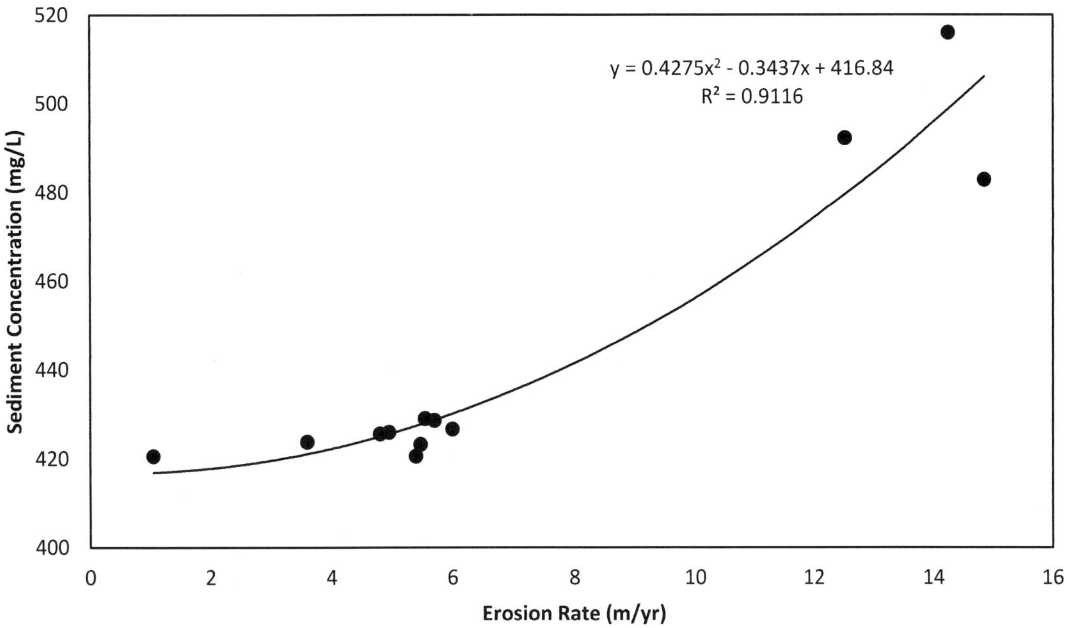

Figure 6: Erosion Rate vs Sediment Concentration for Lower Spencer Creek

4.2 Discussion

The effects of sediment deposition and sediment transport concentrations, both ecological and physical, have become more serious as a result of human activities in close proximity to river channels. The impact of sedimentation processes in a channel can completely alter the morphology and/or ecology of the channel, if extended over several months or years. Excess sediment transport and deposition in a channel is a main cause of habitat degradation and can result in algal blooms, alluvial fans, and deltas. Deposition of fine sediment can reduce light reaching the bed, affecting the food supply in the channel, which will have a lasting effect on fish, benthic organisms, and faunal diversity. Alternatively, limited sediment transport and deposition within a channel can result in bank erosion leading to land loss and/or destruction of nearshore habitat. In addition, excess deposition can diminish vegetation growth, leading to a lack of nutrients in floodplains and marshes (Wood and Armitage, 1997; Fondreist Environmental Inc., 2014; Reiser and White, 1990).

Event 2 showed a large amount of degradation occurring within the channel for all channel configurations. Therefore, when exposed to an event of this magnitude, the ecological habitat within Spencer Creek is at risk of becoming damaged. The Canadian Environmental Quality Guidelines (1999) suggest that a short-term increase in sediment concentration of 25 mg/L could have an adverse impact on aquatic ecosystems. Based on these guidelines, the aquatic ecosystem in Lower Spencer Creek may be adversely impacted in an event of this magnitude.

Based on these results, widening the channel (Modification 1) would be an ideal option for reducing velocity and sediment concentration within the channel. However, due to the resulting decrease in erosion rates with Modification 2, a combination of both options could prove beneficial within the channel. A combination of these modifications could decrease velocity, sediment concentration, and bank erosion within the downstream cross sections.

Although Figures 5 and 6 are based on conditions specific to Lower Spencer Creek, these figures may be used to assist in determining the mean velocity and sediment concentration in a channel based on the permitted amount of erosion. Moreover, these figures could be used to help guide the design of rivers and streams in order to satisfy an erosion limit.

5 CONCLUSION

Models predicting how small, urban streams respond to various rainfall durations and intensities can be useful for future land use planning and river management. When considering alterations to a channel, it is important to consider how the sediment in the channel will be affected, how the flow is altered, and how it will affect erosion of the bed and banks. It is important to understand how the floodplain is affected by rainfall durations and intensities when considering any future construction. Simulations of a 100-year event showed relatively high flow velocities and sediment concentrations in all tested channel configurations. Events of large magnitudes, such as this event, are predicted to occur more frequently in the future. Therefore, it is important to understand the stream hydraulic and morphological response to these events to accurately plan for the future (Takemura and Fukoma, 2014).

The results of this study examined the sediment transport and morphological performance of the Lower Spencer Creek in response to a 100-year precipitation event using the mathematical hydraulic model HEC-RAS. The results from this study are specific to Lower Spencer Creek and should not be used beyond the boundaries of this study. Modification 1 widened the channel by 15%. As expected, this modification decreased the velocity and sediment concentration at cross sections for the averaged base flow event and the 100-year storm event. Modification 2 increased the length between each cross section by 10%, thereby decreasing the slope. There was a decrease in sediment concentration at all cross sections during the averaged base flow event and 100-year event for this modification.

The results of the two modification options suggest that Modification 1, widening the channel, would be overall more successful in reducing the velocity and sediment transport within the channel, which can further benefit the ecology of the channel, and reduce the risk of flooding. However, a combination of the two modifications may better reduce the sediment concentration in the channel. Further analysis, however, is recommended as this study only looked at one large event.

Acknowledgements

This research was financially supported by a Natural Sciences and Engineering Research Council of Canada (NSERC) Discovery Grant. Laboratory experiments and all mathematical modeling was conducted in the Department of Civil and Environmental Engineering at Western University. Additional information was provided by the Hamilton Conservation Authority.

References

Ashmore, Peter, and Michael Anthony Church. *The impact of climate change on rivers and river processes in Canada.* Vol. 555. Geological Survey of Canada, 2001.

Blom, C. W. P. M., and L. A. C. J. Voesenek. "Flooding: the survival strategies of plants." *Trends in Ecology & Evolution*11, no. 7 (1996): 290-295.

Bukaveckas, Paul A. "*Effects of channel restoration on water velocity, transient storage, and nutrient uptake in a channelized stream.*" Environmental science & technology 41, no. 5 (2007): 1570-1576.

Canadian Council of Ministres of the Environment (CCME). Canadian environmental quality guidelines, vol. 2, CCME ed. Winnipeg, Man. 1999.
Citizens of City Hall. *The Floods of Hamilton.* Retrieved from http://www.hamiltoncatch.org/view_article.php?id=1072. 2012
Fondreist Environmental, Inc. *Sediment Transprot and Deposition." Fundamentals of Environmental Measurements.* Retrieved from http://www.fondreist.com/environmental-measurements/parameters/hydrology/sediment-transport-deposition/. 2014.
Hicks, F. E., and T. Peacock. "Suitability of HEC-RAS for flood forecasting." *Canadian Water Resources Journal* 30, no. 2 (2005): 159-174.
Horritt, M. S., and P. D. Bates. "Evaluation of 1D and 2D numerical models for predicting river flood inundation." *Journal of Hydrology* 268, no. 1-4 (2002): 87-99.
Julian, Jason P., and Raymond Torres. "Hydraulic erosion of cohesive riverbanks." *Geomorphology* 76, no. 1-2 (2006): 193-206.
Khattak, Muhammad Shahzad, Faizan Anwar, Tariq Usman Saeed, Mohammed Sharif, Khurram Sheraz, and Anwaar Ahmed. "Floodplain mapping using HEC-RAS and ArcGIS: a case study of Kabul River." *Arabian Journal for Science and Engineering* 41, no. 4 (2016): 1375-1390.
Kroes, Daniel E., and Cliff R. Hupp. "The effect of channelization on floodplain sediment deposition and subsidence along the Pocomoke River, Maryland." *JAWRA Journal of the American Water Resources Association* 46, no. 4 (2010): 686-699.
Marsh, J.H. Hurricane Hazel. Retrieved from http://www.thecanadianencyclopedia.ca/en/article/hurricane-hazel. 2012.
Martin, E.M. Morphodynamic Response of Spencer Creek to Heavy Precipitation Events and Channel Modifications: A Case Study. M.E.Sc. thesis, University of Western Ontario, London, Canada (2017).
Rae, J.G. Chapter 2: Effects of Flooding on Soils. In T.T. Kozlowski (Ed.), Flooding and Plant Growth (pp. 10-44). Orlando, FL: Academics Press. 1987.
Shields Jr, F. Douglas, and Michael R. Palermo. *Assessment of Environmental Considerations in the Design and Construction of Waterway Projects.* No. WES/TR/E-82-8. ARMY ENGINEER WATERWAYS EXPERIMENT STATION VICKSBURG MS ENVIRONMENTAL LAB, 1982.
Sholtes, Joel S., and Martin W. Doyle. "Effect of channel restoration on flood wave attenuation." *Journal of Hydraulic Engineering* 137, no. 2 (2010): 196-208.
Soar, Philip J., and Colin R. Thorne. *Channel restoration design for meandering rivers.* No. ERDC/CHL-CR-01-1. ENGINEER RESEARCH AND DEVELOPMENT CENTER VICKSBURG MS COASTAL AND HYDRAULICSLAB, 2001.
Takemura, Y., and S. Fukuoka. "Effects of channel shape on propagation characteristics of flood flows through a valley." *Journal of Flood Risk Management* 7, no. 2 (2014): 152-158.
Wood, Paul J., and Patrick D. Armitage. "Biological effects of fine sediment in the lotic environment." *Environmental Management* 21, no. 2 (1997): 203-217.

Building Tomorrow's Society
Bâtir la Société de Demain

Fredericton, Canada
June 13 to 16, 2018 / *13 au 16 juin, 2018*

ON THE NATURE OF HIGHWAY LITTER; A METHODOLOGY AND FIELD STUDY

Bolingbroke, Damien[1], Ng, Kelvin T. W.[1,2] and Surendran, Senthuran[1]
[1] Environmental Systems Engineering, University of Regina, Canada
[2] kelvin.ng@uregina.ca

Abstract: Solid waste in Canada is generally well managed but litter still exists. Traditional research in this area has focused on the behavioural aspects of littering, the role of litter legislation, as well as the quantification and classification of litter. Saskatchewan has more highways and roads per capita than any other Canadian province and there exists field evidence of highway litter in this province. This led the authors to develop a new waste quantification approach to study highway or roadside litter. An item-oriented litter survey in highway ditches near Regina was conducted to quantify the variations of litter count with respect to the distance from a Statistics Canada defined Census Metropolitan Area (CMA). A trial test was conducted on July 24th, 2017 to develop the quantification framework and to investigate how the length of roadway studied impacts the total litter count. From this trial, an inverse relationship was found to exist between the length of roadway and the litter count density and a 40m length segments were chosen for the full study conducted on August 13th and 14th, 2017. All visible litter of particle size greater than 2.5 cm as well as cigarette butts were collected using a 2m, by ditch width, grid. Results show some evidence to support the hypothesis that increasing distance from a CMA decreased litter count density. The outcomes may lead to future studies in litter data pattern analysis and the use of the methodology presented for further highway or roadside litter studies.

1 INTRODUCTION

Litter is a persistent issue throughout in Canada. Highway litter or roadside litter piles up on the roadsides and can cause an eyesore for motorist as well as residents who live alongside such highways. Despite there being programs such as the Adopt-A-Highway, where groups volunteer to pick up highway litter, it is still frequently observed in Saskatchewan. In particular, a casual observation of a large amount of highway litter in Saskatchewan and Manitoba during the spring of 2017 lead the authors to question about the quantity and spatial distribution of the highway litter. The objectives of the study are to develop a framework to quantify highway litter in Saskatchewan and to examine the spatial distribution of highway litter with respect to distance from a Census Metropolitan Area (CMA) using both the waste particle count and waste mass. The relationship between mass and count of the litter were examined.

2 LITERATURE REVIEW

There is currently a lack of established standards or guidelines on quantification of highway and roadside litters (Cutter et al. 1991). It appears that most litter surveys and published waste studies have developed and adopted specific methods and procedures according to their study objectives. Among them, waste density (waste quantity divided by the planar area) is commonly used in litter studies. Litter can be quantified by the waste count or waste mass. Waste count is more intuitive and provides more insight on litter studies, but the latter one is commonly used due to the ease of measurement. In this study, both of them are used to quantity the degree of littering.

Due to the heterogeneous nature of roadside litters (Stein, 2010), the size and dimensions of sampling area is an important parameter in the determination of waste quantity and spatial distributions. Table 1 summarized some North American waste quantification studies in chorological order and their selected sampling areas on roadways. Results suggest the sampling area and dimensions were highly site-specific, ranging from 100 m to 210 m. Since different methods and sampling sizes were reported in literature, it is found that direct comparisons of results are difficult.

Table 1: Sampling lengths and areas of various highway litter studies in North America

Site description	Length of Roadway (meters)	Width of Roadway (meters)	Source
31 litter surveys in 16 USA States in 8 different location strata	175 (average)	Various, to edge of right of way (constant feature width)	Syrek, 1986
45 sites across Prince Edward Island, Canada highways and roads	100	10 (max)	Southeast Environmental Association, 2003
94 sites in New Jersey, USA, visible litter survey method	121.9 – 210.0	Various	Gershman, Brickner, & Bratton, Inc., 2005
55 sites in Nova Scotia, Canada of accumulated litter 2.5 cm or larger	100	8 (max)	Smith, 2008
288 visible litter surveys at sites in Maine, New Hampshire, and Vermont, USA	152.4	4.57	Stein, 2010
Highway 6, south of Regina, SK., Canada	200 (Test Section) 40 (Full Study)	Various, to edge of right of way (constant feature width)	This study

The Institute for Applied Research reviewed 31 litter studies across the United States from 1973 to 1986 and systematically evaluated the magnitudes and contributing factors on litter rates (Syrek 1986). The Institute for Applied Research study focused on the similarities and the differences of litter rates among the 31 studies and selected variables to construct a United States standard model. This included reduction factors for traffic volumes, entrapment correction, neighbourhood income, freeway medians, temperature/rainfall, and sampling interval (Syrek 1986). Cutter et al. (1991) conducted a literature review for their field study in New Jersey and reported two commonly used methods for the quantification of litters. The first is an item-oriented surveys, where litter is picked up, counted, and classified. The second method is a visual survey, where litter is only observed and counted, not picked up. Cutter et al. (1991) stated that there is missing information in the methodology for site selection and data extrapolation to statewide waste data.

Some of the studies also focused on the political and social aspects of waste littering, include the role of government comprehensive litter programs on litter reduction, as well as the behavioural links that may cause one to litter (Syrek 1986, Cutter et al. 1991). The present study aimed to develop a framework for the quantification of highway litter from a highway south of Regina, Saskatchewan.

3 METHODOLOGY

3.1 Study Area

Saskatchewan has more highways and roads per capita than any other Canadian province. Regina, is a major metropolitan area of Saskatchewan and western Canada and was chosen as the centre of the study. Statistics Canada defines a CMA as an area where a core city has a population of at least 50,000 with neighbouring municipalities (who must have a high degree of integration such as commuter traffic with the core) totaling a population of at least 100,000 (Statistics Canada, 2017). The Regina CMA has a population in 2016 of 247,224 people (Statistics Canada, 2017). In looking at the core of this CMA (the City of Regina), there exists a number of potential provincial highways include highway's numbered 1, 6, 11, 33, and 46. After accounting for the criteria of this highway litter study, including (i) reach into the core city, (ii) absence of construction, (iii) a clean break between rural and urban city limits, and (iv) no twinned segments, highway number 6 south of Regina was chosen.

3.2 Selection of a 200m Test Section

In reviewing the highway right of way, there are two ditches that run parallel to each side of the main roadway driving lanes. The roadway driving lanes are located roughly within the centre of the right of way. This roadway consists of a two-lane undivided highway with standard shoulders on the outside of each direction of travel. Adjacent to the highway is typically (roughly 90%) various dry land crops that align roughly parallel to the highway right of way. There also exist numerous other features such as dugouts, sections where rural farm yards have approaches and driveways meeting perpendicularly to highway #6, as well as grid roads and agricultural approaches that also access the highway. The alignment of the highway runs directly south with only two curved sections (located between 17.8 km and 18.8 km on the section of roadway) throughout the entire study segment.

Due to the highway #6 orientation in the north and south direction, northbound traffic comes into the CMA and southbound traffic leaves the CMA. Upon observation and subsequent measure in the preliminary study, it was found that the ditches were of a varying width from 14 m to 25 m. As discussed, the selection of the litter sampling area is important due to the heterogeneous nature of highway litter. Two definitions were considered in the present study to determine the width of the sampling areas: (i) to use a constant width of ditch (measured from the edge of the roadway surface), or (ii) to use the variable width between the edge of the roadway and the private land, known as a constant feature width.

After careful considerations of the site characteristics and the possibilities of litter transportation, it was decided that a constant feature width would provide a better definition for this study. This is due to the geographical features and the possibility of the changing wind patterns coming off of the highway and interacting with the crop areas parallel to the ditch. On the contrary, keeping a constant width of ditch would add more complexity such as changing distance of study area from the field area, less area to study, as well as different sloping and runoff considerations in different parts along the roadway. It also leads to a less representative result of the actual highway litter waste that exists per length of roadway.

It was decided that a study of 200 meters of roadway with a constant feature width should be conducted as a test section (Table 1). This length was chosen as it represented a longer distance than all the minimum lengths and 80% of the maximum lengths used in the four studies reviewed. This 200 m length was broken down into 2 m lengths along the roadway and is shown in Figure 1. This length was also chosen to facilitate ease and accuracy of collecting the litter from ditches. The collection method (as shown by the arrows in Figure 1) was be done by having the researchers start at the roadside edge within a 2 m width and walk towards the edge of the right of way. Then, the researchers come back inside the same 2 m width to the roadside, collecting any litter that was visible while looking down as they walked. This meant that a 1 m width of roadway would be collected on each pass the researcher made from the road edge to the edge of the right of way and vice versa. Also, despite varying lighting conditions, researchers ensured accuracy of collection as they travelled in two directions, making certain objects more visible based on their orientation. By observing a 1m width at a time, the researchers had the opportunity to very easily focus and be more likely to find almost all the visible litter, over 2.5 cm in size, present.

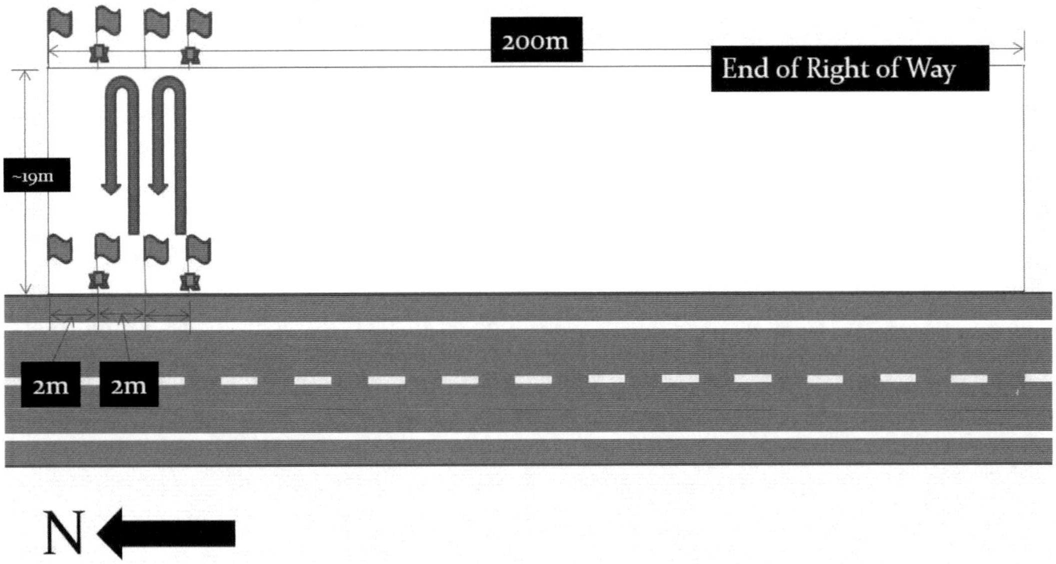

Figure 1: Proposed methodology for a highway litter test section utilizing constant feature width

The lengths of these dimensions were measured out with tape measures and marked with small distinct red flags at 2 m intervals. A handheld GPS device (Garmin eTrex® 10) was used to take points at the corners of the study test area to verify the locations. The locations of the 200 m test sections were selected randomly to avoid bias. All necessary legislation and regulations of the Crown in Saskatchewan, Canada were followed during the sampling. An application was applied and a permit was issued for the study from the provincial Ministry of Highways and Infrastructure.

3.3 Methodology of Highway Litter Trends from a CMA

A length of 40 m was selected for each section to study the 31.52 km highway segment, as discussed in the next section. Further, it was decided that seven study locations in each direction, located 5 km apart, would yield results acceptable to the objectives of the present study. Starting at the beginning of highway #6's most northerly terminus on Township Road 262, the seven locations would occur at increments of 5km as one traveled south of the terminus. This meant that the study locations would be at 31.5km, 26.5km, 21.5km, 16.5km, 11.5km, 6.5km and 1.5km on both sides of the highway. However, upon completing a field survey of the site locations, it was found that the final 1.5km southbound site had a farmyard adjacent to the study location. The ditch area at this location was cleared of all visible litter and the grass was mowed to a very low level. Therefore, it was decided that instead of sampling at the 1.5km distance, the sampling would take place immediately north of the farmyard area at 2.1km.

The field studies were conducted on August 13th and 14th within a 36-hour period. The samples from the litter study were gathered using large plastic bags and transported to the laboratory for storage. The litter samples were allowed to cure and air-dry in a controlled environment for 5 months. Each waste particle with at least 2.5cm diameter was carefully inspected and counted. The mass of the waste samples was measured using electrical balance. This lab work took place from January 15th to January 18th, 2018 in the Geo-Environmental Laboratory at University of Regina.

4 RESULTS AND DISCUSSION

4.1 Results of the Highway Litter Test Section

Following the methodology described in Section 3.2, the trial litter survey occurred on the date of July 24, 2017 and was completed in a single day to minimize the potential impact of weather. The mean count per 2 m length of roadway for both directions was 5.58 with a maximum count of 15 and a minimum count of 0. The litter survey results on highway #6 (with a rough width of 19m) showed great variations in the count of visible litter, with a sample standard deviation of 2.6.

The percentage difference between the maximum 2 m segment and the minimum 2 m segment of roadway in the litter survey was calculated using equation 1 and was found to be 200%. The percentage differences were then further calculated for varying lengths of segments. To identify an acceptable length of segment for a waste count, a plot was prepared and shown in Figure 2.

$$[1]\ Percent\ Difference = \frac{|Max\ count - Min\ count|}{0.5 \times (Max\ count + Min\ count)} \times 100\%$$

A non-linear relationship is observed between the variables in Figure 2. One can see that there is a decreasing percent difference between the samples as the length of roadway segment increases. A highway segment with a length of 40 meters was selected for the full study. This length corresponds to a 35 percent difference between the minimum and maximum values observed.

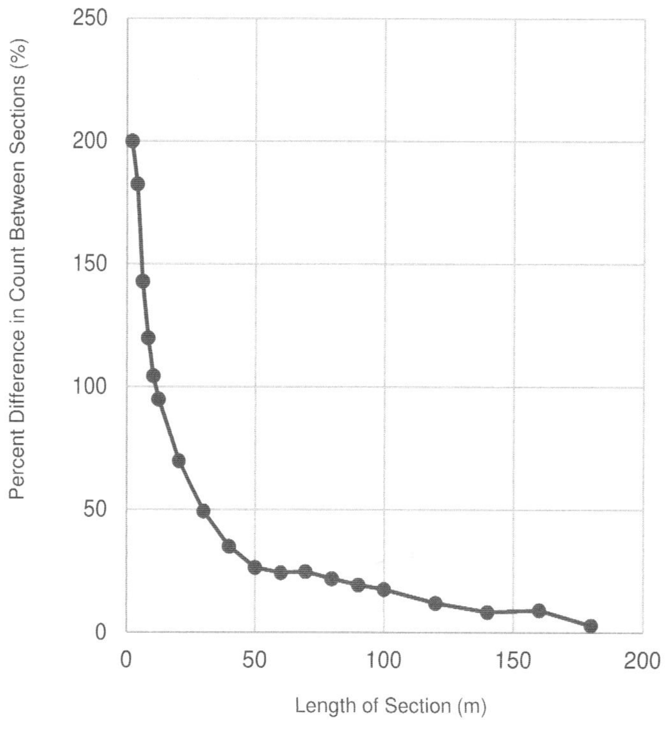

Figure 2: An inverse relationship between the percent difference in count and the section length

4.2 Highway Litter Count and Mass Density Changes with Respect to Distance from a CMA

Figures 3 illustrated the change of count density from the seven roadway segments studied. The distance of 0 km is located furthest from the CMA and a distance of 35 km represents the closest distance to the CMA. It appears that more waste is obtained near the CMA. The dotted lines are the linear trendlines from the data sets. A clear increasing trend is observed in the northbound direction (Figure 3a), although a similar, but less obvious, increasing trend is also observed in the southbound direction. The error bars are derived from the 35% uncertainties associated with the selection of the 40 m segment. The results suggested the CMA may have an observable impact on litter count density, at least for highway #6 considered in this study.

It is not clear why a more significant increasing trend is observed in the northbound direction (Figure 3a) than in the southbound direction (Figure 3b). This could be due to driver litter habits, traffic flow patterns, roadside signage, and/or geographical features. There is also a noteworthy dip in both the northbound and southbound 16.5km sections. This may be due to the impact of an Adopt-A-Highway program. This program allows local volunteer groups collect litter at least once a year from a certain segment of roadway and may have had an impact on the litter count density at section 16.5km, as shown in Figures 3a and 3b. However, evidence is not available at the time of writing to support this claim.

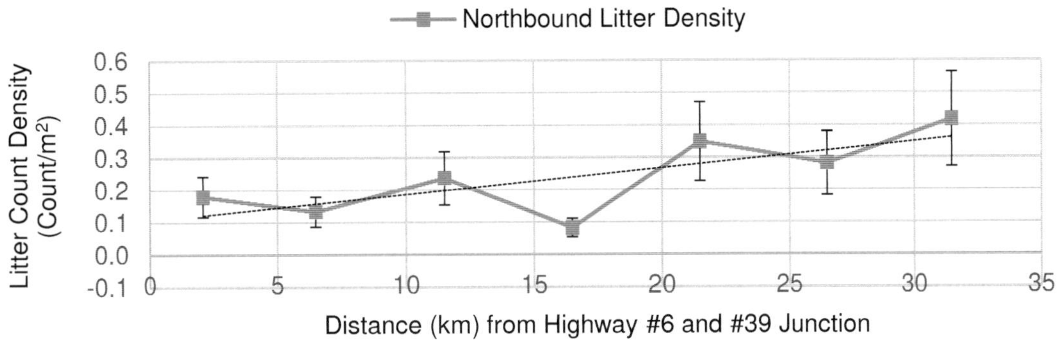

Figure 3a: Northbound litter count density in the study area

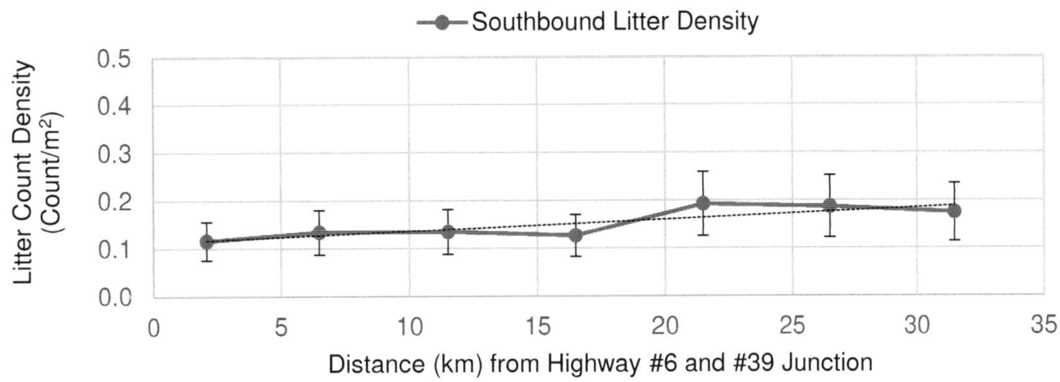

Figure 3b: Northbound Litter Count Density in the Study Area

As discussed, the magnitude and spatial distribution of highway litter can also be quantified by waste mass. Compared to Figure 3, best-fit linear trends depicted in Figures 4a and 4b were less distinctive. The results suggested that waste count may be more appropriate with respect to the study objectives. In the future, an empirical equation will be proposed between waste count and mass using highway litters data from Saskatchewan.

Figures 4a and 4b show the litter mass density changes with distance away from a CMA. Similar to the waste count analysis, a more pronounced increasing trend is observed in northbound direction. This again, may be due to the behavioral and political aspects of highway littering. Also, using all litter samples, it was found that the average mass per highway litter piece was 14.71 g/piece and the average number of highway litter pieces per mass was 68 pieces/kg. This may be of use to further studies looing to estimate the quantity of Saskatchewan highway litter counts from mass, assuming similar study conditions. The compositional changes that occur in the study area is currently being undertaken to investigate the characteristics of highway litters.

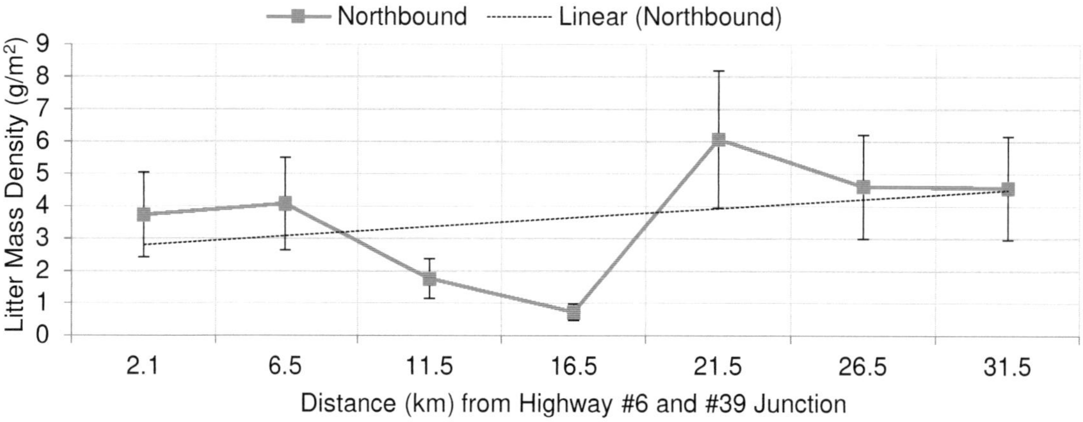

Figure 4a: Northbound Litter Mass Density with Changing Distance from a CMA

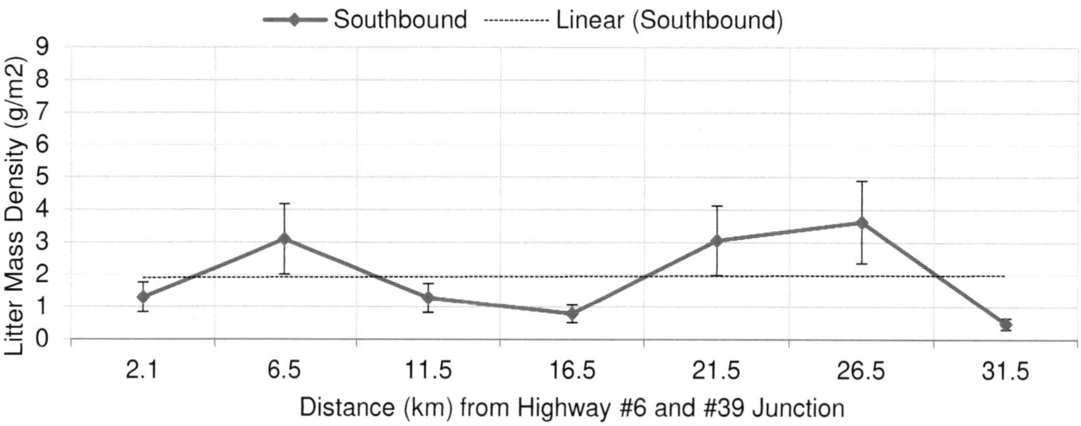

Figure 4a: Southbound Litter Mass Density with Changing Distance from a CMA

5 CONCLUSION

A framework for the collection and quantification of highway litter is proposed in the present study using a trial test and a full field study for a 31.52km highway segment south of Regina, Saskatchewan. Due to the nature of highway litter, the sampling area affects the accuracy and precision of the results. A 40 m segment of highway was selected in the present study for highway litter quantification, with an estimated uncertainty of ±35%. For waste count analysis, a positive relationship is observed between the parameters. For example, it appears that more wastes were littered on highway near the City. The results are more pronounced in the northbound direction. Unlike waste count analysis, definite trends were not observed in waste mass analysis. It was found that the average mass per Saskatchewan highway litter piece was 14.71 g/piece and the average number of Saskatchewan highway litter pieces per mass was 68 pieces/kg. This study provides some of the preliminary work on the quantity and spatial distribution of Saskatchewan highway and roadside litter.

Acknowledgement

This work is financially supported by a research grant (RGPIN-385815) from the Natural Sciences and Engineering Research Council of Canada (NSERC) and an internal FEROF grant from the Faculty of Engineering and Applied Science at University of Regina. The authors are grateful for their support. The views expressed herein are those of the writers and not necessarily those of our research and funding partners.

References

Cutter, S. L., Tiefenbacher, J., Birnbaum, S., Wiley, J, and Solecki, W.D. 1991. Throwaway Societies: A Field Survey of the Quantity, Nature and Distribution of Litter in New Jersey. Applied Geography 11(2): 125-141.

Gershman, Brickner, & Bratton, Inc. 2005. New Jersey Litter Survey: 2004. Retrieved August 28, 2017 from http://www.njclean.org/2004-New-Jersey-Litter-Report.pdf

Smith, H. 2008. A Characterization of Nova Scotian Litter: 2008 Litter Survey. Retrieved on August 28, 2017 from https://novascotia.ca/nse/waste/docs/LitterSurvey_2008.pdf

Southeast Environmental Association. 2003. Roadside Litter Survey Report 2003. Retrieved on August 28, 2017 from http://www.gov.pe.ca/photos/original/ee_roadlitter03.pdf

Statistics Canada. 2017. Table 051-0056 - Estimates of population by census metropolitan area, sex and age group for July 1, based on the Standard Geographical Classification (SGC) 2011, annual (persons), CANSIM (database). Retrieved August 28, 2017.

Stein, S. R. 2010. Northeast 2010 Litter Survey. Retrieved August 28, 2017 from http://www.erplanning.com/uploads/2010_Northeast_Litter_Survey_-_Final_Report_-_Revised.pdf

Syrek, D. B. 1986. American Litter: 1986. The Journal of Resource Management and Technology, 15(2-3): 69-83.

Building Tomorrow's Society
Bâtir la Société de Demain

Fredericton, Canada
June 13 to 16, 2018 / *13 au 16 juin, 2018*

PREDICTIVE MODELLING OF RECREATIONAL WATER QUALITY AT TWO BEACHES IN WINDSOR ESSEX REGION, ONTARIO, CANADA

Madani, Mohammad[1], Seth, Rajesh[1,2]
[1] Department of Civil and Environmental Engineering, University of Windsor, Windsor, ON, Canada
[2] rseth@uwindsor.ca

1 INTRODUCTION

Swimming in contaminated waters may result in gastrointestinal and respiratory diseases. To protect bathers from swimming in such circumstances, microbial water quality is typically monitored for fecal indicator bacteria (FIB). Furthermore, in addition to the health effects, beach closure may have deep financial impacts as well. Annually this ranges from $11.3M to $117M in lost value for the Great Lake recreational swimmers for those days when swimming is banned (Shaikh 2006; Rabinovici et al. 2004). Although E. coli is considered the most suitable indicator for recreational water monitoring according to Guidelines for Canadian Recreational Water Quality, its measurement takes 18 to 24 h before results are available. However, water quality in near shore regions can change over a matter of hours (Myers et al., 1998; Boehm et al. 2002) so concentrations may change between the time of sampling and the reporting of results. Unsafe conditions are frequently announced late due to latencies in the E. coli measurement process. This process results in issuing closures based on previous day data rather than current water conditions. Developing rapid analytical methods which average two hours of laboratory analytical time is a possible solution to address this problem. Although these methods may be operationally available, but they need higher analytical cost than slower culture-based methods (Setty, 2012). Statistical models can be another solution which has been shown to also be an accurate approach (Nevers and Whitman 2005; Feng et al. 2015; Dada and Hamilton, 2016). In the late 1990s the first empirical models developed through statistical techniques such as Multiple Linear Regression (MLR), were used to reduce the risk of infection to users of recreational waters (USEPA, 1999). In Ohio, one-year data at three Lake Erie beaches was used to explore the effectiveness of predictive models (Francy and Darner, 1998). The simplest type of model, namely Rainfall-based alerts, have been used by several communities for a number of years (Francy D., 2009). Later, application of predictive models considering other water quality and hydro-metrological parameters on other beaches became more prevalent and many studies were conducted to develop such models for inland recreational lakes (Dada and Hamilton 2016; Olyphant and Whitman 2004; Nevers and Whitman 2005, 2011; Francy et al. 2013), coastal beaches (Thoe et al., 2012; Boehm et al., 2007) and reservoirs (Francy et al. 2013). Although some deterministic models were developed to predict the water quality and FIB concentration, these models are inferior to statistical models in most situations as they require the determination of FIB sources (Francy D., 2009). In the present study, collected water quality and weather data and their various transformations are examined to develop a multiple linear regression model using the United States Environmental Protection Agency (USEPA) Virtual Beach (VB) toolbox.

2 MATERIAL AND METHOD

Samples were collected on 30 consecutive days between the hours 8:00 and 12:00 (10 August 2010 to 8 September 2010) at Sandpoint beach (Lake St. Clair) and Holiday beach (Lake Erie) in the Windsor-Essex Region. Detailed descriptions of the analysis, field measured parameters and E. coli enumeration techniques were reported in previous work (McPhedran, 2013). Weather data was obtained from the nearest Environment Canada station for each beach. Data included air temperature (℃), daily rainfall (mm) and daily 10:00 am averages of hourly measurements of wind speed (m/s) and wind direction (degree). Minitab and XLSTAT were used to analyse the results. Briefly, to improve linear relations between E. coli concentrations and explanatory variables and also to take care of wide range of expected values, concentrations of E. coli were log10-transformed before any statistical testing and modeling (Francy et al., 2013). The candidate explanatory variables included: 24h and 48h antecedent cumulative rainfall, wind direction, wind speed, turbidity, conductivity, air and water temperature, dissolved oxygen, pH and conductivity. Model performance was examined by determination of metrics such as Root Mean Square Error (RMSE), Accuracy, Sensitivity and Specificity which are defined as:

$$[1]\ RMSE = \sqrt{\sum_{i=1}^{N} \frac{(\log_{10} P_i - \log_{10} O_i)^2}{N}}\ ,\quad Sensitivity = \frac{TP}{(TP+FN)}\ ,\quad Specificity = \frac{TN}{(TN+FP)}\ ,\quad Accuracy = \frac{(TP+TN)}{N}$$

Where N is number of observations, $\log_{10} P_i$ is the log10-transformed predicted model value, $\log_{10} O_i$ is the log10-transformed observation, and TP, TN, FP, and FN are numbers of true positives, true negatives, false positives (Type I error), and false negatives (Type II error) respectively.

3 RESULTS AND DISCUSSION

Time series illustration of observed and modeled E. coli count for the two beaches are shown in Figure 1 (a-b). Also, Figure 1 (c and d) show observed vs. model values of the advisory threshold for Sandpoint beach and Holiday beach respectively. During the study time, for Holiday beach, there were five EC exceedances (>100 CFU/100mL) while at Sandpoint beach, there were three EC exceedances. The non-parametric Mann-Kendall trend test reveals that there is no trend in the time series data (P-value of 0.28 and 0.97 for Sandpoint beach and Holiday beach respectively, $\alpha = 0.05$). Also, using the Durbin–Watson statistic ($\alpha = 0.05$), the selected variables were not autocorrelated ($D = 2.7$ and 2.3 for Sandpoint beach and Holiday beach respectively which are greater than $D_u = 1.93$). This check is important to make sure serial-correlation conditions which lead to improper estimation of the model are satisfactory to conduct MLR analysis (Ge and Frick, 2007; WYMER, 2007). The obtained RMSEs of 0.27 and 0.26 logCFU/100mL in Sandpoint beach and Holiday beach respectively are lower than the common range for MLR models reported in previous studies which were in the range of 0.4-0.5 logCFU/100 mL (Thoe et al., 2014). Virtual beach provides the top 10 best fit correlations, which based on cross-validation, the optimal parameters are those that minimize the mean squared error of prediction (MSEP).

Table 1: Summary of the model parameters

Beach Name	Sandpoint Beach	Holiday Beach
R2	0.78	0.89
RMSE	0.27	0.26
Accuracy (%)	96	96
Sensitivity	0.75	1.0
Specificity	1.0	0.96
Cross Validation MSEP	0.26-0.33	0.24-0.35
Correct Exceedance	3	5
Type I Error rate (%)	0	3.3
Type II Error rate (%)	3.3	0
Total Error rate (%)	3.3%	3.3%

Results show that although turbidity was the most important variable in VB models for both beaches, different combinations of other variables were found to be significant for each of the two beaches. While polynomial transformed turbidity, 24h rainfall and wind direction along shoreline were significant parameters in Sandpoint beach VB-Model, LOG(turbidity), wind direction, and INVERSE(wind_direction) were the most important variables for Holiday beach. Note that although EC levels at the study beaches can be impacted by a variety of other variables (humidity, cloud cover, animal and human sources etc.), these selected variables were reasonably satisfactory for building regression equations. Good correlation between the fitted VB models and observations were observed for both beaches (Figure 1). For the Sandpoint beach VB Model, the R^2 value was observed to be 0.78, sensitivity of 0.75, specificity of 1.0 and accuracy of 96%. Holiday beach VB model had higher R^2 of 0.89. Sensitivity, specificity and accuracy of model for Holiday beach are observed to be 1.0, 0.96 and 96% respectively. A summary of model evaluations for Sandpoint beach and Holiday beach, is shown in Table 1. This model resulted in 8 (22%) Type I and 5 (14%) Type II errors out of 30 testable outcomes at Sandpoint beach, and 3 (6%) Type I and 3 (6%) Type II errors out of 30 testable outcomes at Holiday Beach over the same time periods. 25% of the sample size (~ 8 days) and 1000 trails was set for cross-validation purpose. The best fit equation with the lowest MSEP is selected between top 10 results that provided by VB. Results of MSEP range for selected model are shown in the Table 1 for both beaches.

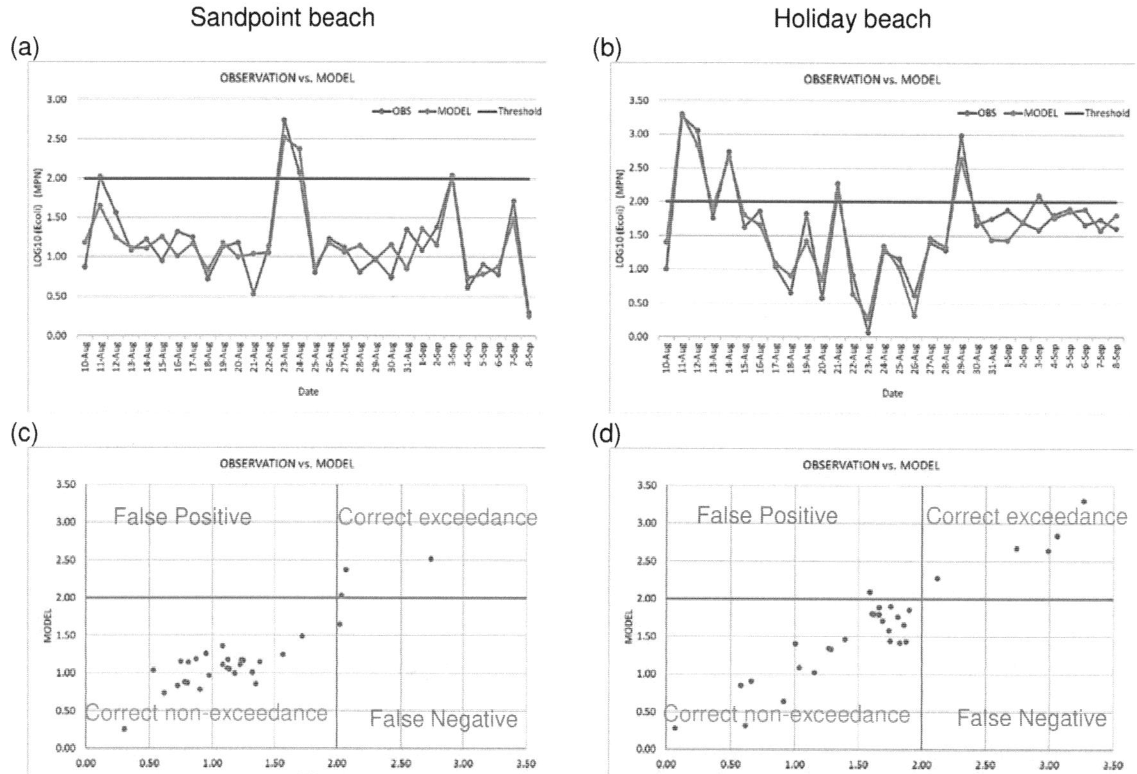

Figure 1: Actual vs predicted Log10-E. coli concentrations for Sandpoint beach and Holiday beach.

4 CONCLUSION

As a brief model interpretation, the structure of this model implies that approximately 78% of the variation of the E. coli concentration is accounted for by the variations of 24h rainfall, turbidity and wind direction for Sandpoint Beach. In the case of Holiday Beach, 96% of the variation is covered by explanatory variables such as turbidity, wind direction and water temperature. Since the variables are transformed, their effects are all nonlinear. Rainfall, which affects the transport of microbial pollution and spikes in turbid conditions, has an independent impact, while turbidity may be associated with the microbial build-up processes as a result of sediment resuspension that could influence E. coli concentrations into the water column. Storm events that usually coincide with high wind speed results in sediment resuspension as a result of wave/ current motions. In the case of Holiday Beach, rainfall is found to not be as effective as usually expected for these particular variables, which might be due to the beach geographical condition or sampling locations which need to be explored in more detail in future studies.

References

Boehm, A.B. 2007. Enterococci concentrations in diverse coastal environments exhibit extreme variability. *Environ. Sci. Technol.* **41**(24): 8227–8232.

Boehm, A. B.; Grant, S. B.; Kim, J. H.; Mowbray, S. L.; McGee, C. D.; Clark, C. D.; Foley, D. M.; Wellman, D. E. 2002. Decadal and shorter period variability of surf zone water quality at Huntington Beach, California. *Environ. Sci. Technol*, **36**: 3885–3892.

Dada, A. C. and Hamilton, D. P. 2016. Predictive Models for Determination of E. Coli Concentrations at Inland Recreational Beaches. *Water, Air, & Soil Pollution*, **227**(9): 347

Feng, Z.; Reniers, A.; Haus, B. K.; Solo-Gabriele, H. M.; Wang, J. D.; Fleming, L. E. 2015. A predictive model for microbial counts on beaches where intertidal sand is the primary source. *Marine Pollution Bulletin*, **94**(1–2): 37–47.

Francy, D. S. 2009. Use of predictive models and rapid methods to nowcast bacteria levels at coastal beaches. *Aquatic Ecosystem Health and Management*, **12**(2): 177–182.

Francy, D.S.; Brady, A.M.G.; Carvin, R.B.; Corsi, S.R.; Fuller, L.M.; Harrison, J.H.; Hayhurst, B.A.; Lant, J.; Nevers, M.B.; Terrio, P.J.; and Zimmerman, T.M. 2013. Developing and implementing predictive models for estimating recreational water quality at Great Lakes beaches: *USGS Scientific Investigations Report* 2013–5166, 68

Francy, D.S.; Darner, R.A., 1998. Factors affecting Escherichia coli concentrations at Lake Erie public bathing beaches: *U.S. Geological Survey Water-Resources Investigations Report* 98-4241, 41

Ge, Z., Frick, W.E. 2007. Some statistical issues related to multiple linear regression modeling of beach bacteria concentrations. *Environ. Res.* **103**(3): 358–364.

McPhedran, K.; Seth, R.; Bejankiwar, R. 2013. Occurrence and predictive correlations of Escherichia coli and Enterococci at Sandpoint beach (Lake St Clair), Windsor, Ontario and Holiday beach (Lake Erie), Amherstburg, Ontario. *Water Quality Research Journal of Canada*, **48**(1): 99.

Myers, D.N., Koltun, G.F., and Francy, D.S. 1998. Effects of hydrologic, biologic, and environmental processes on sources and concentrations of fecal bacteria in the Cuyahoga River, with implications for management of recreation waters in Summit and Cuyahoga Counties, Ohio: *USGS Water-Resources Investigations Report* 98: 4089

Nevers, M. B. and Whitman, R. L. 2005. Nowcast modeling of Escherichia coli concentrations at multiple urban beaches of southern Lake Michigan. *Water Research*, **39**(20): 5250–5260.

Olyphant, G. A. and Whitman, R. L. 2004. Elements of a predictive model for determining beach closures on a real time basis: The case of 63rd Street Beach Chicago. *Environmental Monitoring and Assessment*, **98**(1–3): 175–190.

Rabinovici, S.J.M.; Bernknopf, R.L.; Wein, A.M.; Coursey, D.L.; Whitman, R.L. 2004. Economic and health risk trade-offs of swim closures at a Lake Michigan beach. *Environ. Sci. Technol.* **38**(10):2737–2745.

Setty, K. 2012. Adapting and adopting rapid molecular methods for beach water quality monitoring. *Urban Coast* **3**:48- 53.

Shaikh, S.L. 2006. Fact Sheet: The value of Chicago beaches, University of Chicago, Chicago, IL.

Thoe, W.; Gold, M.; Griesbach, A.; Grimmer, M.; Taggart, M. L. L.; Boehm, A. B. B. 2014. Predicting water quality at Santa Monica Beach: Evaluation of five different models for public notification of unsafe swimming conditions. *Water Research*, 67C, 105–117.

USEPA. 1999. Review of Potential Modeling Tools and Approaches to Support the BEACH Program. EPA-823-R-99-002. U.S. Environmental Protection Agency, Office of Science and Technology, Washington, DC.
WYMER, L. J. 2007. Statistical Framework for Recreational Water Quality Criteria and Monitoring. Presented at 7th Annual Great Lakes Beach Association Meeting, Cincinnati, OH, October 03 - 05,

Building Tomorrow's Society
Bâtir la Société de Demain

Fredericton, Canada
June 13 to 16, 2018 / *13 au 16 juin, 2018*

REMOVAL OF SULFOLANE FROM WATER USING ACTIVATED CARBON ADSORPTION

Yiqiao, Yang[1], Linlong, Yu[1,3], Sobhan, Iranmanesh[1], Ian, Keir[2], and Gopal, Achari[1]
[1] Department of Civil Engineering, University of Calgary, Canada
[2] Bonavista Energy Corporation, University of Calgary, Canada
[3] corresponding author linyu@ucalgary.ca

Abstract: The objective of this research is to determine the effectiveness of two commercially available activated carbons to remove sulfolane from aqueous media. Batch adsorption experiments were conducted to determine each activated carbon's maximum sulfolane adsorption capacity. The activated carbon demonstrating the highest sulfolane adsorption capacity was selected for further investigation, where the impact of pH, temperature, water quality and diethanolamine (DIPA) on sulfolane adsorption capacity was investigated. In addition, a regenerative method based on UV/H_2O_2 was developed to treat and recycle the exhausted activated carbons. The results showed the coconut shell based activated carbon performed better than coal based activated carbon. It was also found that the adsorption capacity was significantly impacted by DIPA being present in the water matrix. The regeneration study revealed the UV/H_2O_2 process can be successfully employed to regenerate the exhausted activated carbon by degrading the sulfolane desorbed from activated carbon.

1 INTRODUCTION

Activated carbon adsorption is a well-known physical separation method featured by the accumulation of contaminants onto the surface of activated carbon. It is widely used in water and wastewater treatment processes to remove organic matters (Marsh et al 2006). In past several decades, the activated carbon adsorption has been successfully employed to remove industrial contaminants including dyes, surfactants, perfluorinated compounds, phenolic compounds and others (Karabacakoğlu et al. 2008, Al-Degs et al. 2008, Wu and Pendleton 2001, Hansen et al. 2010). Activated carbon's adsorption capacity depends on multiple factors including temperature, type of contaminants, surface properties, solution ionic strength and pH. The activated carbon adsorption process can be driven by electrostatic or non-electrostatic interactions (Moreno-Castilla 2004). Electrostatic interaction takes place when adsorbate is an electrolyte and in dissociated or protonated form. The charged adsorbate and carbon surface develop attractive and repulsive forces, which controls the overall adsorption performance. Conversely, non-electrostatic interactions between activated carbon and adsorbate involve van der Waals force of attraction, hydrogen bonding and hydrophobic interactions (Moreno-Castilla 2004).

Activated carbons are produced from carbonaceous materials such as coal, lignite, wood, peat, nut shell and coconut shell (Cecen et al. 2011). Activated carbon derived from different raw materials usually differ in pore structure and surface area distribution. It is also available in different particle sizes and classified as granular activated carbon (GAC) or powdered activated carbon (PAC). GAC has diameter between 1.2 mm to 1.6 mm, is commonly used in continuous water treatment process and is easy to regenerate. In contrast,

PAC has finer particle size with diameter less than 0.1 mm. The PAC exhibits a faster adsorption kinetics but hard to regenerate, leading to high operation cost.

In this study, GACs were investigated to treat an emerging industrial contaminant, sulfolane, from contaminated water. Sulfolane is an organosulfur compound mainly used to extract aromatics from hydrocarbon mixtures and sweeten sour gas in the Sulfinol® process. It has been frequently detected in the soil and groundwater near refineries and gas plants in North America. With a high water mobility, sulfolane can pose risk for off-site contamination.

So far, the study on sulfolane removal by using activated carbon adsorption is very limited and there is only one publication (Coggeshall and Price 2013) evaluated the feasibility of adopting activated carbon adsorption to remove sulfolane from aqueous media. In this research, the removal of sulfolane under various experimental conditions using activated carbon adsorption is investigated. Two commercially available activated carbons prepared from coal and coconut shell were selected as adsorbents. In addition, the impact of temperature, pH, water quality and diethanolamine (DIPA) on the performance of activated carbon adsorption is determined. Furthermore, a UV/H_2O_2 based method is developed to regenerate the exhausted activated carbons.

2 METHODS AND MATERIALS

2.1 Materials

Sulfolane with 99% purity, sodium hydroxide with 98% purity, hydrogen peroxide (ACS grade, ~30%), hydrochloride (ACS grade, ~37%) and diethanolamine (DIPA) with 99% purity were obtained from Sigma Aldrich, Canada. Sodium bicarbonate with 99% purity was purchased from VWR. Ultrapure water was used to prepare all chemical solutions. Groundwater samples were collected from sulfolane contaminated site. Two GACs derived from coal (Type A) and coconut shell (Type B) were obtained from confidential sources. The characteristics of activated carbons are presented in Table 1.

Table 1. Physical Properties for Type A and Type B activated carbon

Physical Properties	Type of activated carbon	
	A	B
Raw Material	Coal	Coconut Shell
Iodine Number (mg/g)	900	1100
BET surface area (m^2/g)	660.3	895.2
Total Pore Volume (cm^3/g)	0.407	0.483

2.2 Batch equilibrium adsorption study

Different amount of GACs (20 mg ~ 400 mg) were added into 20 ml solution containing 100 ppm sulfolane. All samples were then tightly sealed and shaken vigorously for 24 hours at room temperature. After shaking, activated carbons were separated from water samples using a no. 80 mesh (180 µm) prior to the extraction. The amount of sulfolane adsorbed by activated carbon at equilibrium, q_e (mg/g), was determined by Equation [1]:

[1] $q_e=(C_0-C_e)V/W$

where C_0 and C_e (mg/L) are initial and equilibrium sulfolane concentration in the water, V (L) is the volume of solution and W (g) is the mass of dry activated carbon.

2.3 Interference study

For interference study, 20 ml of solution containing 100 ppm sulfolane were mixed with 0.05 g activated carbon for one hour under different experimental conditions. The sulfolane concentration in the bulk solution before and after adsorption experiments were determined. Three pH levels (4, 7 and 10) and two temperature levels (6 °C and 22 °C) were investigated. pHs of solution were adjusted by using hydrochloric acid and sodium hydroxide, while the temperature was controlled by a fridge with digital temperature setting. Two types of water including groundwater and ultrapure water were used in this study. 100 ppm DIPA was also introduced into the solution to determine its impact on sulfolane adsorption by activated carbon.

2.4 Regeneration study

0.3 grams of activated carbons were added into 100 ml solution containing 100 ppm sulfolane. The prepared sample was mixed vigorously by shaker for 1 hr to reach adsorption equilibrium. After adsorption, the activated carbon was separated from the solution using a no. 80 mesh. The activated carbons were then placed in a quartz beaker filled with 100 ml ultrapure water and exposed to UVC irradiation in presence of 3000 ppm H_2O_2 for 1.5 hours. The UVC irradiation is provided by a Luzchem photoreactor equipped with 10 UVC lamps. The light intensity measured in the quartz beaker was determined to be 5.3×10^{17} photon/s. Magnetic stirrer was used to enhance the contact between activated carbon and water. The regenerated activated carbons were then reused to adsorb sulfolane from water following the procedures outlined before. The sulfolane concentration in bulk solution and surface of activated carbon were analyzed to determine the effectiveness of regeneration.

2.5 Analytical procedure

Liquid-liquid phase extraction procedure was used to extract sulfolane from the aqueous phase. Each 5 ml of water sample was extracted with 3 ml of dichloromethane (DCM). After extraction, 1 ml of DCM layer was transferred into GC vial using a pasturing pipette. The sulfolane concentration was then analyzed by Agilent 6890 gas chromatography (GC) equipped with auto-sampler and Flame Ionization Detector (FID). A fused silica capillary column (ZB 5MSI, Phenomenex) was used to separate sulfolane from other compounds. High purity helium was used as a carrier gas. The GC was operated on splitless mode with an injection volume of 1.0 μL. Temperature of injection port was held at 160 °C. The oven temperature was programmed as follows: initial temperature was set at 120 °C, then increased to 300 °C at a rate of 10 °C/min, the final temperature was maintained at 300 °C for 3 min. The FID temperature was held at 330 °C.

3 RESULTS AND DISCUSIONS

3.1 Adsorption Isotherm

The adsorption isotherms of sulfolane onto Type A (coal based) and Type B (coconut shell based) activated carbons are presented in Figure 1. The results showed that the adsorption isotherm for both activated carbons followed a concave curvature trend and are increased gradually. No definite plateau was observed with the investigated equilibrium sulfolane concentrations. It can also be seen from Figure 1 that Type B activated carbon adsorbed more sulfolane than Type A activated carbon for the same equilibrium sulfolane concentration in water. In order to quantitatively compare the adsorption capacity of two activated carbons, the experimental data were fitted into two classical isotherm models, namely Langmuir and Freundlich. The Langmuir isotherm model depicts monolayer adsorption process and suggest there is a limited number of adsorption sites available on adsorbent surface. Once an adsorbate adhered to a specific homogenous site, it assumes no further molecules can occupy that site (Allen et al. 2004). The linearized form of Langmuir isotherm can be represented as:

[2] $C_e/q_e=1/(Q_m b)+C_e/Q_m$

where qe is the amount of solute adsorbed per weight of adsorbent (mg/g), Q_m and b are the Langmuir constant represents the adsorption capacity and energy, and C_e is the equilibrium solution concentration of solute (mg/L).

The Freundlich isotherm model describes a multilayer non-ideal adsorption process and assumes heterogeneous surface energies (Karri et al. 2017), in which the adsorption activation energy is dependent on surface coverage. The linearized form of Freundlich isotherm can be expressed as:

[3] $\log(q_e)=\log(K)+1/n[\log(C_e)]$

where K and n are the Freundlich constants represents adsorption strength and intensity. Other terms are as defined previously.

Table 2 summarizes the Langmuir and Freundlich constants along with the coefficient of determination (R^2) for Type A and Type B activated carbon. It was calculated both R^2 are greater than 0.9, indicating the adsorption process of sulfolane can be adequately described by both Langmuir and Freundlich model. Comparing the model parameters, Type B activated carbon has a higher adsorption capacity (Q_m=28.82 mg/g) and considerable adsorption strength (K=5.33). The higher adsorption capacity of Type B activated carbon can be explained by its better surface properties. As shown in Table 1, coconut shell based Type B activated carbon has a larger iodine number (1100 mg/g), BET surface area (895.2 m^2/g) and total pore volume (0.48 cm^3/g) than Type A activated carbon. As reported by Hashimoto et al. (1975), activated carbon derived from coconut shell had a larger ratio of micropores (higher surface area), whereas coal based activated carbon had a macroporous structure (lower surface area).

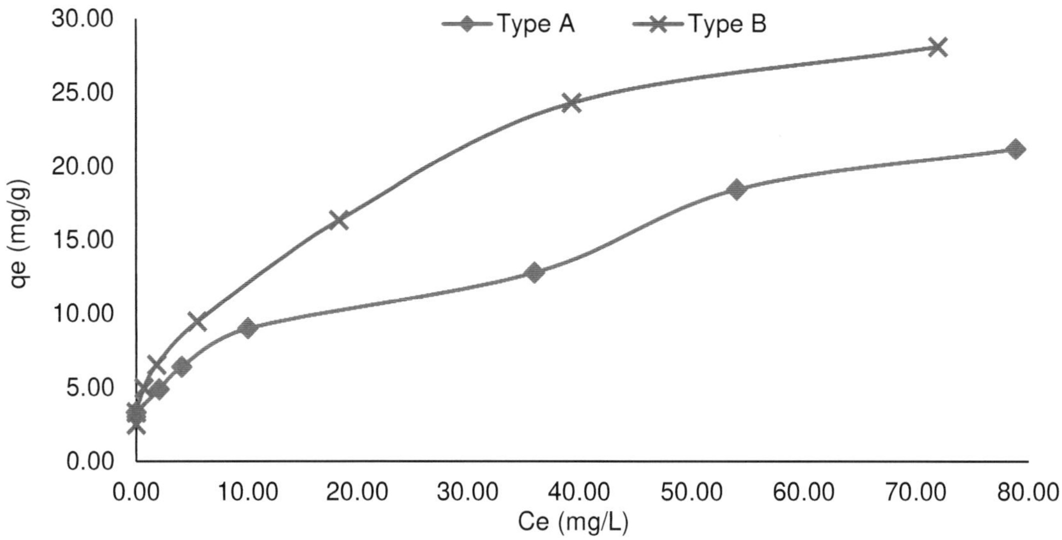

Figure 1. Adsorption isotherm of sulfolane onto Type A and Type B activated carbon

Table 2. Fitting parameters of Langmuir and Freundlich models for Type A and Type B activated carbon

	Langmuir Equation			Freundlich Equation		
	Q_m	b	R^2	K	n	R_2
Type A	21.79	0.10	0.931	3.63	2.55	0.984
Type B	28.82	0.18	0.962	5.33	2.56	0.991

3.2 Adsorption Interferences

Type B activated carbon (coconut shell based) was selected for further optimization study due to its high sulfolane adsorption capacity. The optimization study investigated the impact of pH, temperature, water quality and DIPA on sulfolane adsorption by activated carbon. The degree of impact was assessed by comparing the sulfolane removal percentage under different conditions with the control experiments and the result is summarized in Table 3. It is observed that pH had a negligible influence on sulfolane adsorption, whereas, the temperature, water quality, presence of DIPA have relatively higher impact. Less than 1% variations between pH adjusted samples and control samples shows a possible non-electrostatic interaction between sulfolane and activated carbon. Lower temperature (6 °C) benefited the activated carbon adsorption by a 7.8% increase on sulfolane removal, which can be attributed to the decreased sulfolane solubility as temperature decreased (Tan et al. 2008). The reduced solubility indicates a weakened bonding force between sulfolane and water, leading to more sulfolane to be adsorbed by activated carbon. The sulfolane removal through activated carbon adsorption in the groundwater was 7.1 % lower than ultrapure water; presence of DIPA reduced the sulfolane removal percentage by 25.8%. All of these impacts are caused by competition of different interferences for the available adsorption sites on the activated carbon surface (Worch 2012, Matsui et al. 2003). The humic matters in the groundwater samples can possibly compete with sulfolane for activated carbon's available adsorption sites and inhibit the sulfolane removal rate. Zeinali and Najafpour (2010) reported that DIPA can be effectively removed from water by using activated carbon, indicating that DIPA occupies the activated carbon's adsorption sites and reduces sulfolane adsorption capacity.

Table 3. The impact of pH, temperature, groundwater and DIPA on percent sulfolane removal

Sample No.	Experimental condition*		Variation from control (%)
Control	pH: 7 Ultrapure water matrix	Temperature: 22 °C DIPA Conc.: 0 ppm	± 0.00
A1	pH: 4		+ 0.67
A2	pH: 10		- 0.83
B1	Temperature: 6 °C		+ 7.80
C1	Groundwater matrix		- 7.10
D1	DIPA Conc.: 100 ppm		- 25.80

*The parameter listed is the only experimental condition changed compared to control experiment.

3.3 Feasibility Study of Using UV/H2O2 to Regenerate Exhausted Activated Carbon

The regeneration method used in this study follows a two-step process, where sulfolane is firstly desorbed from activated carbon in ultrapure water and furtherly degraded by UV/H_2O_2 treatment. Table 4 presents the amount of sulfolane on activated carbon before and after UV/H_2O_2 treatment. It is observed the amount of sulfolane on activated carbon decreased from 4.32 mg to 2.46 mg after 1.5 hours UV/H_2O_2 treatment, indicating the two-step process can remove and degrade the sulfolane adsorbed onto activated carbon, and release the number of active adsorption sites. Yu et al. (2016) reported that more than 99% of sulfolane can be degraded after 30 minutes of exposure to UV/H_2O_2 system. However, the sulfolane removal rate (44% in 1.5 hours) in our study is considerably low. This can be explained by the following two possible mechanisms: (a) the overall sulfolane removal rate from activated carbon is governed by both desorption process and UV/H_2O_2 degradation. Sulfolane desorption process is relatively slower than UV/H_2O_2 degradation, leading to a lower overall degradation rate; (b) The presence of activated carbon in the water can significantly reduce the UV transmission and result in an inefficient photon utilization. The results show that a 1.5 hour regeneration can only partially remove sulfolane from activated carbon. It is expected that a longer treatment period can completely desorb and degrade sulfolane from exhausted activated carbon.

Table 4. Amount of sulfolane on activated carbon before and after 1.5 hours UV/H_2O_2 treatment

Adsorption Stage	Sulfolane on activated carbon (mg)	Sulfolane in water (mg)
Before UV/H_2O_2 treatment	4.32	0
After UV/H_2O_2 treatment	2.46	0

The sulfolane adsorption performance after multiple adsorption-regeneration cycles was investigated and the results were summarized in Table 5. It is observed that the regenerated activated carbon can be reused to adsorb sulfolane from water. Initially, the fresh activated carbon can remove 21.6 mg/g sulfolane from water. After one, two, three and four adsorption-regeneration cycles, the amount of sulfolane adsorbed by the activated carbon are 14.27 mg/g, 11.43 mg/g, 12.23 mg/g and 11.30 mg/g, respectively. The decreased sulfolane removal after regeneration can be explained by the reduced available adsorption sites resulted by the incomplete regeneration. It is hypothesized that completely regenerated activated carbon (longer UV/H_2O_2 exposure) can obtain similar sulfolane adsorption as fresh activated carbon. As shown in Table 5, the amount of sulfolane adsorbed on activated carbon was not significantly changed after four adsorption-regeneration cycles, indicating that UV/H_2O_2 treatment did not alter the treated activated carbon's surface chemistry and pore structure. The total amount of sulfolane removed from water by using activated carbon adsorption increased with the number of adsorption-regeneration cycles. The cumulative mass of sulfolane removed from water through four adsorption-regeneration cycles achieved 21.25 mg, which is three times greater than the mass of sulfolane removed by the fresh activated carbon, indicating that the four cycles regeneration can reduce 70% of activated carbon usage.

Table 5. Adsorption performance after consecutive batch adsorption cycles

Regeneration Cycle	Mass of sulfolane removed from water (mg/g)	Total amount of sulfolane adsorbed on activated carbon (mg/g)*	Cumulative mass of sulfolane removed from water (mg)
0 (Fresh)	21.60	21.60	6.48
1	14.27	-	10.76
2	11.43	-	14.19
3	12.23	-	17.86
4	11.30	21.80	21.25

*The total amount of sulfolane adsorbed includes the previously unremoved and newly adsorbed sulfolane mass.

4 CONCLUSION

In this study, activated carbon was shown to be a promising adsorbent to remove sulfolane from contaminated water. Specific conclusions were summarized as below:

- Coconut shell based activated carbon achieve a better sulfolane adsorption performance than coal based activated carbon.
- pH did not impact the sulfolane adsorption efficiency.

- Presence of DIPA can decrease the sulfolane adsorption on activated carbon.
- UV/H2O2 process can be used to regenerate exhausted activated carbon.
- Multiple adsorption-regeneration cycles can reduce the usage of activated carbon.

Acknowlegment

The authors would like to thank Bonavista Energy Corporation and NSERC for providing the financial supports for this research.

References

Al-Degs, Y. S., El-Barghouthi, M. I., El-Sheikh, A. H., and Walker, G. M. 2008. Effect of solution pH, ionic strength, and temperature on adsorption behavior of reactive dyes on activated carbon. *Dyes and pigments*, **77**(1), 16-23.

Allen, S. J., Mckay, G., and Porter, J. F. 2004. Adsorption isotherm models for basic dye adsorption by peat in single and binary component systems. *Journal of colloid and interface science*, **280**(2), 322-333.

Coggeshall, M., and Price, S. 2013. U.S. Patent Application No. 13/196,697.

Hansen, M. C., Børresen, M. H., Schlabach, M., and Cornelissen, G. 2010. Sorption of perfluorinated compounds from contaminated water to activated carbon. Journal of Soils and Sediments, **10**(2), 179-185.

Hashimoto, K., Miura, K., Yoshikawa, F., and Imai, I. 1979. Change in pore structure of carbonaceous materials during activation and adsorption performance of activated carbon. Industrial & Engineering Chemistry Process Design and Development, **18**(1), 72-80.

Karabacakoğlu, B., Tümsek, F., Demiral, H., and Demiral, İ. 2008. Liquid phase adsorption of phenol by activated carbon derived from hazelnut bagasse. J. Int. Environmental Application & Science, **3**(5), 373-380.

Karri, R. R., Jayakumar, N. S., and Sahu, J. N. 2017. Modelling of fluidised-bed reactor by differential evolution optimization for phenol removal using coconut shells based activated carbon. *Journal of Molecular Liquids*, **231**, 249-262.

Marsh, H., Rodríguez-Reinoso, F., and ebrary, I. 2006. *Activated carbon.* 1st ed, Amsterdam: Elsevier.

Matsui, Y., Fukuda, Y., Inoue, T., and Matsushita, T. 2003. Effect of natural organic matter on powdered activated carbon adsorption of trace contaminants: characteristics and mechanism of competitive adsorption. Water Research, **37**(18), 4413-4424.

Moreno-Castilla, C. 2004. Adsorption of organic molecules from aqueous solutions on carbon materials. Carbon, **42**(1), 83-94.

Tan, I. A. W., Ahmad, A. L., and Hameed, B. H. 2008. Adsorption of basic dye on high-surface-area activated carbon prepared from coconut husk: Equilibrium, kinetic and thermodynamic studies. Journal of hazardous materials, **154**(1-3), 337-346.

Worch, E. 2012. *Adsorption technology in water treatment: fundamentals, processes, and modeling.* Walter de Gruyter.

Wu, S. H. and Pendleton, P. 2001. Adsorption of anionic surfactant by activated carbon: effect of surface chemistry, ionic strength, and hydrophobicity. Journal of colloid and interface science, **243**(2), 306-315.

Yu, L., Mehrabani-Zeinabad, M., Achari, G., and Langford, C. H. 2016. Application of UV based advanced oxidation to treat sulfolane in an aqueous medium. Chemosphere, **160**, 155-161.

Zeinali, F., Ghoreyshi, A. A., and Najafpour, G. D. 2010. Adsorption of dichloromethane from aqueous phase using granular activated carbon: isotherm and breakthrough curve measurements. Middle-East J. Sci. Res., **5**(4), 191-198.

Building Tomorrow's Society
Bâtir la Société de Demain

Fredericton, Canada
June 13 to 16, 2018 / *13 au 16 juin, 2018*

SEASONAL AND TEMPORAL VARIATIONS OF CHLORIDE LEVEL IN RIVERINE ENVIRONMENT

Situ, Qianru[1], Akther, Musa[1,2] and He, Jianxun[1]
[1] Department of Civil Engineering, Schulich School of Engineering, University of Calgary, Alberta, Canada
[2] akthermusa@gmail.com

Abstract: Chloride is essential but it can also have detrimental impacts on ecosystem at elevated level. Chloride in natural water bodies often presents both intra- and inter-annual variations which could result from both natural and anthropogenic factors. Among many anthropogenic activities, the use of road salts for de-icing in winters is a common practice across Canada and has been demonstrated to elevate chloride level of rivers/streams. Besides, flow, a hydrologic factor, appears to be closely associated with chloride level regardless of the impact of anthropogenic activities. This paper aimed to further understand the variations in chloride and the possible impact of the use of road salts on chloride in rivers through statistically analyzing data collected across Canada. The results showed that the seasonal (intra-annual) variation of chloride presented in general and chloride concentration was negatively correlated with flow at most stations. The seasonal variation pattern and the dependency of chloride on flow however appeared to be different between large and small rivers. At the inter-annual scale, the increase of chloride concentration was qualitatively linked to the increase of the use of road salts in general; however the Code of Practice for the Environmental Management of Road Salts issued in 2004 appeared to help in reducing the increase trend of chloride level. Furthermore, the results from the locally weighted scatterplot smoothing (LOWESS) analysis showed that the temporal trend observed in chloride at the inter-annual scale could be primarily ascribed to the effect of either anthropogenic activities, flow, or their combination.

1 INTRODUCTION

Chloride is essential to all organisms including humans. Although chloride exists ubiquitously in environment, it can pose detrimental impacts on environment and human health when its level is elevated. Chloride is also inert and thus not commonly involved in geochemical reactions in aquatic environment. Therefore chloride has been often used as one of indicators of pollution and monitored in surface water bodies. Although chloride is conservative in aquatic environment owing to its inert chemical property, there are diverse factors influencing its level or concentration in surface water bodies. Among the influencing factors on chloride level in surface water bodies, the anthropogenic interference has been ascribed as the most significant cause of increasing chloride level (Eneji et al., 2012). Anthropogenic activities interfere on chloride level in many different ways, for example through releasing effluents of wastewater treatment plants and treated wastewater from industries which use large quantities of sodium chloride in many processes (point source pollution), and snowmelt contaminated by road salts (non-point source pollution) in cold regions.

A large number of studies have identified the impact of the use of road salts on various surface water bodies including rivers/streams and lakes as well as groundwater. The use of road salts has been ascribed to cause the elevation of chloride concentration, for example in the Mohawk River in New York State (Godwin et al., 2003), streams in New Hampshire (Daley et al., 2009), urban lakes in the Twin Cities Metropolitan Area of Minnesota (Novotny et al., 2007), and streams in snow-affected urban watersheds (Corsi et al., 2015) in U.S. Besides, the use of road salts has been observed to affect chloride level of groundwater, which would also affect the chloride level of surface water but at a delayed time (Kelly et al., 2012; Meriano et al., 2009; Coris et al., 2015). Therefore, the impact of the use of road salts could occur on a short time scale (during snow melting season) and a long time scale (during a year or years) as road salts can be transported into water bodies through both surface and subsurface flows.

Besides the effect of anthropogenic activities, natural factors (especially hydro-meteorological variables) could influence chloride level of surface water bodies. In riverine environment, the temporal variations/changes in hydrogeological variables such as precipitation, surface runoff, interflow, groundwater flow, and pumped in and outflows in general influence the river flow and subsequently the pollutant levels in streams (Vega et al., 1998; Rani et al., 2011). Variations in water quality and quantity have been widely investigated across the world (e.g., Gupta and Chakrapani, 2005; Rani et al., 2011; Eneji et al., 2012). It is unexceptional that flow, as one of natural factors, can largely affect chloride concentrations in rivers by either diluting chloride or transporting chloride into water bodies from both point and non-point sources. A large number of previous studies have documented that chloride level is in general negatively dependent on flow (e.g., Meriano et al.; 2009, Kelly et al., 2010; Coris et al., 2015; Boyer et al., 1999), which suggests that chloride is diluted by high flows. In such a circumstance, chloride concentration is usually higher in low flow than in high flow.

Many hydro-metrological variables such as flow show prominent seasonal (intra-annual) variations. Some anthropogenic activities such as the use of road salts also have seasonal nature, namely their applications occur in a specific time period in a year; whereas other anthropogenic activities such as releasing effluent from wastewater treatment plants and industries are relatively constant over a year. Therefore, the change and/or variation of chloride level can be very complicated as natural factors and anthropogenic activities drive the change/or variation of chloride level in surface water bodies differently and at different time scales. As many water quality parameters including chloride demonstrate strong seasonal variation in surface water, the characterization of their seasonal variations is very important to evaluate temporal changes of river pollution from natural or anthropogenic point and non-point sources (Ouyang et al., 2006). In addition at the inter-annual time scale, the understanding of the temporal change or trend in water quality is also required to formulate effective management strategy. Although multiple influencing factors could potentially affect the temporal variation/change of chloride at the intra- and inter-annual time scales, some influencing variables might play more important roles than other variables. Thus the role of one variable could be masked by other variables. Therefore, it might be necessary to study the potential impact of one variable after removing the impact of other variables.

In Canada, on average five million tonnes of de-icing salts have been used annually on roadways (Environment Canada, 2012). The existing body of knowledge has provided substantial evidence that the increase of chloride level results in detrimental effects on aquatic environment, terrestrial vegetation, wildlife mortality and soil chemistry. Road salts thus have been put on the second Priority Substances List of the Canadian Environmental Protection Act. The increase of chloride level in both groundwater and surface water has been linked to the increase of the use of road salts. Especially, surface runoff quality affected by the deicing operation has been blamed to cause the degradation of the water quality of freshwater bodies including streams and lakes (Environment Canada, 2012). To effectively mitigate the negative impacts posed by the application of road salts, the Code of Practice for the Environmental Management of Road Salts (referred as Code throughout this paper) was published in 2004. According to the above review, the objectives of this paper were to: (1) investigate the seasonal variation of chloride level; (2) study the temporal trend of chloride level and the effect of the Code; and (3) identify the influencing factor(s) primarily driving the temporal trend of chloride in rivers.

2 METHODOLOGY

2.1 Study Area and Data Collection

Due to the temporal and spatial nature of the road salt application, it leads to the change/variation of chloride level in water bodies at both temporal and spatial scales (Ramakrishna and Viraraghavan, 2005). Thus this paper did not target specific locations and collected data from rivers/streams across Canada. A total of 4288 datasets of chloride were collected from six provinces including Alberta, Saskatchewan, Manitoba, Ontario, Prince Edward Island (PEI), and Nova Scotia. Due to requirement for statistical analysis described below, datasets collected from 103 water quality stations in four provinces including Alberta, Manitoba, Ontario, and PEI were used.

From statistical analysis perspective, longer datasets are always desirable; whereas data availability is always the primary limiting factor for conducting statistical analysis. Schertz et al. (1991) stated that the water quality data should be collected according to a regular and fixed agenda and that a minimum record length of 5-year monthly data or longer for less sampling frequency is needed for monotonic trend analysis. In addition, Vecchia (2000) also mentioned that the data duration should not be less than 5 years to identify a true trend (including simple monotonic trends, compound monotonic trends, and step trends) in the mean because of the effect of autocorrelation in trend analysis. In addition, data both before and after the implementation of the Code in 2004 is required to investigate the potential effect of the Code on the variation of chloride concentration. Therefore, the selection of datasets was primarily conducted based on the following criteria:

- Sampling frequency is approximately monthly or less than monthly. The datasets that cover a wide range of stream flow conditions (from base flow to high flow) are preferred. Note that the datasets, which provide chloride concentration in non-winter season (e.g., at many water quality stations in Ontario), are also selected considering that the use of road salts can have prolonged impact (not only in snow-melting season) on chloride concentration in surface water bodies.

- Datasets consist of at least 5-year data before and after 2004, respectively, as the Code has been implemented since 2004.

Flow data corresponding to water quality monitoring locations were collected from Water Survey of Canada. If the water quality monitoring locations are not coincident with flow gauge stations, flow gauge stations in the proximity of the water quality monitoring locations (either upstream or downstream) were used. Therefore, flow data are not available for all water quality stations. The use of road salts was collected from municipal, provincial, and federal governments across Canada, and a total of 304 datasets were collected. Note that the data on the use of road salts are often available after the implementation of the Code in 2004.

2.2 Statistical Analysis

To investigate the presence of seasonal variation in chloride concentration, the analysis of variance or its non-parametric alternative, the Kruskal-Wallis test, was adopted. The analysis was employed to investigate if the mean(s)/or median(s) of a group(s) data is statistically different from other group(s). In the analysis, the dataset was grouped by months, thus there were in general 12 groups each of which corresponds to a month. If the statistical difference in medians or means is detected in the analysis, it is concluded that seasonality presents in the dataset. To investigate the presence of temporal trend in chloride concentration and the annual use of road salts, the parametric approach (regression approach) and the non-parametric trend test, the Mann-Kendall (MK) test, were employed. As shown in the section of results, seasonality commonly show in chloride data, thus the seasonal MK was primarily used; while the MK test was applied in the absence of seasonality.

To investigate if the Code issued in 2004 helps in effectively reducing the use of road salts and consequently the negative effect of the use of road salts on chloride level. Each dataset of chloride concentration was split into two sub-datasets, one for the time period before 2004 and the other for time period after 2004. The effectiveness of the Code was studied by comparing the identified temporal trends in chloride concentration and their means/medians in these two time periods (before 2004 and after 2004) using the parametric test (t-test) or the nonparametric test (Wilcoxon rank sum test).

To distinguish the role of flow from the role of the combination of all other influencing factors (except flow) in the observed temporal trends in chloride concentration, the Locally Weighted Scatterplot Smoothing (LOWESS) method was applied to remove the confounding effect of changing streamflow on chloride concentration. In the LOWESS, the smoothing factor of 0.5 was used. The smoothing factor defines the fraction the data is used to fit each local polynomial. The trend test was then applied to detect the trends in the flow-adjusted chloride concentration, which is residual of chloride concentration after the effect of flow on the variation of chloride concentration is removed, and predicted chloride concentration, which is modeled using flow as the only independent variable in the LOWESS. In addition, Pearson or Spearman's correlation coefficient between chloride concentration and flow was calculated to demonstrate their dependency.

All statistical analyses except the seasonal MK and MK tests were conducted using MATLAB. The selection of parametric and non-parametric tests in the analyses was conducted according to the normality of a dataset, which was tested using the Kolmogorov–Smirnov (K-S) test. If a dataset was normally distributed, parametric tests were applied otherwise non-parametric testes were adopted. The seasonal MK and MK tests were conducted using the computer program of USGS for trend analysis (Helsel et al., 2005). All statistical tests were conducted at a significance level of 5%.

3 RESULTS AND DISCUSSION

3.1 Seasonal Variation of Chloride

The seasonal variation of chloride concentration was statistically detected at most stations in the time periods of both before 2004 and after 2004. In the time period of before 2004, seasonal variation was detected at 82% stations (84 out of 103); while in the time period of after 2004, seasonal variation was observed at 87% stations (90 out of 103). Three different seasonal variation patterns were seen from the studied stations. Figure 1 displays the monthly boxplots of chloride concentration at three selected stations and illustrates the three different seasonal variation patterns. Chloride concentration is often lower in spring/summer months than that in other months (Figure 1(a) and (b)); while at some stations, relatively higher chloride concentration was measured in late summer and fall (Figure 1(c) and (d)). At several stations the seasonal variation was statistically absent (Figure 1(e) and (f)), although relatively high chloride concentrations were measured during summer/fall months. In general, chloride level tends to be low when flow is high. In a river, high flow is often observed in warm months while low flow is measured in winter months. The low level of chloride during high flow show that the high flow dilutes the concentration of chloride. However relatively higher chloride concentration in late summer and fall or no seasonality detected are more likely to be seen in small rivers. Small rivers are more prone to local hydrological events, such as snow melt and rain events and thus often behave differently from large rivers.

3.2 Temporal Trend of Chloride

Different temporal behaviors of chloride at the inter-annual scale, namely significant and insignificant upward and downward trends, and no trend, were statistically detected in the both time periods of before 2004 and after 2004 from the studied stations. Among the 103 stations, upward trend (both significant and insignificant) of chloride level was identified at the majority of stations (about 85%) in the time period of before 2004. In the time period of after 2004, chloride level at about 43% of stations remained upward trend; whereas chloride level at 57% of stations showed downward trend (both significant and insignificant) or no trend. Figure 2 presents the trend analysis results at three sample stations. The temporal trend at a station can be resulted from various causes including anthropogenic activities, natural variations/changes (e.g., hydro-meteorological variations/changes), and the combination of the both. The anthropogenic activities such as urbanization, the use of road salts, anthropogenic point-sources (e.g., effluent from wastewater treatment plants), and anthropogenic non-point sources (e.g., snowmelt and stormwater runoff) can have immediate and/or prolonged impacts on the chloride in surface water bodies. Among the hydro-meteorological variables, river flow has often been the primary driver of seasonal variation of chloride as discussed previously. In addition, the hydrologic cycle and thus hydrologic variables are subjected to variations/changes due to the variations/changes in meteorological condition as well as anthropogenic activities. Therefore it is crucial to investigate that either anthropogenic or natural factor leads to the temporal trend of chloride in order to more effectively manage riverine water quality.

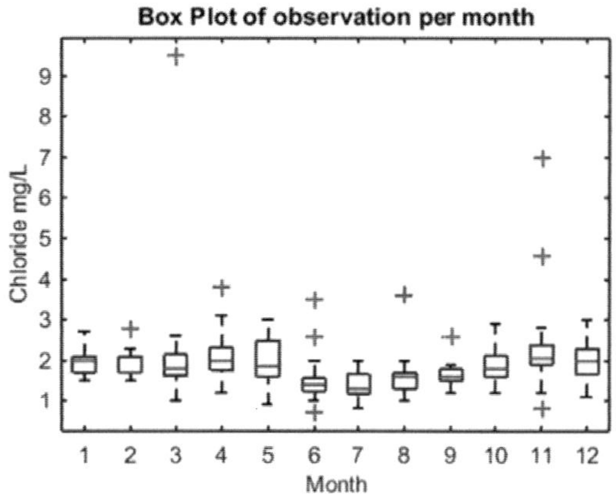

(a) Before 2004

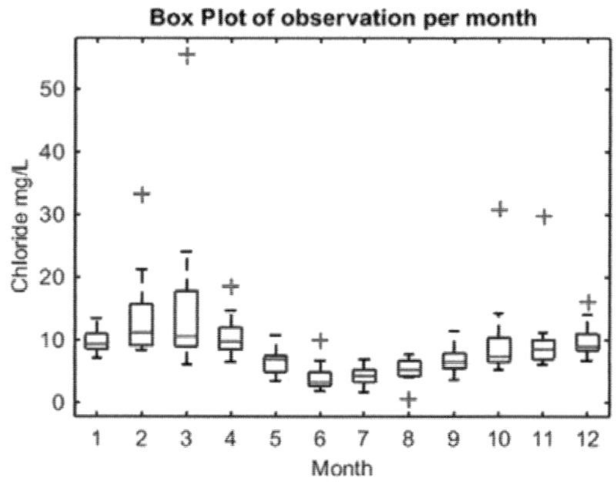

(b) After 2004

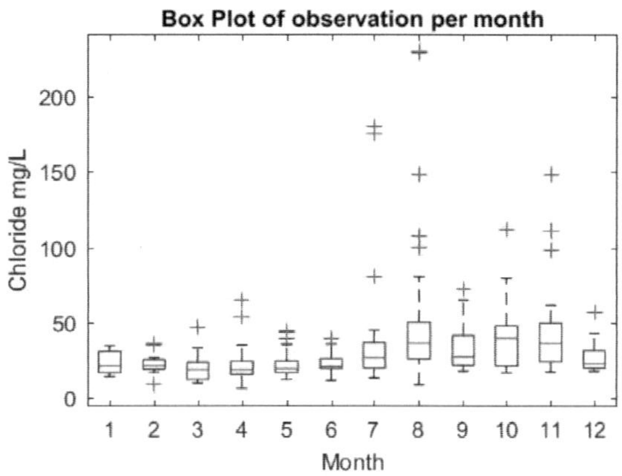

(c) Before 2004

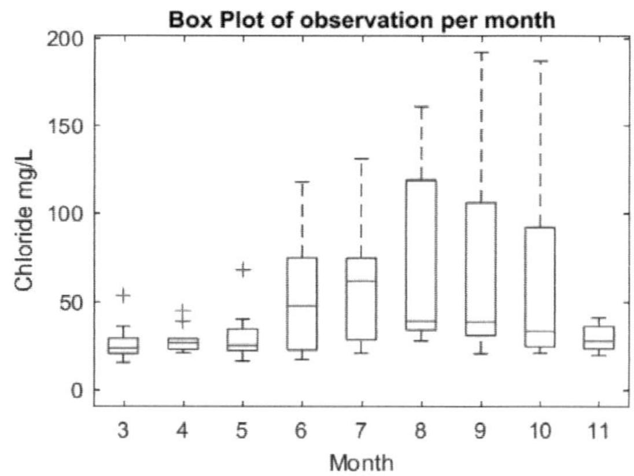

(d) After 2004

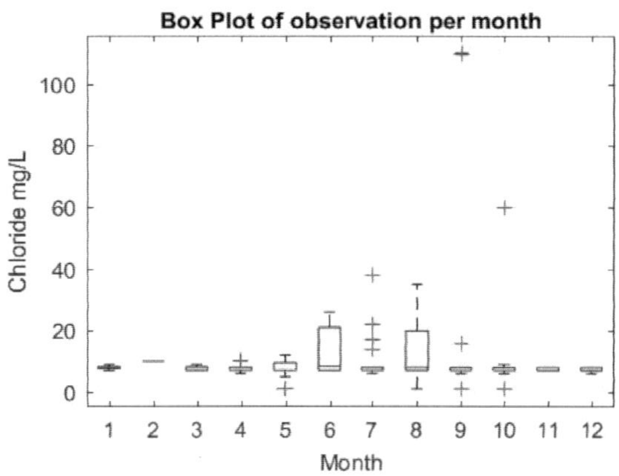

(e) Before 2004

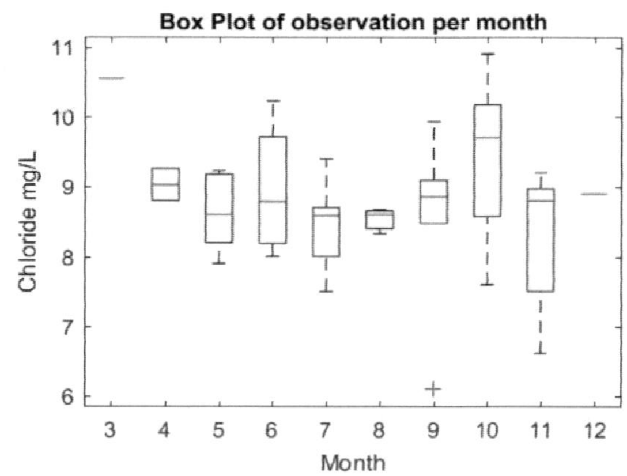

(f) After 2004

Figure 1: Monthly Boxplots of chloride concentration in the time periods of before 2004 and after 2004: (a) and (b) at the Bow River below Carseland Dam in Alberta; (c) and (d) at the Ausable River in Ontario; and (e) and (f) at the West River at Bolger Park Road in Prince Edward Island

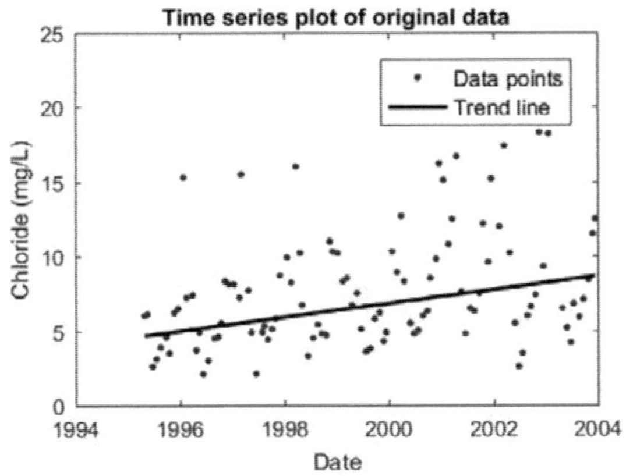

(a) Before 2004 (significant upward trend)

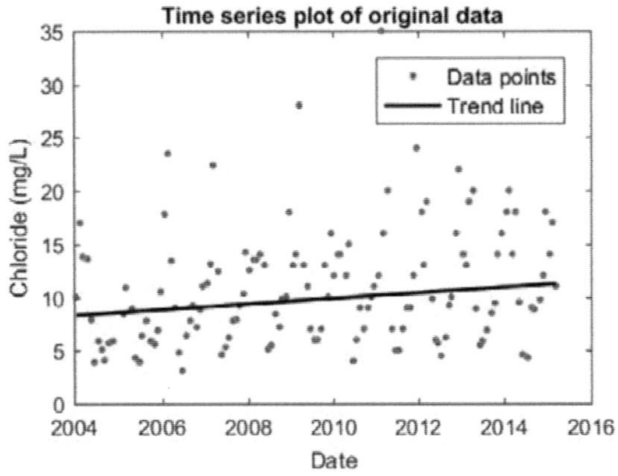

(b) After 2004 (significant upward trend)

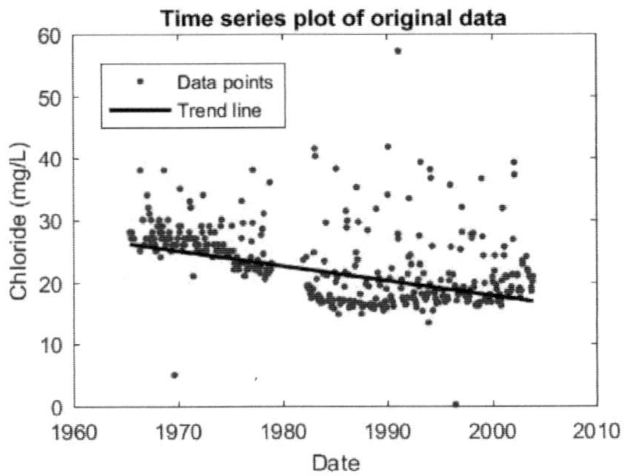

(c) Before 2004 (significant downward trend)

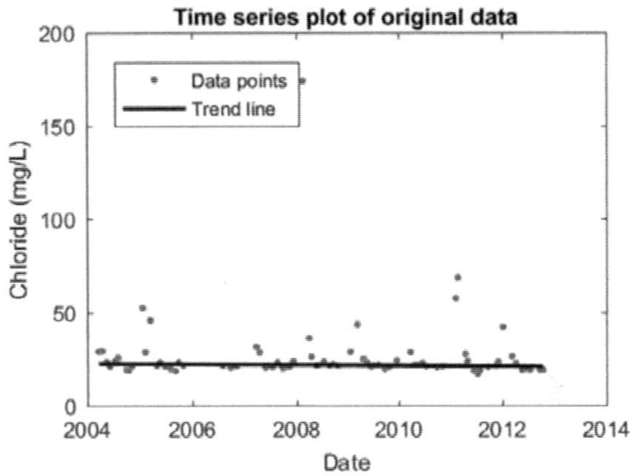

(d) After 2004 (significant downward trend)

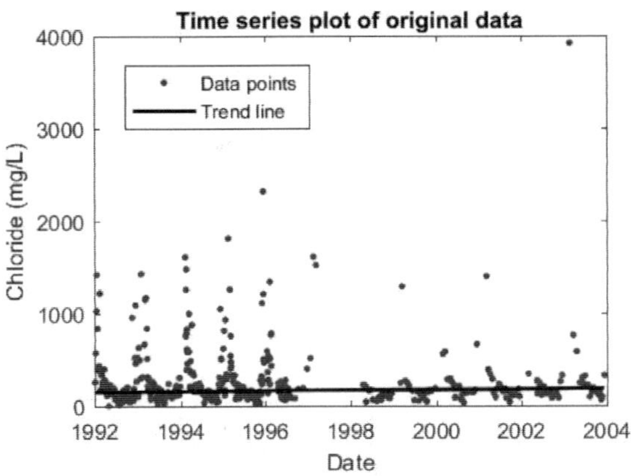

(e) Before 2004 (insignificant upward trend)

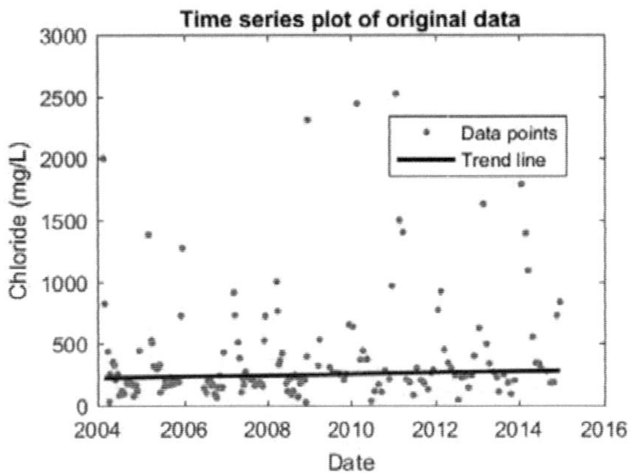

(f) After 2004 (insignificant upward trend)

Figure 2: Time series plots and trend lines of chloride concentration in the time periods of before 2004 and after 2004: (a) and (b) at South Saskatchewan River above Medicine Hat in Saskatchewan; (c) and (d) at Welland Ship Canal in Ontario; and (e) and (f) at the Don River in Ontario

3.3 Primary Factor Driving the Temporal Trend of Chloride

Acknowledging that many potential factors including both anthropogenic and natural factors could lead to the temporal change at the inter-annual scale of chloride in surface water bodies, it is vital to differentiate their roles in order to formulate efficient and effective management strategy to reduce chloride level in surface water bodies. In the following analysis, it is assumed that flow is the primary natural factors driving the variation of chloride as the variation of other hydro-meteorological variables is either reflected into flow (such as precipitation) or negligible. The LOWESS was applied to identify the primary driving factor (either flow or combined anthropogenic factors) causing the temporal variation of chloride. In the LOWESS, flow was used as the only explanatory variable, thus the predicted chloride concentrations are those that are explained by flow. The residuals in the LOWESS represent the chloride concentrations that can be explained by all other factors (especially anthropogenic factors). The trend analysis was then applied to investigate the trends of predicted chloride concentrations and the residuals, which reflect the trend caused by flow and anthropogenic activities, respectively. The analysis was conducted for four selected stations, at which chloride concentration is associated with flow differently (significant negative correlation at two stations, no significant correlation at one station, and significant positive correlation at one station). Note that chloride concentration was found to be in general negatively correlated with flow at most stations.

The trend analysis results for original data, predicted data, residuals, and correlation coefficient (significant) are summarized into Table 2. Furthermore, the time series of predicted chloride concentration and residual from LOWESS and their trend lines are presented in Figure 3 at two stations as examples. As the results shown, the temporal trends detected in chloride concentration are primarily caused by either flow at the station on the Canagagigue Creek or by anthropogenic activities at the stations on the Elbow and Bighead Rivers. At the station on the Duffins Creek, both flow and anthropogenic activities significantly contribute to the increase of chloride concentration, whereas the role of anthropogenic activities is more prominent than that of flow. In addition, the anthropogenic activities appear to always result in the temporal upward trend in chloride; whereas flow can lead to significant increase (station on the Duffins Creek) or decrease (station on the Canagagigue Creek) of chloride concentration. Comparing the four streams/creeks, the magnitude of flow in Duffins Creek and Canadagigue Creek is much lower than that in Elbow and Bighead Rivers. Small rivers/creeks are likely more prone to local hydrological events, such as snow melt and rain events, which could largely change their flow and also transport pollutants into water bodies. This might explain the different hydrologic and water quality behavior of small and large rivers.

3.4 The Effect of the Code

From six datasets of the annual use of road salts for the time period of before 2004 collected across Canada, upward trend (both significant and insignificant) was detected from four out of six datasets; while insignificant downward trend was found from two datasets. After 2004, upward trend was found from 70% of datasets. Thus it can be concluded that the annual use of road salts increases significantly or insignificantly in general after the implementation of the Code in 2004.

To further illustrate the change in chloride level after the implementation of the Code, the medians of chloride concentration were compared between the two time periods of before 2004 and after 2004. The results of the comparison are shown for the southern part of Ontario where all water quality stations are located in in Figure 4 (as example). The median of chloride level was elevated at the majority of stations (about 83% of the studied stations) after 2004; while at the remaining stations, the medians of chloride concentration either decreased or were remained same. Although chloride level in terms of median did not decrease statistically in general, chloride concentrations were found to decrease after 2004 at some stations where the trend of chloride concentration changed from upward trend before 2004 to downward or no trend after 2004. The results suggest that the effect of the Code at a certain degree, however the implementation of the Code has not reversed the increasing trends in chloride in general. The previous results demonstrate that both natural and anthropogenic factors could result in the temporal variation in chloride. In addition, urbanization can lead to the increase of the use of road salts, which can possibly be masked the effectiveness of the Code in reducing the use of road salts. A more elaborate research, which takes all potential influential factors on chloride into account, is recommended to assess the effect of the Code.

Table 2: Trend analysis results of the original data, predicted chloride concentration, residual, and significant correlation coefficient between chloride concentration and flow at four selected stations.

Station number (station name)	Time	Data set	Trend			Correlation
			Significant or Insignificant	Upward or Downward	Slope	
AB05BJ0450 (Elbow River at 9th Ave Bridge)	1988-2015	Original data	Significant	Upward	0.48	-0.37
		Predicted data	Insignificant	Upward	0.01	
		Residual	Significant	Upward	0.51	
03003000202 (Bighead River)	1975-2014	Original data	Significant	Upward	0.09	--
		Predicted data	Insignificant	Downward	-0.00	
		Residual	Significant	Upward	0.09	
06010400102 (Duffins Creek)	1989-2014	Original data	Significant	Upward	1.11	0.38
		Predicted data	Significant	Upward	0.19	
		Residual	Significant	Upward	0.91	
16018401602 (Canagagigue Creek)	1966-2014	Original data	Insignificant	Downward	-0.08	-0.64
		Predicted data	Significant	Downward	-0.32	
		Residual	Insignificant	Upward	0.09	

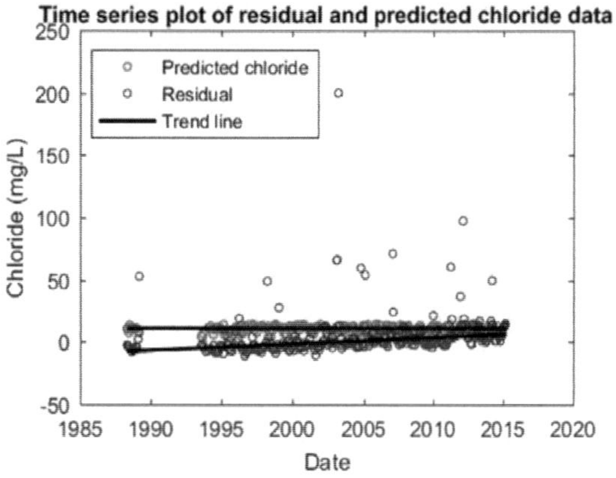

(a)

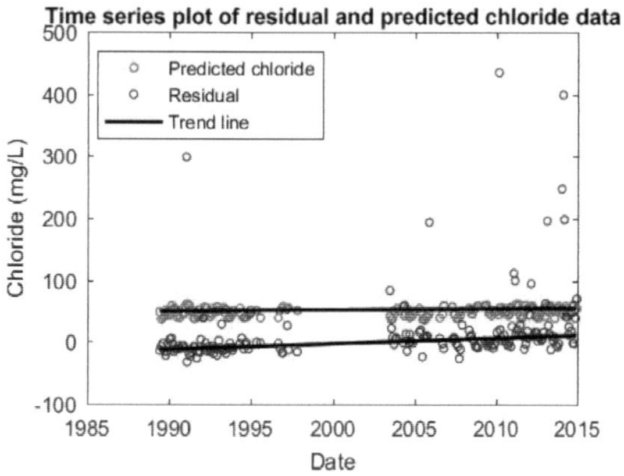

(b)

Figure 3: Time series of predicted chloride concentration and residual with their trend lines at: (a) the station on the Elbow River in Alberta; and (b) the station on the Duffins Creek in Ontario

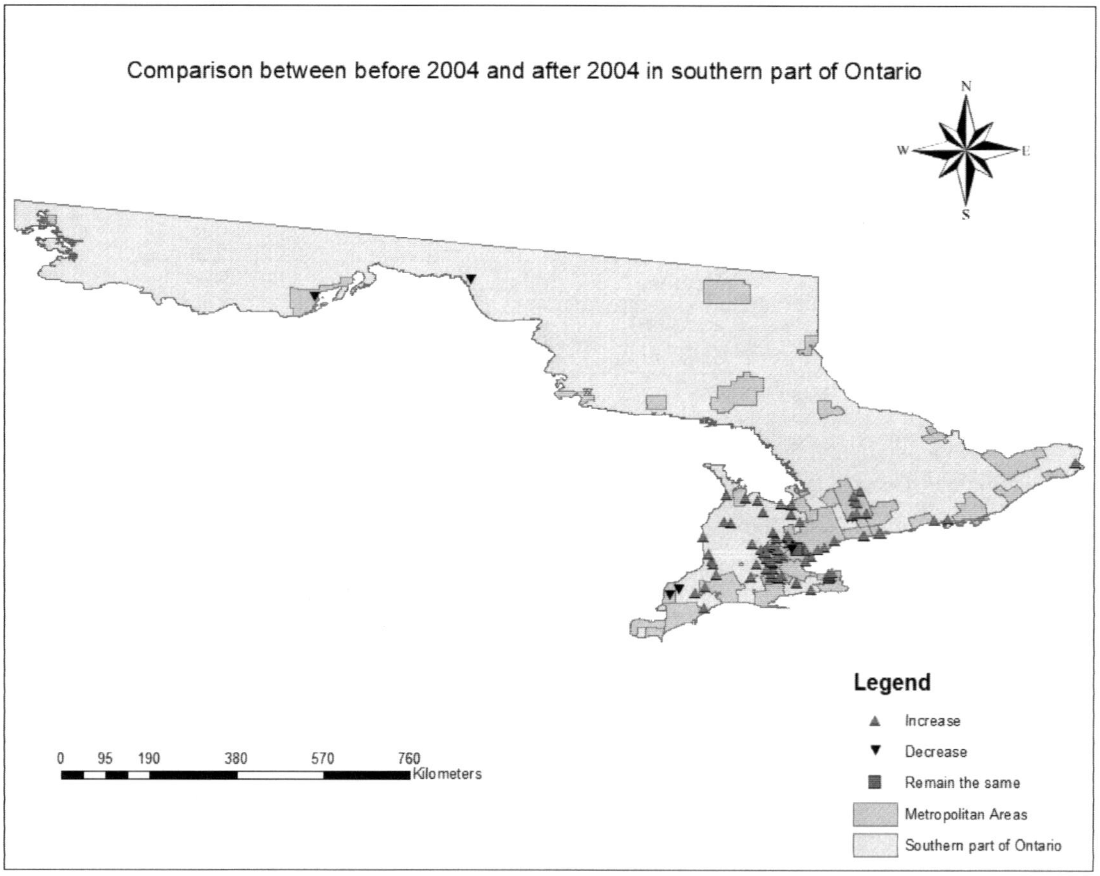

Figure 4: Comparison of the medians of chloride concentration between the two time periods of before 2004 and after 2004 for southern part of Ontario

4 CONCLUSION

In general, prominent seasonal variation in chloride concentration was observed at the majority of studied stations. Low concentrations were often measured in warm months, while high concentrations were measured in cold moths when the road salts are applied. Thus chloride concentration was often negatively dependent on flow. However different seasonal variation patterns and dependence of chloride on flow were also identified at some stations (especially on small rivers). In the time period of before 2004, significant upward trend of chloride was detected at many stations; while in the time period of after 2004, no trend and downward trend were detected at many stations, especially in Ontario. Although compared to the time period before the implementation of the Code, the chloride level (in terms of median of chloride concentrations) was elevated and the use of road salts also increased, the change from upward trends to downward/or no trends might imply the effectiveness of the Code. Furthermore, the analysis results confirmed that the anthropogenic activities often result in the increase of chloride in general; while flow can also be the primary or the secondary driving force of the detected temporal trend in chloride, especially in small rivers/creeks. Therefore to further investigate the role of the Code and formulate more efficient management strategy, more elaborate research in which the effects of various potential influential factors on chloride as well as the river physical characteristics (flow magnitude, river watershed, and anthropogenic activities in watershed) are taken into consideration is recommended.

Acknowledgements

This work was funded by Natural Sciences and Engineering Research Council of Canada (NSERC) and Environment Canada. The authors would like to acknowledge Environment Canada, and the federal, provincial, municipal governments who provide data used in this study.

References

Boyer, J. N., Fourqurean, J. W., and Jones, R. D. 1999. Seasonal and long-term trends in the water quality of Florida Bay (1989–1997). Estuaries, 22(2): 417-430.

Corsi, S. R., De Cicco, L. A., Lutz, M. A., and Hirsch, R. M. 2015. River chloride trends in snow-affected urban watersheds: increasing concentrations outpace urban growth rate and are common among all seasons. Science of the Total Environment, 508: 488-497.

Daley, M.L., Potter, J.D. and McDowell, W.H., 2009. Salinization of urbanizing New Hampshire streams and groundwater: effects of road salt and hydrologic variability. Journal of the North American Benthological Society, 28(4), pp.929-940.

Environment Canada, 2012. Five-year Review of Progress: Code of Practice for the Environmental Management of Road Salts. Environment and Climate Change Canada. https://www.ec.gc.ca/sels-salts/default.asp?lang=En&n=45D464B1-1 (accessed on December 2, 2017)

Eneji, I. S., Onuche, A. P., and Sha'Ato, R. 2012. Spatial and temporal variation in water quality of River Benue, Nigeria. Journal of Environmental Protection, 3(08): 915.

Godwin, K. S., Hafner, S. D., and Buff, M. F. 2003. Long-term trends in sodium and chloride in the Mohawk River, New York: the effect of fifty years of road-salt application. Environmental Pollution, 124(2): 273-281.

Gupta, H., and Chakrapani, G. J. 2005. Temporal and spatial variations in water flow and sediment load in Narmada River Basin, India: natural and man-made factors. Environmental Geology, 48(4-5): 579-589.

Helsel, D. R., Mueller, D. K., and Slack, J. R. 2006. Computer program for the Kendall family of trend tests (No. 2005-5275).

Kelly, W.R., Panno, S.V., Hackley, K.C., Hwang, H.H., Martinsek, A.T. and Markus, M., 2010. Using chloride and other ions to trace sewage and road salt in the Illinois Waterway. Applied Geochemistry, 25(5): 661-673.

Kelly, W.R., Panno, S.V. and Hackley, K.C., 2012. Impacts of road salt runoff on water quality of the Chicago, Illinois, region. Environmental & Engineering Geoscience, 18(1): 65-81.

Meriano, M., Eyles, N. and Howard, K.W., 2009. Hydrogeological impacts of road salt from Canada's busiest highway on a Lake Ontario watershed (Frenchman's Bay) and lagoon, City of Pickering. Journal of Contaminant Hydrology, 107(1-2): 66-81.
Novotny, E., Murphy, D., and Stefan, H. 2007. Road Salt Effects on the Water Quality of Lakes in the Twin Cities Metropolitan Area. Project report No. 505, Local Road Research Board (LRRB), Minnesota Department of Transportation.
Ouyang, Y., Nkedi-Kizza, P., Wu, Q. T., Shinde, D., and Huang, C. H. 2006. Assessment of seasonal variations in surface water quality. Water Research, 40(20): 3800-3810.
Ramakrishna, D. M., and Viraraghavan, T. 2005. Environmental impact of chemical deicers – A review. Water, Air, and Soil Pollution, 166(1-4): 49-63.
Rani, N., Sinha, R. K., Prasad, K., and Kedia, D. K. 2011. Assessment of temporal variation in water quality of some important rivers in middle Gangetic plains, India. Environmental Monitoring and Assessment, 174(1-4): 401-415.
Schertz, T. L., Alexander, R. B., and Ohe, D. J. 1991. The computer program Estimate Trend (ESTREND), a system for the detection of trends in water-quality data. US Department of the Interior, US Geological Survey.
Vecchia, A.V., 2000. Water-quality trend analysis and sampling design for the Souris River, Saskatchewan, North Dakota, and Manitoba. US Geological Survey.
Vega, M., Pardo, R., Barrado, E., and Debán, L. 1998. Assessment of seasonal and polluting effects on the quality of river water by exploratory data analysis. Water Research, 32(12): 3581-3592.

Building Tomorrow's Society
Bâtir la Société de Demain

Fredericton, Canada
June 13 to 16, 2018 / *13 au 16 juin, 2018*

MODELLING SEDIMENT MOVEMENT FOR A MULTI-OPTION LARGE-SCALE HYDROELECTRIC DAM REMOVAL PROJECT

Haralampides, Katy[1,3], Yamazaki, Gordon[1,2], Ndong, Mouhamed[1,2], and Curry, R. Allen.[1,2]
[1] University of New Brunswick, Canada
[2] Canadian Rivers Institute, Canada
[3] katy@unb.ca

Abstract: The Mactaquac Generating Station (MGS), a 672 MW run-of-the-river hydroelectric facility on the main stem of the Saint John River, New Brunswick, Canada was built in 1968 by New Brunswick Power. Since 2013, the Canadian Rivers Institute has been undertaking the Mactaquac Aquatic Ecosystem Study (MAES), a planned, whole-river multidisciplinary ecosystem study and manipulation of the MGS, in support of a decision-making process regarding the future of the aging facility. Future options under consideration range from in-situ refurbishment to full removal and river restoration. Prediction of the potential transport and fate (or long-term retention) of post-inundation sediment is a crucial aspect for decision-makers. Extensive bathymetric surveys, sediment surveys, core sampling, and laboratory studies were conducted to estimate the volume of sediment upstream, sediment bed composition, thickness, particle size diameter, erosion rate and threshold shear stress. These results were integrated into the hydrodynamic model Delft3d that was calibrated and validated using water level and flow data. The model results predicted the downstream fate of the sediments under various hydrodynamic regimes, including the case of complete removal of the hydraulic structure. Given drawdown scenarios were simulated, and sediment movement was quantified. The sediment transport studies have helped to identify areas of interest for future research focus, and have furthered the field and laboratory techniques undertaken by the team. The results of the MAES project will help to advance the science of dam renewal decision-making processes using a holistic, multidisciplinary approach.

1 INTRODUCTION

The Mactaquac Generating Station (MGS), a 672 MW run-of-the-river hydroelectric facility on the main stem of the Saint John River, New Brunswick, Canada was built in 1968 by New Brunswick Power (the "power company"). The Canadian Rivers Institute is undertaking the Mactaquac Aquatic Ecosystem Study (MAES), a planned, whole-river multidisciplinary ecosystem study and manipulation of the MGS, in support of a decision-making process (the "Mactaquac Project") regarding the future of the aging facility. Future options under consideration range from in-situ refurbishment to full removal and river restoration. The Mactaquac Project is the largest dam removal ever taken under consideration and the MAES is the largest coordinated study of its type ever undertaken in support of a potential dam removal project.

The MAES is a multi-year assessment of the structure and function of a large river ecosystem and the predictive analyses of the proposed manipulations, i.e., what are the environmental challenges and opportunities for either replacing or removing the dam? The highly-integrated research establishes baseline environmental conditions, appropriate metrics for biomonitoring, and predicts the water, sediment, thermal regimes, and future habitats for a dam removal scenario. In the dam refurbishment scenario, the research examines the consequences for species at risk and fish passage, and models future management options for environmental flows.

The MAES is an investment by the power company in the research and development required to sustain its hydropower business model in a changing world of new green economies. The premature ageing of the MGS and the pressing consideration of future options for the dam and the river's ecosystem goods and services is a precedent setting case for Canada. The MAES was intended to create a template of approaches and methods that facilitate the incorporation of aquatic ecosystem science into informed decision making and management for this and many future hydropower projects and additionally, the science of river restoration. As a consequence, the MAES will also set new standards for the future of Environmental Impact Assessments in Canada. This paper focuses on one of the keystone objectives of the MAES: understanding the current and future sediment environment under the dam removal scenario.

1.1 Study Area

With a catchment area of 55,000 km^2 and overall length of 618 km, the Saint John River (Figure 1) is one of the largest river systems on the eastern seaboard of North America. At the MGS, the river discharge averages ~1000 m^3/s with river widths and depths averaging 750m and 2m, respectively (Kidd et al. 2011). Average annual air temperature is 5°C and there is an average of 1,100mm of precipitation annually. The river supports the greatest diversity of aquatic plants and animals in eastern Canada, including 15 species at risk (McAlpine and Smith 2010). Human development began in the watershed in the 1600s (logging, fisheries, and later heavy industries), but the watershed remains 80% forested with a human population of ~500,000. There are >200 dams across the river network with 11 hydroelectric generation stations (first built in 1890) and of these, the MGS (Figure 2) located approximately 20km upriver from the City of Fredericton, New Brunswick, is the largest with a nameplate hydroelectric generating capacity of 672MW (6 Kaplan-type 112 MW units). The earthen and concrete structures of the dam span ~1,100m and it is 55m high. The MGS headpond covers an area of 84km^2 and extends 100km upstream to just above the Town of Woodstock, New Brunswick.

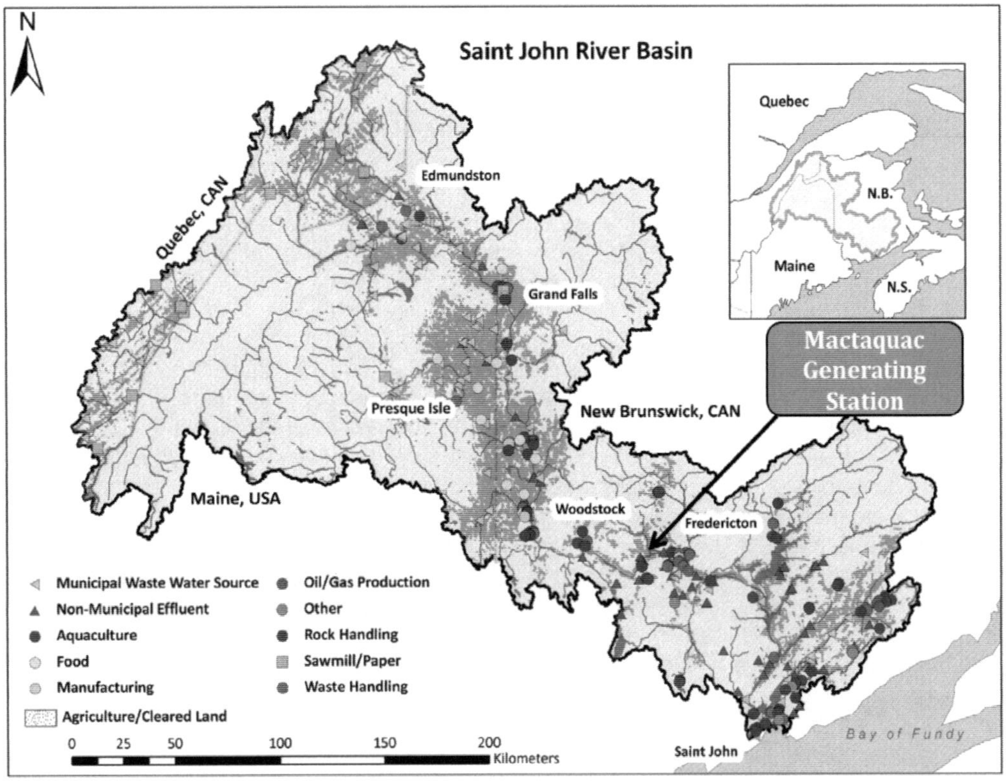

Figure 1: Saint John River Watershed and MAES Study Area

Figure 2: The Mactaquac Generating Station, Fredericton, NB

1.2 Objective

There are two levels of analysis for a drawdown programme associated with a large dam removal: the actual drawdown and the physical and biological conditions that will remain post-drawdown. During the drawdown phase, the key environmental challenges to be modeled are 1) sediment re-suspension in the headpond and in the water being discharged downstream, and 2) subsequently understanding the potential environmental impacts associated with this activated sediment. To do this required first answering a series of questions, including: what is the current state of the river morphology and sediment environment (i.e., prior to the headpond drawdown); how will the sediment relocate during and after drawdown; and how does the post-drawdown equilibrium-state river morphology and sediment environment differ from the current conditions?

The broad objective of this sediment component of the MAES can be defined as: collecting field and historical data to establish the baseline sediment and other necessary conditions in the environment to develop a computer-based mathematical model to simulate various drawdown scenarios that could be used to estimate the changes to the sediment environment caused by potential Mactaquac Project options. There are similar questions and objectives relating to flow dynamics, water quality, and other abiotic and biotic factors that were also incorporated into and/or addressed via the model outputs that are not included within the scope of this paper. These questions included, but were not limited to: what is the resulting morphology of the restored river within the headpond area; what are the effects of sediment relocation on downstream infrastructure and aquifer recharge zones; and how are fish habitat and fish passage conditions changed under the various options? It should also be noted that this paper does not contain the specific results of the model or any subsequent studies as these are not publically available at this time. The primary purpose of this paper is to present the methods applied (and created) to develop the sediment transport model to the point at which these questions could be answered.

2 METHODS

Working on such a large, actively regulated river system that includes substantial seasonal flow differences, a tidal range that exceeds 7m in the estuary, a limited fieldwork sampling window because of the Canadian winter, and with varied water depths from very shallow island habitat to a deep and lake-like headpond has presented challenges in data collection. It is fair to say that our 'lessons learned' list is ponderous in length. Developing a 2D/3D hydrodynamic and sediment transport model of this magnitude is an exercise in collecting, organizing and manipulating many types of big data. Relatively recent technologies (e.g., Acoustic Doppler Current Profiler (ADCP), Light Detection and Ranging (LiDAR)) allow for the collection of big data in short increments of time while modern computers allow for computational models with greater refinement then previously possible. Deciding on the metrics and the appropriate means of data acquisition required consideration of environmental properties, availability of tools, cost of data acquisition and/or processing, and working within the Mactaquac Project schedule. To this end, the MAES team – comprised of nine professors of varying engineering and biological disciplines, multiple Post-Docs and Research Assistants, and many graduate-level students – combined their expertise, along with that of other experts consulted around the world to arrive at a set of metrics necessary to meet the objective.

2.1 Morphology

2.1.1 Geomorphological Conditions

The first step in establishing the current state of the river morphology and sediment environment was to review existing information and assess the headpond geomorphology and estimate sediment distribution processes (O'Sullivan et al., 2016). The geomorphological assessment concluded that sediment deposited within the headpond may not be significant in relation to other large dam removals (e.g. Elwha, Glines Canyon, and processes based experiments such as the Cougar Reservoir drawdown). This was based on the observation that the relief surrounding the Model Study Area (Figure 3) is gently sloping with gradients ranging from 1-10° (with anomalous occurrence of slopes ≥10° also present) with the proximal surficial geology comprising of shallow morainal deposits constituting various till deposits and sands and gravels. Thus, the Model Study Area lies in a low energy environment and the sediment yield supplied to the headpond is likely to have been relatively low even when erosion from weathering and anthropogenic forcing are considered (soil erosion is a function of relief, surficial geology/soil erodibility, rainfall intensity, and land use). It is predicted that a high energy event, e.g., heavy rainfall, snow melt runoff, etc., would be required to entrain coarser grained sediments in addition to relief alone. It was further hypothesized that the presence of the Beechwood Dam (located approximately 140km upstream of MGS) and its associated headpond acts to trap most of the bedload sediment travelling downstream from the majority of the Saint John River basin above MGS.

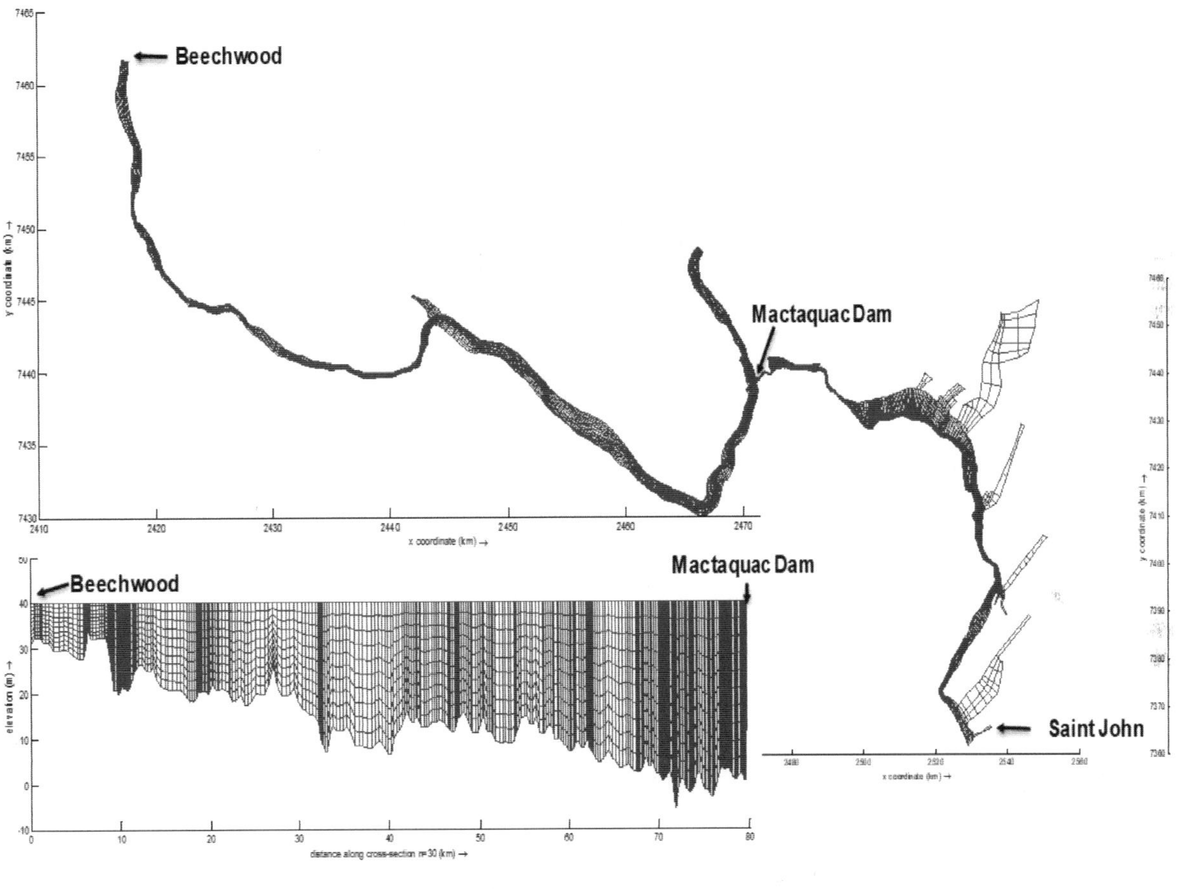

Figure 3: Model Study Area with Model Curvilinear Grid

2.1.2 Bathymetry

Previously-derived bathymetry, recorded in the first few years after the MGS was commissioned, was available for the headpond and for the river downstream of the City of Fredericton. However, these data were nearly 45 years old at the start of the study and were based on soundings taken at 100m intervals. It was determined that a new and reliable bathymetry for all areas incorporated into the model was required. Reliable updated bathymetric data for the headpond allows for water/sediment volumes to be calculated for use in hydrodynamic, sediment transport, and water quality models.

To obtain maximum coverage of the Mactaquac Headpond, the bathymetric survey (Bremner et al., 2016) used a multibeam system in water depths greater than approximately 5 m following prescribed methods to meet or exceed the CHS Order 1a standard (based on International Hydrographic Service standards). The power company provided a vessel (the 'Sea Truck'; Figure 4) for use as the platform for all multibeam work. In shallower depths (<5m), such as occur in near-shore areas of the headpond and throughout the fluvial portions of the Model Study Area, single-beam bathymetry was collected using Lowrance HDS-series sonar units deployed from small flat-bottom boats (Wallace and Gautrea, 2016). This relatively inexpensive set of tools allowed for a small fleet operated by students to collect large quantities of reliable bathymetric data in difficult-to-access locations over a single season. Logged single-beam sonar data were converted to a GIS-compatible format using CIBioBase software (www.cibiobase.com).

Figure 4: The Sea Truck Survey Vessel

The bottom elevation of some areas, particularly marginal island habitats and riparian flood zones that emerge in low-water conditions, were not accessible using sonar-based methods. For these areas, the surficial elevations were obtained from LiDAR images and corrected using standard GPS-based survey methods then merged with the sonar-based bathymetry into a single layer.

2.2 Hydrodynamic Conditions

Water level data were obtained from existing gauges maintained by Environment Canada and the power company within the Saint John River. The power company was also able to provide discharge data from the MGS (since 1969) and the upstream Beechwood Generation Station (since 1957) for the duration of their operation.

Flow data in the island reach between MGS and the City of Fredericton were obtained via an Acoustic Doppler Current Profiler (ADCP) mounted to a remote controlled boat. The ADCP used was a nine-beam system and the raw data were processed using the Velocity Mapping Toolbox software from the US Geological Survey to extract the streamwise, transverse, and vertical velocities, and velocity magnitude by taking account the secondary flow vector. Transect measurements were taken at 14 stations including within the main channel and at major channel bifurcations. Discharge at MGS was regulated during three ADCP events to represent late spring fish migration, summer low, and fall storm flow conditions. ADCP data were also collected in the headpond at three transect locations during summer low flow.

2.3 Sediment Quantity

2.3.1 Post-inundation Sediment Layer Thickness

In order to understand the distribution of sediment that has accumulated in the headpond since the emplacement of the MGS in 1968, a four-tiered approach was applied which included: acoustic sub-bottom profiling, coring, bathymetry comparison, and suspended sediment mass calculations.

Acoustic sub-bottom profiling surveys were undertaken over a two-year period (Grace and Butler, 2016). Initial attempts using a 28 / 3.5 kHz chirp sonar survey proved ineffective for imaging the relatively thin layer of sediment that has accumulated in the headpond. Subsequent attempts using a higher resolution Seistec system was successful for the lower one-third portion of the headpond where the sediment layer is relatively thin (average thickness of approximately 26cm) and comprised entirely of sediment deposited from suspension. In the middle one-third of the headpond, gas was present in the sediments at the sediment/water interface which interfered with the acoustic penetration such that reliable sediment thickness estimates were not obtained. A small-scale ground penetrating radar survey was undertaken to test this method for estimating thickness in gas-charged sediments and the results were encouraging. The gas is likely the result of organic waste from a nearby pulp mill buried within a post-inundation deposited sandy bedload. The portion of the headpond upstream of the Village of Nackawic remains relatively fluvial and the sub-bottom profiles do not reveal any soft sediment in excess of the minimum resolvable 12 cm. This conclusion is also supported by the multibeam bathymetry in which boulders and rocks are clearly visible, indicating the bottom composition is much larger grain size, and the thin silty sediments accumulated downstream in the headpond have only sparsely accumulated in this reach of river.

A coring program was undertaken to truth and support the sub-bottom profiling program. A weighted and powered vibracore was deployed from an A-frame crane fitted to the Sea Truck. The vibracore was determined to be ineffective at extracting clean cores from the relatively thin sediment layer in the lower one-third of the headpond and in the fluvial reach above Nackawic. It is believed that the vibration was not needed to penetrate in these areas and that it inadvertently caused disturbance to the loosely compacted semi-cohesive sediments. Deploying the vibracore with the vibrator not engaged, thus using it as a gravity corer, produced much more useable cores. Additional cores were obtained through the winter ice using a hand-deployed weighted corer. Results from the coring program appear to coincide well with the mapped thicknesses based on the sub-bottom profiles.

Deriving the sediment thickness layer by comparing the historical bathymetry (1969) with the current reservoir bathymetry may be possible. This has been done for individual cross sections; the method for doing this is as whole GIS layers is currently being worked out. Back calculating the sediment deposition in the headpond may also be possible using recently collected total suspended solids (TSS) data fitted to a flow curve and matching this to historical flow conditions from 1969-present. The TSS data have been collected (Section 2.3.2) but the process for doing the calculations is still in progress.

2.3.2 Total Suspended Solids (TSS)

To establish baseline conditions, a combined approach of historical data review and field data collection was applied (Yamazaki et al., 2017). MAES requested water quality data for stations along the Saint John River (SJR) and its tributaries from the New Brunswick Department of Environment and Local Government (NBDELG) to identify typical values and understand how suspended sediment moves through the River in time and space. Two historical data sets were provided: SJR Stations 1994-1998 and SJR and Tributary Stations 2003-2014. To understand typical turbidity and suspended sediment values at stations located on or near the Saint John River, the data were organized to demonstrate spring freshet averages, spring freshet peaks, fall storm peaks as well as summer, fall, and spring averages. Additional recent TSS data were collected by MAES under various flow and precipitation conditions throughout the Model Study Area.

2.4 Sediment Quality

For the purpose of this paper, 'sediment quality' is limited to the information required to model its transport, specifically grain size and shear stress. Other metrics like sediment chemistry were also collected but are not discussed herein. Material for determining grain size was taken from within the cores which were sliced in 1cm intervals. Grain size distributions for the samples were determined (Yamazaki et al., 2016) using a laser diffraction particle size analyzer.

Laboratory experiments were conducted in an erosion flume to determine the resuspension and transport properties of the sediment, including the shear stress needed to initiate movement. Using an adjustable vertical gate at the downstream end of the flume, and a variable speed pump, a constant water level can be achieved for an increasing pump setting. A core can be inserted into the floor of the flume so that only the top is exposed to the moving water. Once the top layer has eroded away, the core is pushed upwards to expose a new layer to the flow. As the volume flow rate was increased, velocity profiles were recorded for each setting. The velocity follows a logarithmic curve near the sediment/water interface, and by plotting the velocity as a function of the log(depth) in the water column, a relationship was derived to calculate the shear stress present. The variation of critical shear stress and erosion rate with depth were observed and recorded for the length of each core.

2.5 Hydrodynamic and Sediment Transport Model

The selection of the model platform was accomplished in a similar manner to that described for the model metrics presented above. Delft3D was selected for its demonstrated successes in similar (though smaller) undertakings, for its modular flexibility to address the stated and other objectives, and because it is open source. As described by the developer (Delft3D-FLOW, 2013), the FLOW module is a multi-dimensional (2D or 3D) hydrodynamic and sediment transport simulation programme which calculates non-steady flow and transport phenomena resulting from tidal and meteorological forcing on a curvilinear, boundary fitted grid. In 3D simulations, the vertical grid is defined following the Z-layer approach. The MOR module computes sediment transport (both suspended and bedload) and morphological changes for an arbitrary number of cohesive and non-cohesive fractions. For the suspended load this module connects to the advection-diffusion solver of the FLOW module and density effects may be taken into account. An essential feature of the MOR module is the dynamic feedback between modules, which allows for simulations on any time scale from days (e.g., drawdown stages) to centuries (e.g., long-term effects of sediment entrainment).

Early consultations with the software developer resulted in dividing the Model Study Area into two reaches: A) from the Beechwood Generating Station to the MGS; and B) from the MGS to just above the Port of Saint John where the river enters the Bay of Fundy. This dual-model approach allows for the discharge data from Beechwood, located at the furthest upstream extent of the Model Study Area, to be used as the input to the upstream portion of the model while the MGS discharge serves as the input to the downstream portion. One of the primary benefits of having and using discharge data from these facilities is that the contribution from upstream tributaries at both points is inherently included. Another benefit is that the data sources are longstanding (from dam commissioning to the present) and reasonably accurate. The approach also breaks the Model Study Area up into two approximately 100km river reaches which results in reduced computation time with improved accuracy.

Prior to working in the Delft3D platform, a 1D, steady-state model of the Saint John River between the Town of Woodstock and the City of Saint John was developed (NATECH, 2015) to calculate average current speeds at eleven locations (four sections in the headpond, and seven downstream of the MGS). The purpose of the 1D model was to establish predictions and a working understanding of general flow conditions in the Model Study Area to guide and check the development of the 2D model. The Delft3D model was first established in 2D (depth averaged) using rectangular grid elements. Sensitivity tests were carried out with two physical parameters (i.e., roughness and eddy viscosity) and calibration was accomplished using water level data from a year with a large spring freshet (2008). The hydrodynamic model output was then compared to that of a more normal spring freshet (2014) for verification. Predicted depth-averaged water surface elevations and velocities, as well as associated sediment dynamics, were then simulated under various hydroclimatic and seasonal flow conditions, in addition to given drawdown scenarios. TSS and historic/present bathymetric data were used to assess the sediment transport predictions. A second calibration was later undertaken using the ADCP-generated flow data to improve the model accuracy in the island-rich reach between MGS and the City of Fredericton. The final step was to advance the model to 3D.

3 DISCUSSION

It is important to note that from start to finish, the development of a fully functioning and reasonably accurate 3D hydrodynamic and sediment transport model for the approximately 200km-long Model Study Area took nearly four years to accomplish. Throughout this time, at various steps, the model in its various forms was useful for answering questions posed by the power company and their engineering consultants in support of the Mactaquac Project decision making process. Similarly, early model outputs were used by the MAES graduate students in support of their research. By the time that the model reached its full functionality (March 2018), the decision to refurbish rather than remove the MGS had been made. Thus, the sediment transport component of the model, though useful during the decision making process, may not be the most useful aspect of the model going forward. That said, sediment transport simulations will continue to be used to predict the long-term effects of fine sediment retention in the headpond on downstream sediment-dependent wetlands and of bedload sediment retention on the morphology of downstream islands and bars.

What were once considered as secondary objectives are now considered as primary and the flexibility and robustness of the model is proving worthy of the transition. For example, the hydrodynamic component of the model is already in use to understand flows in near proximity to MGS to meet the power company's objective of improving existing fish passage facilities. Simulations will be run to complement the fish passage studies, giving estimates of velocity distributions and water levels to support the engineering of new fish pass structures and approaches. Similarly, the 2D model is also being used to support ecosystem studies linking biotic and abiotic components in an ecohydraulics context in critical inter-island habitats via the development of a fuzzy-logic based habitat simulation software.

Finally, with the completion of this model, the power company is already getting returns on its investment in the research and development required to sustain its hydropower business model in a changing world of new green economies by using ecosystem-based science to support its decision making process, and through its investment in the local training of students to ensure that the professionals of tomorrow are ready to enact the decisions of today. For the MAES, the model development process has succeeded at creating a template of approaches and methods for this and many future hydropower projects and additionally, the science of river restoration, that are applicable around the world.

Acknowledgements

The authors would like to extend their heartfelt gratitude to their Industry Partner, New Brunswick Power. It was with both surprise and honour (and some suspicion) that the authors were approached, unsolicited, by the power company to undertake an unprecedented multi-year, multi-dimensional study in support of their decision making process. Their unwavering commitment and support throughout this process, including financial as well as in-kind contributions of time and equipment, and their willingness to allow it to unfold without their interference, has contributed greatly to the outcome and was sincerely appreciated.

The Canadian National Science and Engineering Research Council (NSERC) provided matching funding to the MAES and afforded the opportunity to pursue several offshoot research studies including those of more than a dozen graduate students and several dozen undergraduate students.

The authors would also like to thank the following people and organizations for providing data and information used to build the model: Dr. Karl Butler (UNB), Mitch Grace, M.Sc. (UNB), Dr. John Hughes Clarke (University of New Hampshire), Dr. Joshua Kurek (Mount Allison University), New Brunswick Department of Environment and Local Government, and Environment Canada. We would also like to thank the multitude of technicians and students who endured long field/lab days taking samples and doing analyses.

NATECH Environmental Services Inc. were instrumental in the crucial early and middle stages of model development. The authors are sincerely thankful for their expert and diligent contributions in this effort.

References

Bremner, M., G. Keirstead, J. Muggah, B. Wallace, A. Chateauvert, G. Yamazaki. 2016. Determining the Bathymetry of the Mactaquac Headpond and Saint John River. Mactaquac Aquatic Ecosystem Study Report Series 2016-032. Canadian Rivers Institute, University of New Brunswick, 10p.

Delft3D-FLOW, 2013. Delft3D-FLOW User Manual: Hydro-Morphodynamics. Deltares, 3.15 ed. 677p.

Grace, M. and K. Butler. 2016. Sediment Thickness Map for the Mactaquac Headpond Derived from Acoustic (Seistek) Sub-bottom Profiles. Mactaquac Aquatic Ecosystem Study Report Series 2016-040. Canadian Rivers Institute, University of New Brunswick, 28p.

NATECH Environmental Services Inc. 2015. Hydrodynamic Model Setup for the Saint John River at Mactaquac. Mactaquac Aquatic Ecosystem Study report (unnumbered), submitted to the Canadian Rivers Institute, University of New Brunswick, 45p.

O'Sullivan, A., Curry, R.A., and Yamazaki, G. 2016. Assessment of the Mactaquac Headpond Geomorphology and Estimated Sediment Distribution. Mactaquac Aquatic Ecosystem Study Report Series 2016-29. Canadian Rivers Institute, University of New Brunswick, 10p.

Wallace, B. and M. Gautreau. 2015. Methods Paper: Downstream Bathymetry and BioBase Analyses of Substrate and Macrophytes. Mactaquac Aquatic Ecosystem Study Report Series 2015-006. Canadian Rivers Institute, University of New Brunswick, 5p.

Yamazaki, G, A. Chateauvert, B. Wallace, and K. Haralampides. 2016. Fine Surface Sediment Particle Size from Perth-Andover to Fredericton, New Brunswick. Mactaquac Aquatic Ecosystem Study Report Series 2016-037. Canadian Rivers Institute, University of New Brunswick, 18 p.

Yamazaki, G., Wishart, L, Haralampides, H., and Mana Ahmady. 2017. Historical and Observed Suspended Sediment Characteristics of the Saint John River. Mactaquac Aquatic Ecosystem Study Report Series 2016-036. Canadian Rivers Institute, University of New Brunswick, 26p.

Leadership in Sustainable Infrastructure
Leadership en Infrastructures Durables

Fredriction, Canada
June 13 to 16, 2018 / *13 au 16 juin, 2018*

SERVICE CONSTRAINTS ON A SIMPLE GIS BASED CURBSIDE WASTE COLLECTION MODEL

Vu, Hoang Lan[1], Bolingbroke, Damien[1], and Ng, Kelvin T.W.[1,2]
[1]Faculty of Engineering and Applied Science, University of Regina, Canada
[2] kelvin.ng@uregina.ca (corresponding author email)

Abstract: Waste collection expenditure alone made up about 47% of the annual residential waste management budget at City of Regina in 2005. Optimized waste collection routes, such as using Geographic Information System tools, help reduce the waste collection system's collection time, labour cost, fuel consumption, and environmental footprint. For North American cities and towns with a front-of-street curbside collection program, a bin or collection point is typically serviced on the truck's passenger side. However, it was found that this service constraint was not explicitly considered in many GIS-based studies. The objectives of this study are to apply the truck's passenger side constraint and other service constraints to a small study area in Regina, and to investigate their effects on overall travel distance and collection time using different scenarios. A simple GIS-based door-to-door collection model with various service constraints is developed and studied. In some scenarios, the "one-way road" rule is applied in the modeling to force waste collecting vehicles to service the collection points only at the truck's passenger side. In another scenario, a single collection point is assumed to be shared between two individual waste generation sites to increase efficiency. The results suggested that the model approach with the truck's passenger side constraint provides more realistic results. In this study, conventional GIS modelling approaches with no service constraints underestimate the travel distance in collection zones by 60.9%. Also, sharing a collection point among waste generators could reduce collection time by at least 12.5% when compared to the status quo.

1 INTRODUCTION

Waste collection cost in industrial countries accounted for more than 50% of the total expenditure on solid waste management (Sanjeevi and Shahabudeen 2016). In Regina, Canada, waste collection expenditure made up about 47% of the residential waste management budget (City of Regina, 2005). Optimization of residential collection routes increases collection efficiency, and also helps to reduce travel distances, labor costs and the fuel consumption of trucks. Emissions from these collection vehicles, and the associated environmental impacts, will also be reduced.

Geographical Information System (GIS) techniques are widely studied for waste collection optimization (Vijay et al. 2005, Alvarez et al. 2008, Karadimas and Loumos 2008, Zamorano et al. 2009, Zsigraiova et al. 2013, Boskovic and Jovicic 2015, Gallardo et al. 2015, Son and Louati 2016, Nguyen et al. 2017). By applying GIS to identify the shortest routes for waste collection trucks, a 9.9% to 71.8% reduction in total truck travel distances are attainable (Zamorano et al. 2009, Malakahmad et al. 2014, Abdelli et al. 2016; Sanjeevi and Shahabudeen 2016; Son and Louati 2016; Nguyen et al. 2017). A reduction in total collection time, in the range of 14.3% to 62.0%, is also reported (Thanh et al. 2009; Zsigraiova et al. 2013, Son and Louati 2016). It is found that most of these GIS-based studies focused on waste collection systems in larger areas, and the distances between collection points are relatively large. Trucks in these studies are often allowed to access generation sites and collection points using both sides of road without imposing traffic direction constraints. On the contrary, trucks in a door-to-door curbside collection system collect wastes using the passenger's side of the trucks for reasons including ease of operation and lower traffic impacts. As such, traffic direction constraints are necessary for realistic modeling results, especially for smaller residential areas.

Truck fuel consumptions, collection time, and emissions in curbside waste collection were studied and reported by a number of studies (Nguyen and Wilson 2010, Maimoun et al. 2013; Maria and Micale 2013; Edwards et al 2016; Maimoun et al. 2016). Some of the key factors affecting truck fuel consumption and emissions were the distance between two collection points, truck speed, and truck idle time (Nguyen and Wilson 2010, Edwards et al. 2016). The truck speeds reported in curbside collection studies appear to change with distance between stops and traffic density. Edwards et al. (2016) found that the average truck speed between stops in urban area was 7 km/h, whereas the figures for semi-urban and rural areas were 9 km/h and 15 km/h, respectively. Curtis and Dumas (2000) found that the waste collection truck speed was 8 km/h, with an average travel distance of 23 m between stops. Abdelli et al. (2016) suggested similar results, with a collection truck speed of less than 10 km/h in their studies. Boskovic et al. (2016) and Maimoun et al. (2016) both used an average truck speed of 10 km/h. A higher truck speed of 16 km/h - 20 km/h was reported by Maria and Micale (2013) for a longer truck travel distance between stops range of 145 m to 315 m. Idling time (lifting time) per stop reported in waste collection literature were generally between 8.3 seconds to 8.7 seconds (Nguyen and Wilson 2010, Maimoun et al. 2015, Edwards et al. 2016).

Most collection optimization studies using GIS focused on the design and optimization of waste collection with transfer stations in larger areas without traffic direction constraints. In the present study, GIS optimization techniques were applied to a small residential neighborhood with a door-to-door curbside collection system. The objectives of this study are (i) to model door-to-door waste collection by applying various realistic service constraints using GIS, and (ii) to investigate the impact these constraints have on total collection time using different scenarios. The "one-way road" rule was applied in the GIS modeling to force waste collecting vehicles to service collection points at the truck's passenger side only along a given street. A total of 3 scenarios were considered in the study to evaluate the effects of service and operational constraints on collection distance and time.

2 METHODOLOGY

2.1 Study Area, Current Waste Collection Service and Field Measurements

The selected study area is located in the Hillsdale area in south Regina covering an area of 0.88 km^2, with about 675 households (Figure 1). The average number of people per home in the study area was about 2.39 (City of Regina, 2015). The distances between the doors of any two houses, measured parallel to the curb, in the study area range from 9 m to 20 m. Roads in the study area are mainly residential roads with a width ranging from 8.5 m to 9.5 m (OpenStreetMap 2017). The City's only landfill is located in the north east corner of the city and the road distance between the study area and this landfill is about 15 km.

The residential waste generation rate was approximately 16.65 kg/household·week (City of Regina, 2005). Thus, the total waste generation in the study area was about 11.24 tonnes/week. Currently, automated front-of-street collection (curbside collection) service is provided weekly (bi-weekly in the winter) in the study area. A 240 L or 360 L garbage bin and a 320 L blue recycling bin are provided to each household. Homeowners are responsible for taking their bins to the curb on the collection day. The capacity of waste collection truck is 22 m^3, with side-loading at the truck's passenger side for material collection.

According to literature, truck speed and average travel distance between waste generation sites are important parameters in door-to-door collection system modelling. The truck speeds reported in literature were generally from locations with mild climates and no or limited winter driving conditions. A field study during winter driving conditions was conducted within the study area in January 2018 to measure the actual vehicle speeds and travel distances using a passenger car for scenarios with (i) a single collection point at each house, and (ii) a single collection point shared by two houses. A total of 36 potential truck stops were considered and a Garmin GPSMAP 64s handheld was used to determine the geophysical data. For scenarios with a single collection point per house, 9 m to 20 m average travel distances were recorded, with an average truck speed of 9 km/h. For the scenario with shared collection point, 21 m to 40 m average travel distances were recorded, with an average truck speed of 12 km/h. Both parameters agreed well with literature and winter driving conditions appear to have no to little observable impacts on travel distance and truck speed, at least for the conditions encountered in the present study.

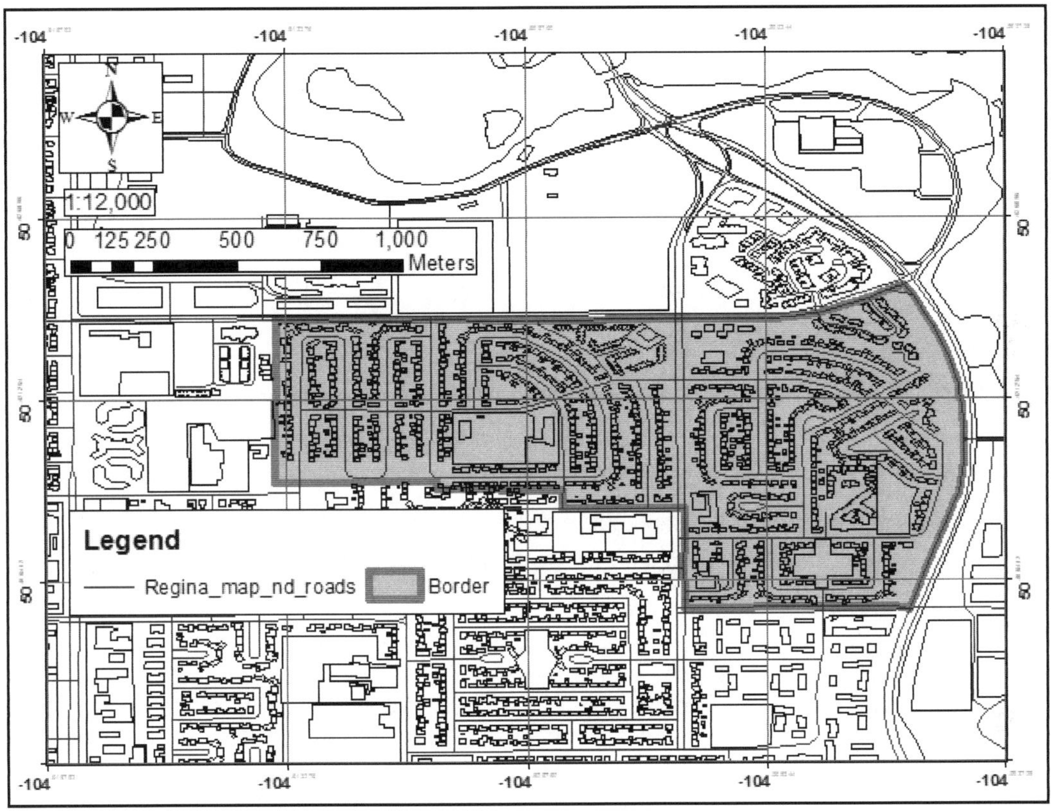

Figure 1: The study area (Open Street Map, 2017)

2.2 Conventional GIS Modelling with no Service Constraints (Scenario A)

Vehicle route problem - Network Analysis (ArcGIS version 10.5) was used to analyze and optimize truck collection routes. Inputs for the model included: (i) a road network, (ii) order locations, and (iii) route class (Figure 2). The road network was derived from OpenStreetMap (2017), and contained details such as the width and length of roads, traffic direction (one way/two way roads), and truck speed and height restrictions. In our conventional modeling approach (Scenario A), the road network was modeled such that trucks could collect waste at either side intermittently along a road, with the system objective of minimizing truck travel distance. Order location data consisted of pickup time at a stop, time windows, and waste volume per stop. Time windows were chosen from 8 am to 8 pm in this study. Waste volume per stop was calculated from the waste generation rate (16.65 kg/household·week), average waste density, and number of households per stop. The route class contained the capacity of trucks (22 m^3), depot and landfill locations, and other constraints.

2.3 Curbside Waste Collection with Service Constraints (Scenario B)

Similar to Scenario A, the Scenario B inputs consisted of the road network, order locations, and route class. The major difference between Scenarios A and B was the modeling of the roads. In Scenario B, each road segment was modelled as two individual one-way roads in the opposite direction to force trucks to collect materials on the passenger side only. Other inputs such as truck speed and height limits, pickup time at a stop, waste volume, time windows, truck capacity, depot/landfill locations, and other constraints were identical to Scenario A. The truck collects waste at every single house and there are a total of 675 orders (houses). Some key input parameters of the model are provided in Table 1. Default values of the GIS vehicle route problem were used for other parameters.

Table 1: Key input parameters

Parameters	Description	Unit	Value	Remarks
Capacity of truck	Maximum volume of waste that a truck can hold	L	22,000	Based on existing capacity of trucks
Pickup time	The time required for a truck to service an order (house)	min	0.15	Selected from literature
Truck traveling speed in Scenarios A and B	Truck speed when collecting waste at every single house	km/h	9	From field study
Truck traveling speed in Scenario C	Truck speed when collecting waste at a shared collection point	km/h	12	From field study
Volume of waste at a household	Waste generation per week at a household	L	120	Based on waste generation rate and waste density

2.4 Curbside Waste Collection with Shared Collection Points and Service Constraints (Scenario C)

Literature review suggests truck speed increases with increasing travel distance between stops. Scenario C was similar to Scenario B except that two houses were served per stop using a common collection point. Order locations of the model were adjusted to allow waste collection from up to two houses per stop. The new arrangement helped reduce the number of stops and thus extended the travel distance between stops and increased average truck speed. The distances between two stops and the speed of collection truck were measured from the field study. A summary of the methodology in this study is presented in Figure 2.

3 RESULTS AND DISCUSSION

Outputs of the models (optimal truck routes and travel times) under different scenarios were compared to investigate the effects of service constraints and shared collection point in a door-to-door collection system in the study area.

3.1 Model with no Service Constraints (Scenario A)

A total of four routes were used to collect waste from 675 houses in the study area. The optimized truck travel distances and times for different scenarios are tabulated in Table 2. For Scenario A, the truck travel distance of each route varied from 27.12 km to 30.59 km while travel time ranged from 66 min to 76 min. The total truck travel distance was 115.93 km, whereas the overall travel time was 278 minutes.

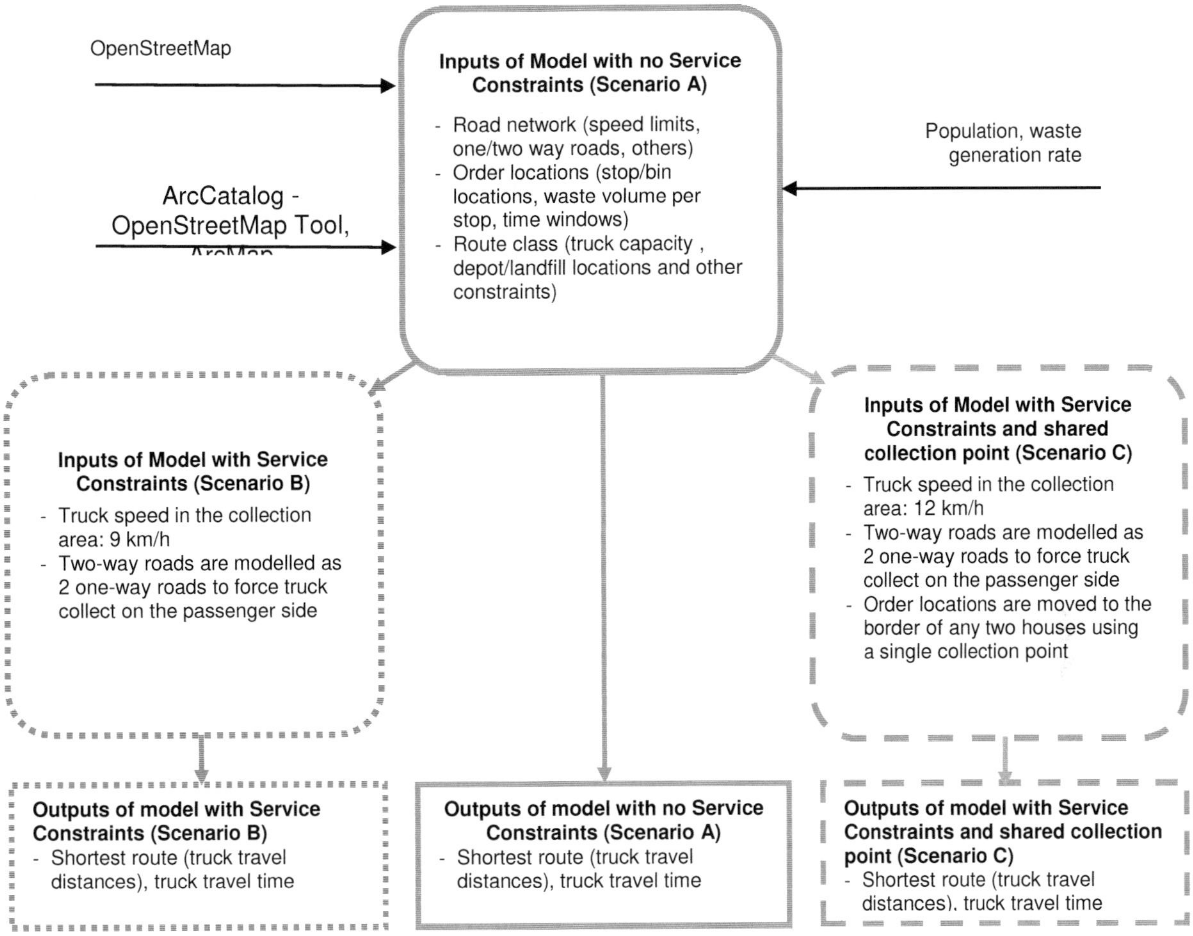

Figure 2: Summary of Methodology of this study

Although travel distance in route 2 (28.78 km) was lower than that of routes 3 and 4 in Scenario A (Table 2), route 2 truck travel time was considerably higher (76 min). This apparent discrepancy is partly due to a lengthier travel distance within the collection zone (5.08 km, Table 3). In this study, the truck speed within the collection zone was limited to 9 km/h in Scenario A, while truck speed outside the collection zone ranged from 50 km/h to 100 km/h. Collection on routes with lengthier travel distance and more customers (number of stops) within the collection zone generally takes more time due to these lower speeds.

Table 2: Comparison of total travel distance and time

	Scenario A – no service constraints			Scenario B – with service constraints			Scenario C – with service constraints and shared collection point		
	Number of orders	Travel distance (km)	Travel time (min)	Number of orders	Travel distance (km)	Travel time (min)	Number of orders	Travel distance (km)	Travel time (min)
Route 1	180	27.12	68	159	30.86	83	135	30.79	69
Route 2	183	28.78	76	176	28.09	79	183	29.81	74
Route 3	162	30.59	68	183	31.94	86	183	27.13	71
Route 4	150	29.44	66	157	27.53	81	174	27.63	74
Total	**675**	**115.93**	**278**	**675**	**118.42**	**329**	**675**	**115.36**	**288**

Compared to Scenarios B and C, Scenario A has the lowest overall truck travel time (278 min, Table 2). However, a closer look of the proposed routes reveals the highly idealized nature of the routes. In Scenario A, the road network was modelled using single line segments from OpenStreetMap, and the network allowed trucks to collect waste on opposite sides of the road uninterruptedly. Figure 3a showed an example of the optimized truck routes in Scenario A. The optimized truck route shows a zigzag pattern, utilizing both sides of the road to minimize overall travel distance and time. The optimized routes in this case are less practical because the proposed operation may severely impact traffic flow and pose considerable road hazards to other road users (Figure 3a). It is believe that this modelling approach underestimated the total travel distance and time than the actual situation.

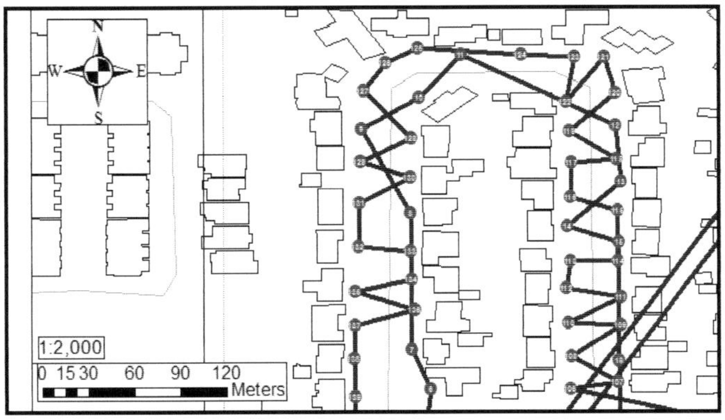

(a)

(b)

(c)

Figure 3: Examples of truck routes under the three scenarios: (a) Scenario A-conventional vehicle route problem with no constraints; (b) Scenario B-vehicle route problem with service constraints; and (c) Scenario C-vehicle route problem with service constraints and shared collection points

Table 3: Comparison of truck travel distances with or without service constraints within the collection zone

	Travel distance in collection zone (km) (Scenario A)	Travel distance in collection zone (km) (Scenario B)	Percentage of increasing distance (%)
Route 1	4.05	6.73	66.1
Route 2	5.08	6.39	25.8
Route 3	4.73	8.32	76.0
Route 4	3.87	7.08	83.2
Total	**17.73**	**28.52**	**60.9**

3.2 Model with service constraints and shared collection points (Scenario C)

The overall truck travel distances with service constraints were slightly higher than in Scenario A, ranging from 27.53 km to 31.94 km (Table 2). The overall truck travel times, on the other hand, were noticeably higher. For example, the total travel distance was increased by 2.49 km (2.2%) between Scenarios A and B, whereas the travel time went up by 51 minutes (18.4%). Table 3 showed that the truck travel distances within the collection zone were considerably higher in Scenario B than in Scenario A, with an average increase of 60.9%. It is interesting to note that Route 2 in Scenario B had a slightly higher distance than that of route 4 (Table 2). However, the corresponding truck travel time of route 2 (79 min) was lower than that of route 4 (81 min). This may be in part due to a relatively low travel distance within the collection area in route 2 (6.39 km, Table 2). In this scenario, the route served the highest number of households, 183 (route 3), and also had the highest overall travel time (86 min, Table 2).

Figure 4 shows a qualitative comparison between the optimized routes in Scenarios A and B. Routes in Scenario A (Figure 4, a1 – a4) contained fewer turns and was simpler due to the lack of service constraints. All optimized truck routes in Scenario B (Figure 4, b1 – b4) had lengthier travel routes than those of Scenario A. Overall, there was a 60.9% increase in truck travel distance in Scenario B when compared to those of Scenario A (Table 3). The percentage increases were lower in routes 1 and 2 (Table 3) because of the decreasing number of customers between Scenarios A and B on these routes (Table 2). Due to the trucks being forced to collect waste at only the passenger's side in Scenario B, higher total travel distances are modelled. Figure 3 highlights the potential benefits of modeling the vehicle route problem using two one-way roads rules, especially for door-to-door curbside collection in residential area. The conventional modeling approach, shown in Scenario A, appears to underestimate the true travel distance and time.

Results of the model with service constraints and shared collection points (Scenario C) are provided in Table 2. Overall truck travel distances varied slightly from 27.13 km to 30.79 km. Truck travel time ranged from 69 min to 74 min. Route 1 had the lowest truck travel time (69 min), partly due to it having the lowest number of customers (135). A larger average travel distance within the collection zone is observed with the establishment of the shared collection points (Figure 3c). Truck travel distances and travel times in Scenario C were lower than those in Scenario B. It is interesting to note that while the travel times were reduced by 12.5%, the travel distances went down by only 2.6%, when comparing Scenario's C and B, which both have service constraints but only Scenario C having shared collection points. The results suggested that the overall travel time is sensitive to the truck speed within the collection zone. It appears that the use of shared collection points may slightly enhance the door-to-door collection efficiency. However, certain social concerns with shared collection point operation such as the clean-up of the shared collection points and a lack of privacy may hinder the use of such a system and should be addressed in future studies.

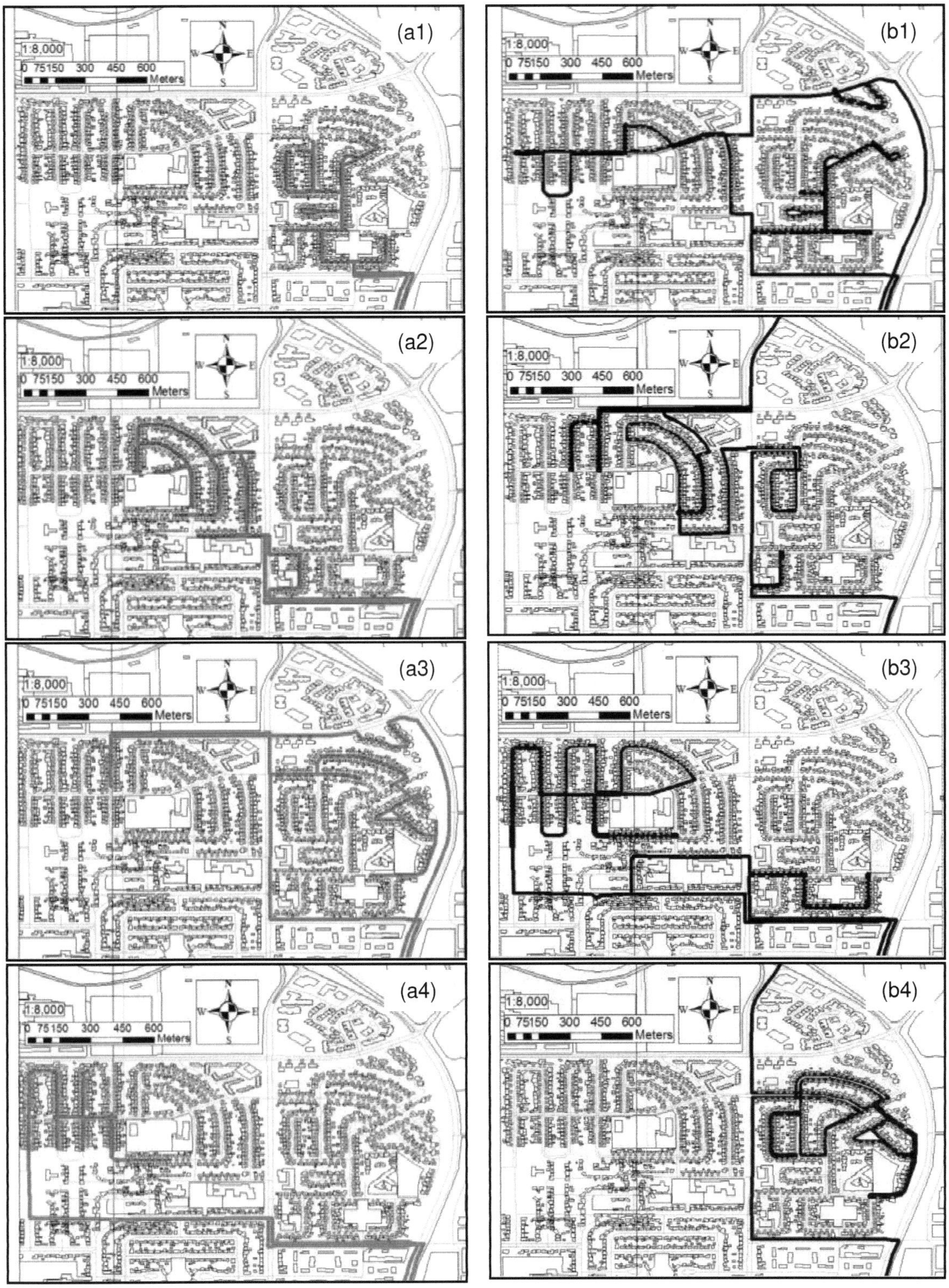

Figure 4: Comparisons of truck routes between Scenario A (a1 - a4) and Scenario B (b1 - b4)

4 CONCLUSIONS

The study examined three different GIS-based curbside waste collection models to estimate the optimal truck collection routes for a small residential area located in Regina, the capital city of Saskatchewan. The trucks in the model with no service constraints (Scenario A) are allowed to collect waste on both sides of the road to minimize overall travel distance. The lowest truck travel time of 278 min is obtained in this conventional modelling approach. The total truck distance and travelling time reported, however, are misleading due to the unrealistic routes this approach creates. By adjusting the road network and applying one-way rules, two models (Scenarios B and C) were developed. A 60.9% increase in the collection zone truck travel distance was observed in Scenario B. Results from this study suggest that Scenario B is a more realistic door-to-door curbside collection system model. To improve the collection efficiency, a model with service constraints and shared collection points was also investigated (Scenario C). The model was created to extend average truck travel distance between stops within the collection zone to increase truck speed. Compared to Scenario B, the overall truck collection time was reduced by 12.5% for the small study area considered in this study. These results highlight the importance of service constraints and average travel distances when modelling a door-to-door collection system using GIS tools.

Acknowledgement

This work is financially supported by a research grant (RGPIN-385815) from the Natural Sciences and Engineering Research Council of Canada (NSERC) and an internal FEROF grant from the Faculty of Engineering and Applied Science at University of Regina. The authors are grateful for their support. The views expressed herein are those of the writers and not necessarily those of our research and funding partners.

References

Abdelli, I.S., Abdelmalek, F., Djelloul, A., Mesghouni, K., and Addou, A., 2016. GIS-based approach for optimized collection of household waste in Mostaganem city (Western Algeria). Waste Management & Research, 34(5) 417- 426. DOI: 10.1177/0734242X16633519.

Alvarez, J.V.L, Larrucea, M.A., Quero, S.F.C., Jime´nez del Valle, A., 2008. Optimizing the collection of used paper from small businesses through GIS techniques: The Legane´s case (Madrid, Spain). Waste Management 28, 282-293.

Boskovic, G., and Jovicic, N., 2015. Fast methodology to design the optimal collection point locations and number of waste bins: A case study. Waste Management & Research, Vol. 33(12) 1094-1102.

Boskovic, G., Jovicic, N., Jovanovic, S., and Simovic, V., 2016. Calculating the costs of waste collection: A methodological proposal. Waste Management & Research, Vol. 34(8) 775-783.

City of Regina, 2005. Service Report. Waste Management Division.

Curtis, E.M., Dumas, R.D., 2000. A Spreadsheet Process Model for Analysis of Costs and Life-cycle Inventory Parameters Associated with Collection of Municipal Solid Waste. Internal Rep., Department of Civil Engineering, North Carolina State University, Raleigh, N.C. http://www4.ncsu.edu/~jwlevis/DST_Collection.pdf (Accessed on 20180121).

Edwards, J., Othman, M., Burn, S., and Crossin, E., 2016. Energy and time modelling of kerbside waste collection: Changes incurred when adding source separated food waste. Waste Management, 56: 454-465. http://dx.doi.org/10.1016/j.wasman.2016.06.033.

Gallardo, A., Carlos, M., Peris, M., and Colomer, F.J., 2015. Methodology to design a municipal solid waste pre-collection system. A case study. Waste Management 36: 1-11.

GENIVAR Consultants LP, 2009. Waste Plan Regina Report. City of Regina.

Ghose, M.K., Dikshit, A.K., and Sharma, S.K., 2006. A GIS based transportation model for solid waste disposal - A case study on Asansol municipality. Waste Management, 26: 1287-1293. doi:10.1016/j.wasman.2005.09.022.

Karadimas, N.V., Loumos, V.G., 2008. GIS-based modelling for the estimation of municipal solid waste generation and collection. Waste Management & Research: 26: 337-346.

Khan, D., and Samadder, S.R., 2016. Allocation of solid waste collection bins and route optimisation using geographical information system: A case study of Dhanbad City, India. *Waste Management & Research*, Vol. 34(7) 666–676.

Maimoun, M.A., Reinhart, D.R., Madani, K., 2016. An environmental-economic assessment of residential curbside collection programs in Central Florida. Waste Management, 54: 27-38.

Maimoun, M.A., Reinhart, D.R., Gammoh, F.T., and Bush, P.M., 2013. Emissions from US waste collection vehicles. Waste Management 33: 1079-1089. http://dx.doi.org/10.1016/j.wasman.2012.12.021.

Maria, F.D., and Micale, C., 2013. Impact of source segregation intensity of solid waste on fuel consumption and collection costs. Waste Management 33: 2170–2176. http://dx.doi.org/10.1016/j.wasman.2013.06.023.

Maria, F.D., Caterina Micale, C., and Morettini, E., 2016. Impact of the pre-collection phase at different intensities of source segregation of bio-waste: An Italian case study. Waste Management 53: 12–21.http://dx.doi.org/10.1016/j.wasman.2016.04.026.

Nguyen-Trong, K., Nguyen-Thi-Ngoc, A., Nguyen-Ngoc, D., Dinh-Thi-Hai, V., 2017. Optimization of municipal solid waste transportation by integrating GIS analysis, equation-based, and agent-based model. Waste Management 59 (2017) 14-22.

Nguyen, T.T.T., and Wilson, B.G., 2010. Fuel consumption estimation for kerbside municipal solid waste (MSW) collection activities. Waste Management & Research, 28: 289–297. DOI: 10.1177/0734242X09337656.

Sanjeevi, V., and Shahabudeen, P., 2016. Optimal routing for efficient municipal solid waste transportation by using ArcGIS application in Chennai, India. Waste Management & Research, Vol. 34(1) 11-21.

Malakahmada, A., Bakria, P.M., Mokhtara, M.R.M., and Khalila, N., 2014. Solid waste collection routes optimization via GIS techniques in Ipoh city, Malaysia. Procedia Engineering, 77: 20 - 27.

Son, L.H., Louati, A., 2016. Modeling municipal solid waste collection: A generalized vehicle routing model with multiple transfer stations, gather sites and inhomogeneous vehicles in time windows. Waste Management 52, 34-49.

Thanh, N.P., Matsui, Y., Ngan, N.V.C., Trung, N.H., Vinh, T.Q., and Yen, N.T.H., 2009. GIS application for estimating the current status and improvement on municipal solid waste collection and transport system: Case study at Can Tho city, Vietnam. Asian Journal on Energy and Environment, 10(02): 108-121.

Vijay, R., Gupta, A., Kalamdhad, A.S., Devotta, S., 2005. Estimation and allocation of solid waste to bin through geographical information systems. Waste Manage Res: 23: 479-484.

Zamorano, M., Molerob, E., Grindlayb, A., Rodríguezc, M.L., Hurtadoa, A., Calvo, F.J., 2009. A planning scenario for the application of geographical information systems in municipal waste collection: A case of Churriana de la Vega (Granada, Spain). Resources, Conservation and Recycling, 54, 123-133.

Zsigraiova, Z., Semiao, V., Beijoco, F., 2009. Optimisation of MSW collection routes for minimum fuel consumption using 3D GIS modeling. Waste Management 29, 1176–1185.

Building Tomorrow's Society
Bâtir la Société de Demain

Fredericton, Canada
June 13 to 16, 2018 / *13 au 16 juin, 2018*

REMEDIATION OF SULFOLANE IMPACTED SOIL BY INTEGRATING SOIL WASHING/FLUSHING AND ADVANCED OXIDATION PROCESS

Yu, Linlong[1,3], Brandão, Mariana[2], Khan, Muhammad Faizan[1] and Achari, Gopal[1]
[1] Department of Civil Engineering, University of Calgary, Canada
[2] Department of Water Resources, Federal University of Sergipe, São Cristóvão, SE, Brazil
[3] corresponding author linyu@ucalgary.ca

Abstract: Sulfolane in soil is an emerging problem in oil and natural gas processing sites. This study investigates remediation of sulfolane contaminated soil through the application of soil flushing followed by UVC/H2O2/O3 and soil washing followed by four different advanced oxidation processes (AOPs) including UVC/H2O2, UVC/O3, alkaline ozonation and neutral Fenton reagent. Results showed that sulfolane can be effectively removed from the soil using water as a flushing/washing solvent. Furthermore, sulfolane in the soil flushing/washing solution was effectively degraded by the following AOP treatment.

1 INTRODUCTION

Remediation of contaminated soils and sites using environmentally sustainable and efficient technologies has been a major challenge. Despite the high cost, ex-situ procedures generally require less time to achieve contaminant cleanup when compared to in situ approaches. It is also easily monitored and uniform, offering greater prospects (Kupuna et al, 2016). Soil flushing/washing has previously been employed as a pre-treatment technique to extract contaminants from polluted soils and further facilitate advanced treatments. Soil flushing/washing is a mechanical process based on desorption of contaminants from soil through the action of water with or without surfactants (Mao et al, 2015). It has been applied to remediate soils contaminated with organic compounds including polychlorobiphenyls, polycyclic aromatic hydrocarbons and pesticides (Trellu et al. 2016). Although soil washing/flushing has shown promising results, this process only transfers the contaminants from the soil phase into aqueous phase, resulting in wash solution still containing contaminants which require further treatment. There are many post treatment options available for soil washing/flushing solutions such as bioremediation, chemical oxidation, advanced oxidation processes (AOPs) and electro-kinetic remediation. Among these options, AOPs have attracted more interests in recent years due to its capacity to degrade a large group of organic contaminants. AOPs are processes designed to degrade recalcitrant organic compounds using highly reactive species such as hydroxyl radicals (•OH). They have been considered as highly efficient processes to treat emerging contaminants including pharmaceuticals and personal care products (PPCPs), pesticides, herbicides, surfactants, steroids and hormones, endocrine disrupting compounds etc (Andreozzi et al. 1999; Esplugas et al. 2007; Parsons 2004).

In this study, soil washing/flushing integrated with advanced oxidation processes was evaluated to treat soils contaminated with an emerging industrial pollutant, sulfolane. Sulfolane ($C_4H_8O_2S$) is a polar organosulfur solvent well known for sweetening sour gas in the Sulfinol® process as well as used to extract aromatics from hydrocarbon mixtures. Due to leaks and spills, soil and underground water have been contaminated with sulfolane. The highly polar nature of sulfolane makes it extremely mobile in water enabling it to easily enter soil and groundwater of nearby gas refinery plants through different routes posing a threat for aquatic life and soil environment.

Previous published works (Yu et al. 2016a; Mehrabani et al. 2016; Izadifard et al. 2017) have shown that AOP is a feasible technology for sulfolane degradation in aqueous medium. However, in contrast to aqueous spiked medium, the composition of flush/wash water is heterogeneous as other compounds are also washed from the soil. Therefore, this paper reports on the effectiveness of AOPs on treating these solutions containing sulfolane. Both soil washing, and flushing were investigated to remove sulfolane from soil. The soil washing solution was treated with UVC/H_2O_2, UVC/O_3, alkaline ozonation and neutral Fenton while the soil flushing solution was treated with UVC/H_2O_2/O_3, which was reported as one of the more effective methods in degrading sulfolane (Yu et al., 2016a).

2 METHOD AND MATERIALS

2.1 Chemicals and Contaminated Soil.

Sulfolane (99% purity) and hydrogen peroxide (30% purity) were purchased from Sigma Aldrich, Canada, sodium hydroxide with 97% purity was purchased from EMD; EDTA chelated iron (III) with 99% purity was purchased from J.T. Baker and milli-Q water was used to conduct soil washing experiments and sample dilutions. Contaminated soil samples were provided by Bonavista Energy Corporation.

2.2 Experimental Procedure

The experimental setup for the integration of soil washing/flushing and AOPs is shown in Figure 1. The soil samples were firstly treated by soil washing or soil flushing with milli-Q water as the solvent. The collected soil washing, and flushing solution samples were then treated with different AOPs.

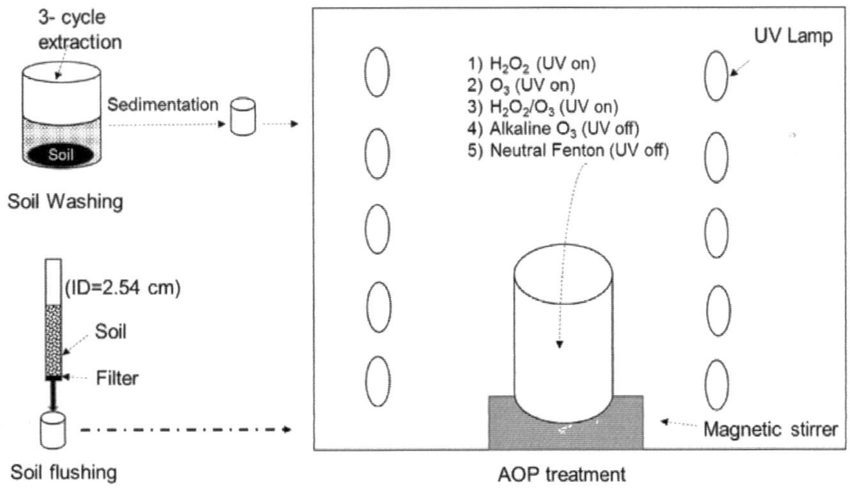

Figure 1. Experimental setup

2.2.1 Soil Washing and AOP

Processed soil weighing 1.5 kg and containing 900 mg/L of sulfolane was washed by water using a 3-cycle procedure. Three litres of water were applied for each cycle. The soil washing solution collected from each cycle was then combined in a settling column. After 24 hr of sedimentation, the supernatant was separated by siphon and kept in the fridge for AOP treatments. The AOP treatments included UVC/H_2O_2, UVC/O_3, alkaline ozonation and neutral Fenton. The details of each AOP experiment including chemical dosage, pH and light condition are listed in Table 1. All experiments were conducted in a 150 ml quartz beaker containing 100 ml of soil washing solution. Luzchem photoreactor (LZC ORG) equipped with 10 germicidal lamps (LZC-UVC) were used to provide UVC irradiation if necessary. The light intensity was determined to be 5.3×10^{17} photon/s through ferrioxalate actinometry. For experiments with ozone, an O_3 generator (A2Z 3-G LAB, A2Z O3 systems Inc.) connected to the beaker through a glass diffuser was used. The ozone flow rate was kept at 3.7 g/h. For alkaline ozonation, the pH of solution was adjusted to 13 by adding sodium hydroxide before ozone exposure. For neutral Fenton experiments, FeEDTA and H_2O_2 were introduced into the solution at a molar ratio of 1:10.

Table 1: Details of AOP experiments

Type of AOP	UVC Light Intensity (photon/s)	$[H_2O_2]$ (mg/L)	O_3 (g/h)	[FeEDTA] (mg/L)	pH
UVC/H_2O_2	5.3×10^{17}	200	-	-	No adjustment
UVC/O_3	5.3×10^{17}	-	3.7	-	No adjustment
Alkaline O_3	-	-	3.7	-	13
Neutral Fenton	-	3400	-	3670	No adjustment

2.2.2 Soil Flushing and AOP

A glass column with 1 inch inside diameter was filled with 45 g of processed soil containing 8000 mg/L of sulfolane. Water was added from the top of column and the leachate was collected from the bottom of soil column at different time intervals. The collected leachate (first 68 mL) was diluted 40 times with milli-Q water. The diluted sample was then treated with $UVC/H_2O_2/O_3$. The H_2O_2 concentration was set at 1000 mg/L, ozone flow rate was controlled at 3.7 g/h and the light intensity was about 5.3×10^{17} photons/s.

2.2.3 Sample Extraction and Sulfolane Analysis

Sulfolane was extracted from water samples by liquid/liquid extraction. A ratio of 5:3 of each water sample to dicholoromethane (DCM) was used. Samples were shaken for 30 mins in a shaker (Model: SI-1100, Roto-Shake Genie) at 35 rpm. The extracted samples were left to stand for 5 mins to allow separation of water and DCM. Extracted sulfolane in DCM was then analyzed with gas chromatography equipped with flame ionization detector (Yu et al. 2016a).

3 3. RESULTS AND DISCUSSION

3.1 Integration of Soil Washing and AOP

Using water as the washing solvent, three consecutive cycles of soil washing resulted in more than 99% of sulfolane removal from the contaminated soil. The solution, obtained from soil washing, containing sulfolane was treated with different AOPs including (UVC/H_2O_2, UVC/O_3, alkaline ozonation and neutral Fenton) and the results are presented in Figure 2. The four AOPs can be classified into two major categories: UV assisted processes and non-UV assisted processes. All four AOPs showed their capabilities to degrade sulfolane in the soil washing solution. Based on the degradation of sulfolane, the four AOPs in descending order were: alkaline ozonation > UVC/O_3 > UVC/H_2O_2 > Neutral Fenton. The percentage of sulfolane removal after 20 mins of retention time was 97%, 83%, 70% and 43% for alkaline ozonation, UVC/O_3, UVC/H_2O_2 and neutral Fenton respectively. The degradation of sulfolane is due to its reaction with the highly reactive hydroxyl radicals generated during AOPs. Each AOP has a distinct pathway for the generation of hydroxyl radicals which are summarized in Table 2.

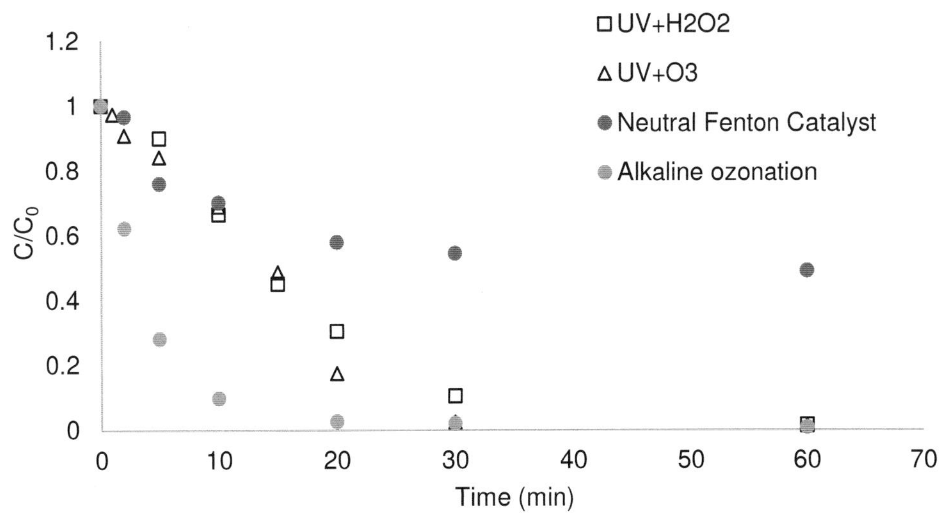

Figure 2. Sulfolane degradation in the soil washing water solution under different AOPs

Table 2: Reaction of different AOPs

Categories	Type of AOPs	Reactions
UV-based	UV/H_2O_2	$H_2O_2 \xrightarrow{hv} HO\cdot + HO\cdot$
UV-based	UV/O_3	$O_3 \xrightarrow{hv} O(D) + O_2$ $O(D) + H_2O \rightarrow H_2O_2$ $H_2O_2 \xrightarrow{hv} HO\cdot + HO\cdot$
Non-UV based	Alkaline Ozonation	$HO^- + O_3 \rightarrow O_2 + HO_2^-$ $HO_2^- + O_3 \rightleftarrows HO_2\cdot + O_3\cdot^-$ $HO_2\cdot \rightleftarrows H^+ + O_2\cdot^-$ $O_2\cdot^- + O_3 \rightarrow O_2 + O_3\cdot^-$ $O_3\cdot^- + H^+ \rightarrow HO_3\cdot$ $HO_3\cdot \rightarrow HO\cdot + O_2$ $HO\cdot + O_3 \rightarrow HO_2\cdot + O_2$
Non-UV based	Fenton process*	$Fe^{2+} + H_2O_2 \rightleftarrows FeO^{2+} + H_2O$ $Fe^{2+} + H_2O_2 \rightarrow Fe^{3+} + OH^- + OH\cdot$ $Fe^{3+} + H_2O_2 \rightleftarrows FeOOH^{2+} + H^+$ $FeOOH^{2+} \longrightarrow HO_2\cdot + Fe^{2+}$

During UVC/H_2O_2 process, each molecule of hydrogen peroxide can generate two moles of hydroxyl radicals under UV irradiation. The UVC/O_3 process produces hydroxyl radicals through initiating the reaction between ozone and UV photons. Both processes achieved more than 99% of sulfolane removal in 60 minutes, which are lower than the sulfolane degradation rate in ultrapure water reported by Yu et al. (2016a). In this study, the residual soil particles in the soil washing solution and other co-washing contaminants reduced the light intensity received by the target chemicals (H_2O_2, O_3 and sulfolane) as well as competed for hydroxyl radicals. In addition, hydrogen peroxide and ozone are strong oxidizers, they can be consumed by other organics. Ozonation with pH = 13 showed that 70% of sulfolane can be degraded within 2 mins. This is due to instant formation of hydroxyl radicals from the reaction between ozone and hydroxide ions as shown in Table 2. For full scale applications, a pH adjustment step is required after this AOP treatment. Neutral Fenton reagent removed 45% of sulfolane in the first half hour but further exposure of 30 mins only improved the removal by 5%. The formation of yellow precipitate was formed during the degradation of sulfolane. As reported by Yu et al. (2016b), the catalytically active iron species precipitated out during the reaction and stopped the degradation of sulfolane. The advantages and disadvantages of these four methods are summarized in Table 3. It is noted that, each method is not optimized in this study. The comparison of these technologies was based on current experimental conditions.

Table 3. Comparison between different AOPs

Type of AOPs	pH Adjustment	UV requirement	Degradation rate	Formation of precipitates
UVC/H_2O_2	No	Yes	Fast	-
UVC/O_3	No	Yes	Fast	-
Alkaline Ozonation	Yes	No	Very fast	-
Neutral Fenton process	No	No	50% removal	Yes

3.2 Soil Flushing and AOP

Soil flushing is a more practical approach than soil washing processes for high conductivity soils as it does not require a soil/water separation step. The pictures of soil flushing solution during different periods are presented in Figure 3. From the images, it is observed that the color of soil flushing solution is dense yellow in the first few samples. The color of the collected leachate in the vials showed in Figure 3 is attenuated as the volume of water flushed accumulates from 2 mL to 152.9 mL. It means the organic matter in the soil, responsible for the strong yellow colour, is being flushed, as well as sulfolane. In that sense, it's important to consider that high organic load in the leachate can potentially affect pollutant degradation efficiency when conducting AOP in the presence of UV irradiation (Bandala et al, 2008). According to Figure 4, sulfolane concentration in the collected leachate dropped from 25000 mg/L to 200 mg/L after 68 mL of water had passed through the soil column. The equivalent sulfolane removal can be calculated from Equation [1].

$$[1]\ X_i(\%) = \frac{\int_0^{V_i} C\,dv}{\int_0^{V_T} C\,dv} \times 100$$

Where C (mg/L) is the concentration of sulfolane in the soil flushing solution; V_i (L) is the volume of water that passed through the soil column, V_T (L) is the total volume of water required to achieve 100% sulfolane removal, X_i (%) is the percentage of sulfolane removal from soil after V_i (L) of water passed through the column.

Figure 5 shows that more than 99% of sulfolane can be removed from the soil after flushing with 67 mL of water. The water/soil mass ratio to achieve 99% sulfolane removal is calculated to be 1.5, which is much lower than soil washing process. However, the flushing process is time-consuming for soils with low conductivity. In this study, it took more than 24 hours for 67 mL of water to pass through the soil column.

Color of water								
Cumulative volume of water flushed	2.0 ml	5.8 ml	8.9 ml	24.2 ml	67.6 ml	112.1 ml	136.9 ml	152.9 ml

Figure 3. Color change of soil flushing water

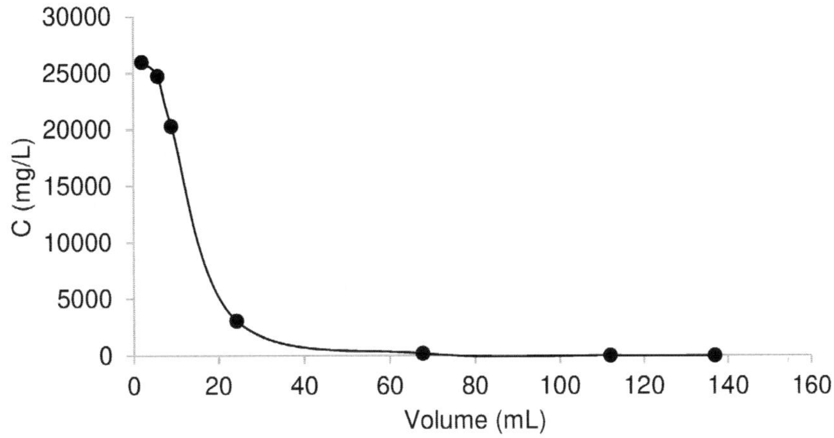

Figure 4. Change in sulfolane concentration in the flush water.

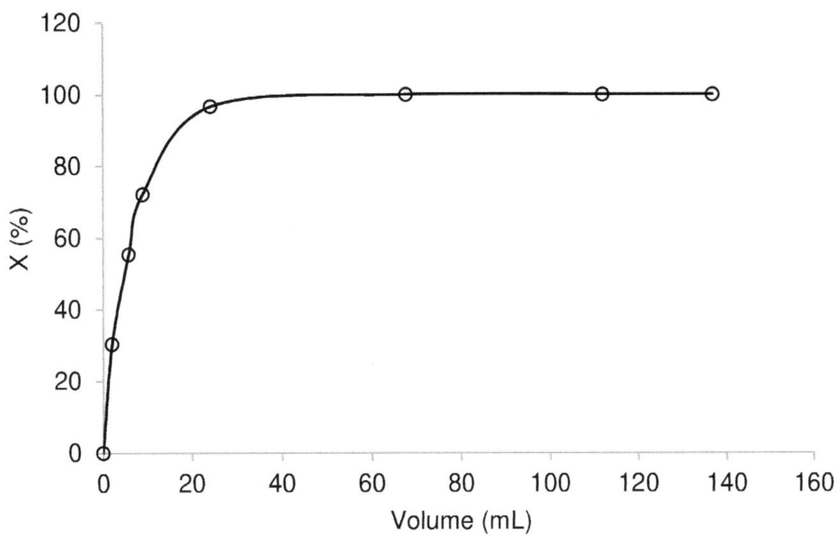

Figure 5. Percentage of sulfolane removal calculated based on Equation [1].

The collected leachate (soil flushing solution) was then exposed to UVC/H_2O_2/O_3 for degrading sulfolane. As a comparison, a similar experiment was conducted in ultrapure water and both results are reported in Figure 6. After 60 mins, more than 99% of sulfolane was degraded in the soil flushing samples. Sulfolane degradation in the ultrapure water was much faster, where it only took 30 mins to achieve 99% sulfolane removal. The soil flushing solution contains not only sulfolane, but all the inorganics and organic compounds found in the soil, such as metals, carbonates, humic substances and others. These co-eluents can reduce the UV light transmittance, consume oxidants and compete with hydroxyl radicals, all these leading to a less efficient sulfolane degradation.

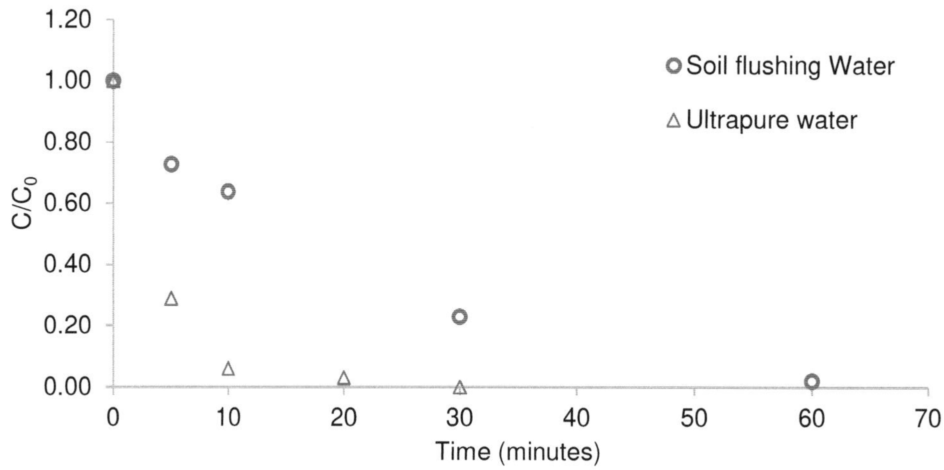

Figure 6. Degradation of sulfolane in the soil flushing solution and ultrapure water by using UV/H_2O_2/O_3; Light intensity = 5.3×10^{17} photon/s; O_3 flow rate = 3.7 g/h; H_2O_2 = 1000 mg/L

4 CONCLUSIONS

The soil washing/flushing processes integrated with AOPs are promising technologies to treat sulfolane contaminated soil. The specific conclusions drawn from this study are provided below:

- Three cycled soil washing can remove more than 99% of sulfolane from soil.
- 68 mL of water can remove more than 99% of sulfolane from 45 grams of soil.
- Alkaline ozonation yield the fastest sulfolane degradation in soil washing solution although it requires pH adjustment before and after treatment.
- UV combined with H_2O_2, ozone or H_2O_2/Ozone can completely degrade sulfolane from wash/flush water within one hour.
- Neutral Fenton reagent was not effective as a post soil washing/flushing treatment.
- Co-contaminants present in the soil washing/flushing samples decrease the AOP treatment efficiency.

Acknowledgements

The authors acknowledge the financial support provided by NSERC, MITACS and Bonavista Energy Corporation to conduct this research.

References

Andreozzi, R., Caprio, V., Insola, A. and Marotta, R. 1999. Advanced oxidation processes (AOP) for water purification and recovery. Catalysis Today, 53(1): 51-59.

Bandala, E.R., Velasco, R. and Torres, L.G. 2008. Decontamination of soil washing wastewater using solar driven advanced oxidation processes. Journal of Hazardous Materials, 160(2-3):402–407.

Esplugas, S., Bila, D.M., Krause, L.G.T. and Dezotti, M. 2007. Ozonation and advanced oxidation technologies to remove endocrine disrupting chemicals (EDCs) and pharmaceuticals and personal care products (PPCPs) in water effluents. Journal of Hazardous Materials, 149(3): 631-642.

Izadifard, M., Achari, G. and Langford, C.H., 2017. Degradation of sulfolane using activated persulfate with UV and UV-Ozone, Water Research, 125: 325-331.

Jafvert, C.T., Chu, W. and Van Hoof, P.L. 1995. A quantitative structure-activity relationship for solubilization of nonpolar compounds by nonionic surfactant micelles." Proc., Surfactant Enhanced Subsurface Remediation: Emerging Technologies, American Chemical Society, Washington, D.C., 24–37.

Juwarkar, A.A., Nair, A., Dubey, K.V., Singh, S.K. and Devotta, S. 2007. Biosurfactant technology for remediation of cadmium and lead contaminated soils. Chemosphere, 68(10):1996-2002.

Khan, J.A., He, X., Khan, H.M., Shah, N.S. and Dionysiou, D.D. 2013. Oxidative degradation of atrazine in aqueous solution by UV/H2O2/Fe2+, UV/Fe2+ and UV/Fe2+ processes: a comparative study, Chemical Engineering Journal, 218:376–383.

Kupuna, S., Palanisami, T., Megharaj, M., Venkateswarlu, K. and De Voogt, R.N. 2016 Ex-Situ Remediation Technologies for Environmental Pollutants: A Critical Perspective. Reviews of Environmental Contamination and Toxicology Volume 236. Springer International Publishing, 117-192.

Legrini, O., Oliveros, E. and Braun, A.M. 1993. Photochemical Processes for Water Treatment. Chemical Review, 93(2), 671-698.

Mao, X., Jiang, R., Xiao, W. and Yu, J. 2015. Use of surfactants for the remediation of contaminated soils: a review, Journal of hazardous materials, 285: 419-435.

Mehrabani, M., Yu, L., Achari, G., Langford, C.H., 2016. Mineralization of sulfolane by UV/O3/H2O2 in a tubular reactor, Journal of Environmental Engineering and Science, 11: 44-51.

Mulligan C.N., Yong R.N. and Gibbs B.F. 2001. Remediation Techonologies for metal-contaminated soils and groundwater: an evaluation. Engineering Geology, 60(1-4):193-207.

Parsons, S. 2004. Advanced Oxidation Processes for Water and Wastewater Treatment. IWA Publishing, London, UK.

Trellu, C., Mousset, E., Pechaud, Y., Huguenot, D., van Hullebusch E. D., Esposito, G. and Oturan M. A. 2016. Removal of hydrophobic organic pollutants from soil washing/flushing solutions: A critical review. Journal of hazardous materials, 306: 149-174.

Yu, L., Mehrabani-Zeinabad M., Achari G. 2016a. Application of UV based advanced oxidation to treat sulfolane in an aqueous medium. Chemosphere, 160:155-161.

Yu, L., Achari, G., Langford, C. H. and Ian K. 2016b. A feasibility study on sulfolane degradation in groundwater using neutral Fenton catalysts, CSCE Conference, London, Ontario, Canada, June 1-4, 2016.

Building Tomorrow's Society
Bâtir la Société de Demain

Fredericton, Canada
June 13 to 16, 2018 / *13 au 16 juin, 2018*

SUSTAINABLE ANAEROBIC MEMBRANE BIOREACTOR (ANMBR) TREATMENT OF INDUSTRIAL WASTEWATERS: FOULING CONTROL

Joshua J. Snowdon[1,2], and Kripa S. Singh[1]
[1] University of New Brunswick, Canada
[2] h557d@unb.ca

Abstract: An external anaerobic membrane bioreactor (AnMBR) equipped with a nanofiltration tubular membrane module was operated for 170 days treating a synthetic wastewater simulating a high strength industrial wastewater. The organic loading rate supplied to the AnMBR was in the range of 0.8 – 8.2 kg/m^3/d, with a mixed-liquor suspended solids (MLSS) concentration in the range of 13.6 – 33.9 g/L. The membrane module was operated for 170 days of continuous data collection, where the loading limitations of the reactor were gauged. Following operation, the membrane module underwent various chemical cleaning processes to alleviate membrane fouling and recover its flux. These successive chemical cleanings applied 5 different methodologies in an effort to analyze their efficiencies in membrane fouling removal. General results from the chemical cleanings concluded that a 1% sodium hydroxide and a 1% sodium hypochlorite solution, without backwashing processes, offered the greatest clean water flux recovery over all other chemical cleaning methodologies. The 10% citric acid cleans were found to decrease recoverable clean water flux. A better understanding of chemical cleaning protocols is required to reduce the quantities of chemical agents used and quantify the relationship between sequential chemical cleanings and reversible fouling removal efficiency. As membrane costs will always be an important constraint in AnMBR applications, prolonging the longevity of a given membrane module through efficient chemical cleaning processes is critical to their success.

1 INTRODUCTION

Sustainable wastewater treatment is important to protect environment and public health. Anaerobic membrane bioreactor (AnMBR) technology offers a promising, sustainable wastewater treatment alternative when it may be applied effectively. Combining anaerobic processes for the degradation of organic pollutants with the physical separation capabilities of membrane filtration provides a number of benefits over more traditional anaerobic and aerobic wastewater treatment technologies, and non-membrane integrated systems currently being used. These benefits are allocated in AnMBRs' generally lower operation and maintenance costs (due to zero oxygen requirements, and much lower excess sludge handling costs), very high quality effluent with the potential for recycle or reuse, and the production of value-added products such as biogas and select biochemicals such as volatile fatty acids, among other benefits (Dvorak et al. 2015, Evren Ersahin et al. 2016, Herrera-Robledo and Noyola 2015, Liao et al. 2007).

What have limited AnMBRs' widespread applications is disadvantages with regards to membrane fouling and membrane costs. Membrane costs have decreased significantly in recent years due to advances in new membrane or module materials and manufacturing technologies, however their costing will always remain an important consideration in large-scale wastewater treatment plant applications (Skouteris et al.

2012). Continuing to remain the most significant factor limiting membranes' cost-effectiveness, and ultimately their widespread application, is the issue of membrane fouling (Ramos et al. 2014 and Skouteris et al. 2012). Membrane fouling increases transmembrane pressure (TMP) and decreases flux, while the control of fouling increases manpower and energy requirements, demands costly chemical cleaning (which further requires waste handling), and may result in increased deterioration of the membrane material, affecting its lifespan (Ramos et al. 2014).

Fouling mitigation for a membrane module is a two-pronged approach: first, the membrane-fouling rate should be minimized and second, technologies to restore the flux of a fouled membrane must be implemented (Liao et al. 2007). After a membrane has undergone sufficient fouling to reach its TMP or flux thresholds, further physical and chemical cleaning alternatives may be applied to remove reversible fouling from the membrane module. Chemical cleaning alternatives are generally preferred, as the strong chemical agents within them may remove both surface and internal pore blockages effectively (Dvorak et al. 2015, Liao et al. 2007, Vera et al. 2014). There have been few comparative studies on different cleaning methods and conditions presented within literature (Ramos et al. 2014 and Zhou et al. 2014).

If chemical membrane cleaning processes could be enhanced and better understood they could help further justify AnMBR applications for wastewater processing. A better understanding of chemical cleaning protocols is required to reduce the quantities of chemical agents used (which incur additional operation and maintenance costs) and quantify the relationship between sequential chemical cleanings and reversible fouling removal efficiency. This research project compared the efficiency of 5 different chemical cleaning methodologies applied to a tubular nanofiltration membrane module following operation in a novel AnMBR system. The different chemical cleanings had their efficiencies analyzed with regards to their abilities to recover clean water flux and clean water operational permeability, two important operational parameters for AnMBR systems.

2 MATERIALS & METHODS

2.1 Reactor Configuration

The pilot-scale AnMBR system used for the experiment consisted of a 1000L stainless steel anaerobic continuous stirred tank reactor (CSTR), integrated with an external tubular nanofiltration membrane module. The reactor was maintained, on average, at 36.3°C (± 0.7°C), standard mesophilic conditions, for the duration of all data collection periods by a heat wrap surrounding the reactor. A simplified process flow diagram (PFD) of the experimental setup may be viewed in Figure 1.

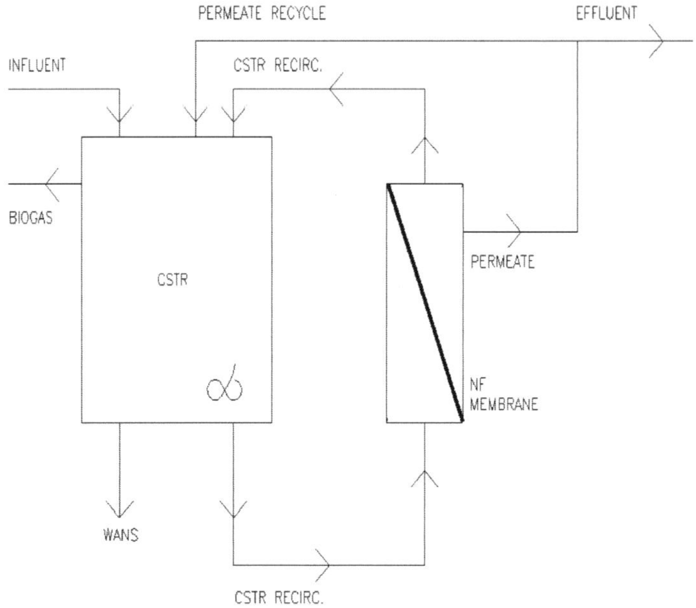

Figure 1: Simplified PFD of reactor configuration.

Influent wastewater was supplied to the reactor through two feed tanks, one containing a sugar-based solution, and the other containing a potato-starch, macronutrient, and alkalinity mixture. The influent wastewater peristaltic pump (Masterflex, Ontario) was set to automatically feed the system whenever the tank level fell below a pre-set threshold level. The tank level decreased through effluent flows and waste anaerobic sludge (WANS) flows. WANS was wasted from the system using an automatic peristaltic pump (Masterflex, Ontario) located at the base of the CSTR (set on a 2h loop, with the duration of waste flow being dependent on the desired mixed-liquor suspended solids concentration).

CSTR mixed-liquor was recirculated through the external nanofiltration membrane module by a progressive cavity pump (Nemo, New Brunswick), which acted as a continuous loop. The progressive cavity pump recirculated the mixed-liquor at a flow rate of 6000 L/h. This flow rate provided a reactor turnover time of 10 minutes, and a cross-flow velocity on the membrane surface of 2.5 m/s. The membrane module was cylindrical in shape, composed of polyvinylidene fluoride (PVDF), with an outer diameter of 50 mm and a total length of 1.436 m. The module contained 13 tubular membrane channels, each 8 mm in diameter, which resulted in a total membrane area of 0.42 m^2. The membrane surfaces had a mean pore size of 30 nm. The membrane module had backwash functionality present. Backwash functionality being present allowed the membrane module to switch from forward filtration flow to backward flow. This created a backwash that rinsed out the pores of the membrane using filtrate, or permeate.

Before passing into the membrane module, the CSTR mixed-liquor was forced through a stainless steel screen (having a pore size diameter of 0.8 mm, and being present to reduce contact between large particulate materials and the membrane surface, especially during initial seeding of the reactor). As the mixed-liquor moved through the membrane module an instantaneous pressure gradient, supplied through a peristaltic pump (Masterflex, Ontario) at the beginning of the permeate flow line, forced a portion of the mixed-liquor to be filtered through the membrane module. This filtrate, henceforth known as permeate, was recycled back into the CSTR through the permeate recycle line or collected for backwashing and effluent flows. Mixed-liquor that was not filtered through the membrane module was recycled back into the CSTR via the CSTR recirculation line.

Biogas exited the CSTR through a gas-trap at the top of the reactor and passed into a plastic gas collection bag, where the flow of biogas was regulated. The regulated outflow of biogas moved through a massflow meter (AALBORG, Orangeburg), where its total mass was measured and logged, before exiting the system. The massflow meter was calibrated with a headspace gas having a composition of 70% methane gas and 30% carbon dioxide gas. There was also a secondary gas trap attached to the system that wasted directly into the environment, in case of emergencies.

2.2 Wastewater Characterization

A potato starch-based wastewater, simulating a high-strength food industry wastewater, was supplied to the reactor through two feed tanks and two respective peristaltic pumps. One feed tank contained sucrose only, accounting for approximately 15% of the total feed volume and approximately 80% of the total feed chemical oxygen demand (COD) (approximately 59 g/L as COD). The second feed tank was continuously mixed and contained waste potato starch obtained from a potato processing plant, macronutrients, and a source of alkalinity. Macronutrient (nitrogen and phosphorous) requirements were calculated using the typical composition of a bacterial cell, $C_5H_7O_2N$, the rate of sludge wastage, and knowing the phosphorous requirement is approximately 1/7th of the nitrogen requirement, as mentioned in pertinent literature (Speece, 2008). Alkalinity dosages were adjusted to maintain reactor pH around a neutral pH. The potato starch mixture accounted for approximately 85% of the total feed volume and approximately 20% of the total feed COD (approximately 15 g/L as COD). A thorough overview of the wastewater composition may be viewed in Table 1. Micronutrients were bulk-dosed to the reactor once weekly on a per-gram COD supplied basis to meet minimum requirements for a high acetic acid utilization rate (Speece 2008).

Table 1: Characteristics of the synthetic industrial wastewater.

Parameter	Unit	Value	
		(Avg.)	(S.D.)
Chemical Oxygen Demand (COD)	g/L	74.3	± 9.7
Biochemical Oxygen Demand (BOD)	g/L	21.5	± 2.8
Total Suspended Solids (TSS)	g/L	6.2	± 2.3
Total Kjeldahl Nitrogen (TKN)	mg/L	630	± 420
Ammonia Nitrogen (NH_3-N)	mg/L	256	± 189
Total Phosphorous (TP)	mg/L	210	± 56
Phosphate Phosphorous (PO_4-P)	mg/L	222	± 55
Total Alkalinity	g-$CaCO_3$/L	5.1	± 1.7
pH	-	8.75	± 0.35

2.3 Seed Sludge Characteristics

The reactor was seeded with a granular sludge acquired from an anaerobic digester treating a fruit-juice processing plant wastewater in Lassonde, Quebec. The seed sludge had an initial total solids (TS) concentration of 23.2 g/L, total volatile solids (TVS) concentration of 19.8 g/L, total suspended solids (TSS) of 21.4 g/L, volatile suspended solids (VSS) of 19.5 g/L, and a specific methanogenic activity (SMA) of 0.18 g-COD_{CH4}/g-VSS/day.

2.4 Analytical Methods

COD, TSS, VSS, biochemical oxygen demand (BOD), and total alkalinity were conducted following Standard Methods (APHA 2005). Total phosphorous (TP), total kjeldahl nitrogen (TKN), phosphate-phosphorous, and ammonia nitrogen were conducted according to HACH Water Analysis Handbook (Hach Company 2002). The methane content of the biogas was measured using a Varian CP-3800 gas chromatograph. The chromatograph was fitted with an Alltech CTR 1 concentric packed column and a TCD detector maintained at a constant temperature of 180°C. Helium was used as a carrier gas operating at a flow rate of 30 mL/min. The temperature of the injection port was set to 120°C. Biogas injections were done in triplicates. SMA tests were conducted in duplicates using batch studies operated at 36°C. The tests were conducted in 500 mL serum bottles with a working volume of 450 mL. The bottles were filled with reactor sludge, acetic acid as a substrate, pH buffer, and mineral bases. Young and Cowan outlined the procedures and mixed solutions used (Young and Cowan 2004). All flow and pressure readings were measured once daily at similar times using inline flow meters and pressure sensors.

2.5 Experimental Design

The chemical cleaning study experimental design was created applying the pretest-posttest control group design methodology. This methodology demonstrated that an effect occurred only following the change in a certain variable from control conditions (Leedy and Ormrod 2005). Using the pretest-posttest control group design methodology, first an initial clean water flux for the membrane module was established on day one of operation, prior to the membrane coming in contact with the mixed-liquor. This was accomplished through filling the CSTR with water and operating the system in forward filtration mode with the membrane module in place, noting down the clean water membrane flux exhibited by the membrane module at various recirculation flow rates.

On day 170, the membrane module loop in the system was isolated from the mixed-liquor flow path, allowing a small 50L tank to be filled with clean water and filtered through the membrane module without requiring the CSTR to be emptied. Chemical solutions were added to this clean water, as required for the particular chemical cleaning methodology in place. A total of 5 chemical cleaning methodologies were applied to the membrane module over 5 days, leading to a total of 5 successive membrane module cleanings. The overview of each chemical cleaning methodology may be viewed in Table 2.

Table 2: Chemical cleaning methodologies for membrane module 1.

Membrane Clean No.	Cycle	Recirculation Time (min)	Soak Time (min)	No. of Backwashes and Duration (No. & min)	Chemicals Used
Clean 1	1	30	N/A	N/A	1% NaOH and Sodium Hypochlorite
	2	N/A	60	N/A	
	3	30	N/A	N/A	
Clean 2	1	30	N/A	N/A	10% Citric Acid
	2	N/A	60	N/A	
	3	30	N/A	N/A	
Clean 3	1	90	N/A	6 and 2 min	1% NaOH
	2	N/A	60	N/A	
	3	30	N/A	1 and 2 min	
	4	48	N/A	3 and 2 min	2% Citric Acid
	5	N/A	30	N/A	
	6	45	N/A	3 and 2 min	
	7	45	N/A	3 and 2 min	1% NaOH and Sodium Hypochlorite
	8	N/A	30	N/A	
	9	45	N/A	3 and 2 min	

Clean 4	1	45	N/A	3 and 2 min	10% Citric Acid
	2	N/A	30	N/A	
	3	45	N/A	3 and 2 min	
Clean 5	1	60	N/A	N/A	1% Sodium Hypochlorite
	2	N/A	45	N/A	
	3	60	N/A	N/A	

In Table 2 the recirculation time is the time that the membrane module was operated in forward filtration mode, where the water-chemical mixtures were filtered through the membrane module in the same forward direction as permeate. The soak time is the time that filtration was ceased, and the membrane module remained in contact with the water-chemical mixtures. The number of backwashes and duration show the frequency and duration at which filtration was switched from forward filtration mode to backwards filtration mode (backwashing) during a cleaning cycle. Finally, the chemicals used column details the exact concentrations of chemicals within the clean water that were used for each cycle of the chemical cleaning.

Following each chemical cleaning, the 50L tank was filled with clean water and operated in forward filtration mode at recirculation flow rates similar to the control conditions. The clean water fluxes and TMPs achieved at each of these recirculation flow rates were recorded, and compared back to the control. This allowed the clean water flux and clean water operational permeability recovery of each chemical cleaning methodology to be quantified to identify which chemical cleaning methodology was the most efficient in recovering clean water flux and operational permeability. Operational permeability is defined as membrane flux divided by instantaneous operational TMP, or in other words, pressure-averaged flux (Hamdan de Andrade et al. 2013). Operational permeability was considered so that the ability of each clean to recover a favourably low TMP could be factored into the flux analysis.

3 RESULTS & DISCUSSION

3.1 Recovery of Clean Water Flux

The average clean water flux recovery for each chemical cleaning methodology applied to the membrane module may be viewed in Figure 2.

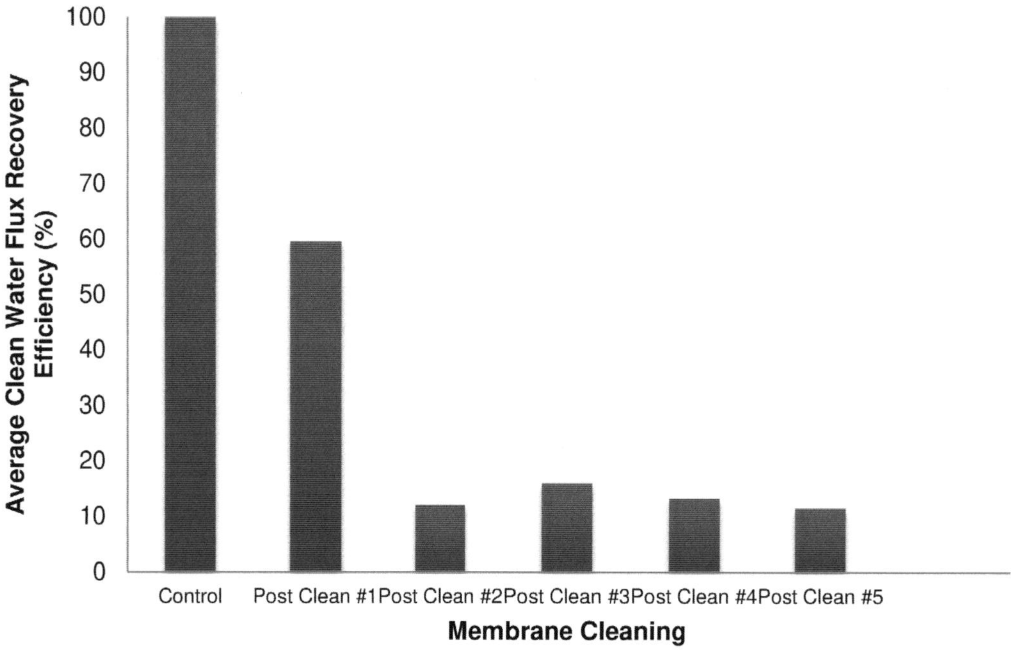

Figure 2: Average clean water flux recovery following each clean for the membrane.

The first chemical cleaning applied to the membrane module offered the only substantial recovery in clean water flux, with a recovery of 60%. Chemical cleaning 1 consisted of a 30 min recirculation phase, a 1h soak, and a second 30 min recirculation phase using a mixed 1% sodium hydroxide and 1% sodium hypochlorite solution. It is likely that the caustic solutions were successfully able to weaken the bonds between foulant materials and the membrane surface. Since the majority of membrane foulants would have most likely been organic in nature (due to a high mixed-liquor volatility of 91% and a high feed wastewater volatility of 99%), sodium hydroxide and sodium hypochlorite would have been highly effective in removing them (Liao et al. 2007). Rinsing with tap water following the completion of the chemical cleaning was then able to remove the loosened foulants from the membrane module effectively.

The second chemical cleaning methodology applied consisted of a 30 min recirculation phase, a 1h soak, and a second 30 min recirculation phase using a 10% citric acid solution. Applying this acidic clean on the day following the first clean lead to a decline in membrane flux recovery, decreasing to 12% of the membranes' initial clean water flux. Citric acid is generally used to remove inorganic fouling, and therefore was not anticipated to provide much more flux recovery than the prior caustic clean, however such a drastic decline in flux recovery was not anticipated. Vera et al. used a combined 0.05% sodium hypochlorite, 0.6% citric acid, and 0.05% sodium hypochlorite chemical cleaning on a PVDF hollow-fibre membrane and found that the citric acid portion of the cleaning did remove a portion of the membrane fouling, leading to an increased recovery (Vera et al. 2014). Considering that cleans 1 and 2 were identical in methodology other than the chemical solutions used, the concentration of citric acid may have been sufficiently high that it damaged the membrane surface, and resulted in the decline in chemical cleaning efficiency. A surface investigation was unable to be performed following clean 2 as it would have required the membrane module be destroyed, which would not have allowed cleans 3 – 5 to be applied.

Following the citric acid cleaning, in an effort to improve the membrane module's clean water flux recovery cleans 3 – 5 were applied, and offered clean water flux recoveries of 16%, 13%, and 12%, respectively. Clean 3, which incorporated sodium hypochlorite, sodium hydroxide, and citric acid cleaning phases as well as backwashing processes recovered only 4% more clean water flux following clean 2. However, if clean 2 permanently damaged the membrane surface then the ability of this cleaning methodology to offer an improvement in clean water flux, overall, lead the authors to believe it could be a viable option for future cleaning studies. Nonetheless, comparing the recovery of clean 3 to that of clean 1, it is apparent that clean 1 outperformed clean 3 significantly. Both of these cleans used similar chemical solutions, in similar concentrations, however clean 1 offered longer soak times, and no backwashing processes, making it a more static cleaning procedure, whereas clean 3 incorporated backwashing and had longer recirculation times. Clean 1 offering a significantly greater clean water flux recovery suggests that a static clean is superior to a more dynamic clean under the given operating conditions and membrane characteristics. Recent results have demonstrated that the soak time, or contact time, between a sodium hypochlorite solution and the membrane surface has a larger impact on clean water flux recovery than adjusting the actual concentration of the cleaning solution used (Wang et al. 2017).

Clean 4, incorporating a 10% citric acid solution once again, resulted in a decline in clean water flux recovery from clean 3. This result again suggests that the citric acid solution may have been sufficiently high in concentration that it had a detrimental impact on the membranes' surfaces. Furthermore, clean 4 utilized the same chemical solution as clean 2, however incorporated a longer recirculation phase and backwashing processes. Since clean 4 offered only a 1% increase in clean water flux recovery over clean 2, it may be concluded that shifting the cleaning methodology from a static clean to a dynamic clean offered no worthwhile improvement in clean water flux recovery. It appears that the actual cleaning solution, with preference being given to caustic solutions in this project, holds a greater influence on cleaning performance than the type of cleaning methodology applied.

Finally, clean 5 applied a 1% sodium hypochlorite cleaning solution with a longer recirculation time and shorter soak time compared to clean 1. Additionally, no backwashing processes were incorporated. Following clean 5 the clean water flux recovery was only 12%, making it one of the lowest performing chemical cleans. Assuming the membrane was not permanently damaged following clean 2, clean 5 performing much worst than clean 1 suggests that the 1% sodium hydroxide solution used in clean 1 may have been the cleaning solution that was the most effective in removing membrane foulants, which is why it had the best performance in terms of clean water flux recovery. Clean 3 also utilized 1% sodium hydroxide, and was found to have the second highest clean water flux recovery. Additionally, these results suggest that an increased soak time is superior to an increased recirculation time. The bonds created between the foulants and the membrane surface may require a longer exposure to the cleaning solutions for them to be broken and removed from the membrane surfaces effectively.

3.2 Recovery of Clean Water Operational Permeability

The average clean water operational permeability recoveries for each chemical cleaning methodology applied to the membrane module may be viewed in Figure 3.

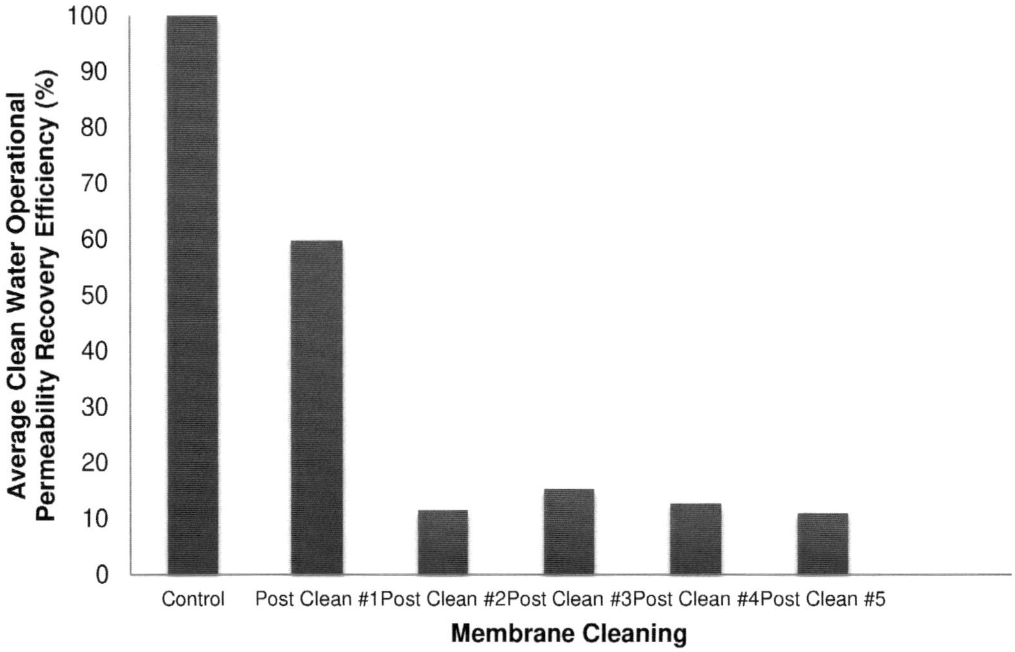

Figure 3: Average clean water operational permeability recovery following each clean for membrane 1.

A similar trend to that of the clean water flux recoveries occurred when examining the operational permeability recovery results whereby clean 1 was the most effective in average clean water permeability recovery (average recovery of 60%), clean 2 resulted in a drastic decline in clean water operational permeability recovery (average recovery of 12%), and clean 3 recovered a small portion of operational permeability (average recovery of 15%). From these results it was concluded that the clean 1 methodology was the most promising, since clean 1 offered a good initial average clean water flux and average clean water operational permeability recovery. Clean 3 showed a promising chemical cleaning alternative, as it was able to increase the membranes' clean water flux recovery following the 10% citric acid clean. Overall, since all operational permeability recoveries were just as effective as their clean water flux recovery counterparts, it may be concluded that all chemical cleanings were effective in recovering TMP as well.

3.3 Reactor Performance

The AnMBR system operated for 170 days prior to the membrane being taken offline treating organic loading rates (OLRs) in the range of 0.8 – 8.2 kg/m^3/d. The OLRs fluctuated greatly during this portion of data collection as the loading capacity of the reactor was being gauged. The mixed-liquor suspended solids fluctuated from a lowest value of 13.6 g/L to a highest value of 33.9 g/L. Even through these large fluctuations in OLR the AnMBR system was able to achieve an average COD removal efficiency of 99.5% (± 0.6%), indicating very good COD removal rates. The TSS removal efficiency remained above 99% throughout the entirety of operation, indicating near 100% solid-liquid separation achieved by the membrane module. The average methane yield remained close to the theoretical value of 0.395 m^3-CH_4/kg-COD throughout this period of operation as well, which signified that the majority of COD removed through biological processes was converted to biogas effectively.

4 CONCLUSIONS

Through this study an AnMBR system integrated with an external nanofiltration tubular membrane module was able to effectively treat a synthetic wastewater simulating a high-strength food processing wastewater. Several chemical cleaning methodologies were applied to the fouled membrane module following 170 days of continuous operation. A summary of the major conclusions is provided:

- A primarily static chemical cleaning using a mixed 1% sodium hydroxide and 1% sodium hypochlorite solution was able to achieve the highest clean water flux and clean water operational permeability recovery efficiencies.
- A chemical cleaning using a 10% citric acid cleaning solution may cause permanent damage to the membranes' surfaces. Even if permanent damage was not done, the membrane module's clean water flux recovery efficiency dropped by 48% between cleans after being exposed to the citric acid cleaning solution, and was not able to recover more than 16% clean water flux after exposure.
- For chemical cleanings, in general, it appears that a higher soak time is more favourable than a higher recirculation time or the incorporation of backwashing processes.
- Overall, it appears that the 1% sodium hydroxide cleaning solution was the most effective cleaning solution for removing membrane foulants.

Acknowledgements

The authors would like to acknowledge and thank the following funding agencies and industrial collaboration: the Natural Science and Engineering Research Council (NSERC), the New Brunswick Innovation Foundation (NBIF), and ADI Systems Inc. They would also like to thank Gustavo Zanatta, Jonathan Dargavel, and Dr. Dennis Connor for their help and contributions with data collection and testing throughout the duration of the study.

References

American Public Health Association (APHA). 2005. *Standard methods for the examination of water and wastewater*, American Public Health Association, Washington, DC, USA.

Dvorak, L., Gomez, M., Dolina, J., and Cernin, A. 2015. Anaerobic membrane bioreactors – A mini review with emphasis on industrial wastewater treatment: Applications, limitations, and perspectives. *Desalination and Water Treatment,* **57**(2016), 19062 – 19076.

Evren Ersahin, M., Tao, Y., Ozgun, H., Gimenez, J. B., Spanjers, H., and van Lier, J. B. 2016. Impact of anaerobic dynamic membrane bioreactor configuration on treatment and filterability performance. *Journal of Membrane Science,* **526**(2017), 387 – 394.

Hach Company. (2002). *Water analysis handbook.* Hach Company, Loveland, CO, USA.

Hamdan de Andrade, L., dos Santos Mendes, F. D., Cawettiere Espindola, J., and Santos Amaral, M. C. 2013. Internal versus external submerged membrane bioreactor configurations for dairy wastewater treatment. *Desalination and Water Treatment,* **52**(2014), 2920 – 2932.

Herrera-Robledo, M., and Noyola, A. 2015. The evolution of pore-blocking during the ultrafiltration of anaerobic effluent-like mixtures. *Separation and Purification Technology,* **147**(2015), 172 – 178.

Leedy, P., and Ormrod, J. 2005. *Practical research. Planning and design.* Pearson, New Jersey, USA. Chap. 10, 217 – 244.

Liao, B. Q., Kraemer, J. T., and Bagley, D. M. 2007. Anaerobic membrane bioreactors: Applications and research directions. *Critical Reviews in Environmental Science and Technology,* **36**(6), 489 – 530.

Ramos, C., Zecchino, F., Ezquerra, D., Diez, V. 2014. Chemical cleaning of membranes from an anaerobic membrane bioreactor treating food industry wastewater. *Journal of Membrane Science,* **458**(2014), 179 – 188.

Skouteris, G., Hermosilla, D., Lopez, P., Negro, C., Blanco, A. (2012). Anaerobic membrane bioreactors for wastewater treatment: A review. *Chemical Engineering Journal*, 198-199, 138 – 148.

Speece, R. E. (2008). *Anaerobic biotechnology and odor/corrosion control for municipalities and industries.* Archae Press, Nashville, TN, USA. Chap. 16, 414.

Vera, L., Gonzalez, E., Diaz, O., and Delgado, S. 2014. Application of a backwash strategy based on transmembrane pressure set-point in a tertiary submerged membrane bioreactor. *Journal of Membrane Science,* **470**(2014), 504 – 512.
Wang, Q., Zeng, H., Wu, Z., and Cao, J. 2017. Impact of sodium hypochlorite cleaning on the surface properties and performance of PVDF membranes. *Applied Surface Science*, **428**(2018), 289 – 295.
Young, J., and Cowan, R. 2004. *Respirometry for environmental science and engineering.* SJ Enterprises, Springdale, AR, USA. Chap. 4, 4-17 – 4-18.
Zhou, Z., Meng, F., Lu, H., Li, Y., Jia, H., and He, X. 2014. Simultaneous alkali supplementation and fouling mitigation in membrane bioreactors by on-line NaOH backwashing. *Journal of Membrane Science*, **457**(2014), 120 – 127.

Building Tomorrow's Society
Bâtir la Société de Demain

Fredericton, Canada
June 13 to 16, 2018 / *13 au 16 juin, 2018*

THE EFFECT OF AN ACTIVATED CARBON (GAC) FILTER ON THE MICROBIAL QUALITY OF WATER

Behboud ahani, Mojtaba[1,2]
[1]Bandar-Anzali Branch, Islamic Azad University, Gilan, Iran
[2]mojtababehboudahani@gmail.com

Abstract: Water is a component of the environment, that its quality is influenced by the health of environment. On the other side, improving environmental health is also directly related to the quality and quantity of water. A healthy water resource must be able to supply safe and sufficient water for the society. For this purpose, the resources have to be under monitoring and control of health responsible regularly; so that by being aware of the presence of any potential contaminants in water, control measures should be applied. Provision of safe water and proper management of wastewaters play the major role in reducing water-related infectious diseases. A set of measures on improving the environment and the health, safe and adequate water supply can reduce 40 to 100% of water related communicable diseases. In this study, the effect of active carbon, as an actual absorbent, on the elimination of turbidity, color, taste and smell has been investigated in order to improve the quality of drinking water. The results of the study indicated that active carbon, as a strong absorbent with high porosities and broad lateral surface is effective to eliminate turbidity, color, and smell.

Keyword: Water, Activated carbon, Microbial Quality, Drinking Water Treatment, Adsorption processes

1 INTRODUCTION

Water is a component of the environment, that its quality is influenced by the health of environment. On the other side, improving environmental health is also directly related to the quality and quantity of water. A healthy water resource must be able to supply safe and sufficient water for the society. For this purpose, the resources have to be under monitoring and control of health responsible regularly; so that by being aware of the presence of any potential contaminants in water, control measures should be applied. Provision of safe water and proper management of wastewaters play the major role in reducing water-related infectious diseases. A set of measures on improving the environment and the health, safe and adequate water supply can reduce 40 to 100% of water related communicable diseases. Seemingly healthy and clear water may contain a variety of pathogenic live microorganisms and therefore we must ensure the health of water before drinking (Crittenden, Weber 1978).

Water treatment is needed to produce drinking water from sources contaminated with pathogenic micro-organisms, anthropogenic chemicals and/or containing an elevated concentration of natural organic matter (NOM) and/or inorganic compounds impacting drinking water quality, both from a public health point of view and/or its aesthetic appreciation, e.g. color, taste, odor and turbidity (Clifford 1983).

Supplying healthy and sufficient water in war zones and emergency situations is also very important; in this regard, proper selection of water resources and making necessary plans to protect water resources and distribution network as well as conducting refinery ways and disinfection methods on water as required process of refinement are among the measures of effective prevention of transferring diseases related to the water and providing the conditions for eliminating possible pathogenic factors (Femogarcia, Utrina 1988).

GAC is typically used in a water treatment plant after the coagulation and sedimentation processes and, commonly, following preliminary disinfection steps during which chemical reactions can occur. Moreover, water is often disinfected before it passes through the GAC adsorbers in order to prevent nuisance biological growths. In many instances, the activated carbon functions as a granular filter medium for removing particulates, although in a few cases in the United States and in most instances in Europe the GAC adsorbers are preceded by filters for particulate removal. Water is usually passed downward through packed beds of GAC. Regeneration of GAC is not generally practiced at water plants in the United States as it is in Europe. If the objective of GAC use is to include the removal of organic compounds in addition to those that cause taste and odor, regeneration is likely to become more common in the United States. The type of contactor selected for the GAC will be influenced by the frequency of regeneration. After treatment of a water supply with GAC, post disinfection is generally used to reduce the total number of bacteria, some of which may be present because of the microbial growths in adsorbers. Sufficient disinfectant is usually applied to ensure a residual in the distribution system to prevent contamination of the water (Hassler 1963).

1.1 Problem Statement

Reference to water treatment methods in medical documents of the past times is indicative of the fact that there is a direct relationship between cleanliness of water and human health. Hippocrates who is considered as the father of medical science says: the person who intends to competently investigate and search in medicine must consider the water consumed by residents of an area since water is of high importance in humans' health. Historical resources and images obtained show that since 2000 B.C. water treatment for drinking has been common (Hutton, Cheremision1978).

GAC filtration is applied in water treatment in the world for the removal of organic micro pollutants and for the reduction of color, and in combination with advanced oxidation for the improvement of the biological stability of finished water. The efficiency of GAC filtration is limited, and GAC regeneration is needed when breakthrough of a target compound is reached22 Microbial activity in GAC filters in drinking water treatment [7]. Thermal regeneration of GAC is expensive, and the costs of the annual regeneration of GAC at location Weesperkarspel of Amsterdam Water Supply comprise 46% of the total operational costs for GAC filtration, and 15% of the drinking water production costs (Mckinny, Roos 2004).

After several decades of research, the relationship between microbial activity in GACFs and the surface characteristics of GAC is still not elucidated. A research project was defined aiming at collecting such knowledge in a series of investigations including observations in full-scale treatment plants, laboratory experiments, and pilot plant studies. For this purpose, both adsorption processes (R. van der Aa, TU Delft) and microbiological processes were studied. The investigations described in this thesis aim at analyzing the microbiological processes occurring in GACFs used in water treatment. In particular the contribution of bacteria to the removal of organic compounds is studied. To elucidate the contribution of the bacterial activity to the effects of GAC filtration on the removal of NOM from the water the study focused on the quantification and the characterization of the bacterial community in GACFs (Nawwar 1989).

The following goals of this investigation have been defined:

1. Development and evaluation of a method for measuring the concentration of active microorganisms in GAC filter beds;
2. Characterization of the bacterial activity in GACFs, which includes:
    a. Isolation and identification of predominating bacterial species;
    b. Elucidation of physiological properties of the predominant species

3. Analysis of the interactions between process conditions and the microbiological processes during GAC filtration.

1.2 Characteristics of Activated Carbon

The ability of activated carbon to remove organic compounds from water is based on the properties of this material. These properties include: (i) a high specific surface area and (ii) specific chemical surface characteristics including hydrophobicity and presence of functional groups.

The specific surface area varies between 400 and 1500 m2 g-1 and is a result of the high porosity of GAC. The shape, size, volume and surface area of the pores and their spatial distribution within the carbon particle characterize the porosity of GAC (Sarai, 2006).

1.3 Adsorption

Adsorption is a process in which a molecule (adsorbate) from the liquid or gas phase is bound to the surface of a solid phase (adsorber) that is in contact with the liquid or gas. In case of *physical adsorption* van der Waal's and electrostatic interactions are responsible for the partitioning of the adsorbate between two phases. Non-polar compounds or hydrophobic parts of complex molecules are attracted to the lipophilic surface of GAC and regions of molecules with electrostatic charge are attracted to surface oxides with the opposite charge. Although physical adsorption is generally considered a reversible process, desorption of complex molecules is unlikely to occur due to the multiple binding by different mechanisms (Simsek 1970).

1.4 Powder of Activated Carbon (PAC)

Manufactured by direct activation, powdered activated carbons have been specifically developed for the removal of a broad range of organic contaminants from potable, waste and process waters. Powdered activated carbon (PAC) has a relatively smaller particle size when compared to granular activated carbons and consequently, presents a large surface to volume ratio.

As such, PAC is not commonly used in a dedicated adsorber vessel, due to the high head loss that would occur. Instead, PAC is generally added directly to other process units, such as raw water intakes, rapid mix basins, clarifiers, and gravity filters.

PAC is made from organic materials with high carbon contents such as wood, lignite and coal. PAC typically has a diameter less than 0.1 mm and an apparent density ranging between 23 and 46 lb/ft3, depending on the material used and manufacturing process. Iodine and molasses numbers are typically used to characterize PAC. These numbers describe the quantity of small and large pore volumes in a sample of PAC. A minimum iodine number of 500 is specified for PAC by AWWA standards.

PAC is used by water treatment plants on either a full time basis or as needed for taste and odor control or removal of organic chemicals. PAC is can be fed as a powder using dry feed equipment or as a slurry using metering pumps. PAC can also be mixed with water and fed as a slurry. Slurry systems are normally used when PAC is frequently added and the required dosages are high. Slurry systems usually include a storage tank, day tank, and a chemical feeder (either a diaphragm pumps or rotary feeders) (Smethurst 1979).

1.5 Removal of Pesticides

Treatment with the activated carbon is the only treatment method effective for removal of pesticides such as aterazin and other artificial organic chemical compounds of water and based on the view of EPA, GAC is the best technology available for removal of fugitive and artificial organic materials from water. Using the 12-18 mg/l dose, the amount of atrazin can be reduced from 8-9 ppb (g/lm) to the amount of 3 ppb (MCL or maximum level of aterazin concentration)

1.6 Factors that Affect the Performance of Activated Carbon

1.6.1 Molecular Weight

As the molecular weight increases, the activated carbon adsorbs more effectively because the molecules are lea soluble in water. However, the pore structure of the carbon must be large enough to allow the molecules to migrate within. A mixture of high and low molecular weight molecules should be designed for the removal of the more difficult species.

1.6.2 pH

Most organics are less soluble and more readily adsorbed at a lower PH. As the pH increases, removal decreases. A rule of thumb is to increase the size of the carbon bed by twenty percent for every pH unit above neutral (7.0).

Contaminant concentration: The higher the contaminant concentration, the greater the removal capacity of activated carbon. The contaminant molecule is more likely to diffuse into a pore and become adsorbed. As concentrations increase, however, so do effluent leakages. The upper limit for contaminants is a few hundred parts per million. Higher contaminant concentration may require more contact time with the activated carbon. Also, the removal of organics is enhanced by the presence of hardness in the water, so whenever possible, place activated carbon units upstream of the ion removal units.

1.6.3 Particle Size

Activated carbon is commonly available in 8 by 30 mesh (largest), 12 by 40 mesh (most common), and 20 by 50 mesh (finest). The finer mesh gives the best contact and better removal, but at the expense of higher pressure drop. A rule of thumb here is that the 8 by 30 mesh gives two to three times better removal than the 12 by 40, and 10 to 20 times better kinetic removal than the 8 by 30 mesh.

1.6.4 Flow Rate

Generally, the lower the flow rate, the more time the contaminant will have to diffuse into a pore and be adsorbed. Adsorption by activated carbon is almost always improved by a longer contact time. Again, in general terms, a carbon bed of 20 by 50 mesh can be run at twice the flow rate of a bed of 12 by 40 mesh, and a carbon bed of 12 by 40 mesh can be run at twice the flow rate of a bed of 8 by 30 mesh.

1.6.5 Temperature

Higher water temperatures decrease the solution viscosity and can increase die diffusion rate, thereby increasing adsorption. Higher temperatures can also disrupt the adsorptive bond and slightly decrease adsorption. It depends on the organic compound being removed, but generally, lower temperatures seem to favor adsorption (Snoeyink 1987).

2 2. MATERIALS AND METHODS

2.1 Method Development

Method development was conducted in four phases: (i), determination of the effects of ultrasound on the ATP concentration and cultivability with selected pure cultures of aquatic bacteria (Spirillum sp., strain NOX and Pseudomonas fluorescens, strain P17) and suspensions of mixed bacterial communities detached from GAC samples. The suspensions were treated (10 times) with ultrasound and after each treatment ATP and cultivability were determined. Untreated suspensions served as a control; (ii), determination of the effect of the adsorptive properties of GAC and the release of fines on the concentration of free ATP. Free ATP was added to a suspension of virgin GAC or GAC with developed biomass and incubated for two hours. A solution of free ATP in autoclaved tap water was used as control (iii), testing of the efficiency of the sonication treatment to remove attached bacteria with bacterial biofilms on plasticized polyvinylchloride (PVCp), GAC and sand. The materials were subsequently treated up to 20 times with ultrasound at different power inputs and after each treatment biomass removal was measured as the ATP concentration. The ATP yield was plotted against the number of treatments; (iv), from the results of these experiments an optimal method for the removal and measurement of active biomass was defined and applied to a series of samples from different filter beds.

2.2 Samples from Filter Beds

Samples of GAC, sand and anthracite were obtained from nine water treatment plants in The Netherlands. Filter material (10-100 g) was collected from below the filter bed surface (2-10 cm) with a properly cleaned multisampler (Eijkelkamp) and stored in 100 cm3 screw-capped borosilicate flasks in the filtrate of the sampled unit at 4°C for a maximum period of 36 hours before the analysis. The specific density of the filter material was calculated from the weight of 50 cm3 of GAC, sand or anthracite that had been dried for four hours at 105° C. The volume is determined with a measuring cylinder (100 cm3).

2.3 Calculation of External Surface of GAC

Data for the external surface of the GAC particles assuming spherical shapes were derived from the literature 1The estimation of the surface of the pores with a diameter > 1 µm is based on data provided by the manufacturer about GAC porosity as based on penetration of butan at increasing pressure and the assumption that pores had a cylindrical shape.

2.4 Test Strains, Mixed GAC Community, and Stationary Bioflm

Spirillum sp. strain NOX and Pseudomonas fluorescens strain P17 were obtained from the stationary growth phase of AOC tests in treated water as described earlier (Snoeyink 1987).

A mixed microbial community as present on GAC was obtained with a two minutes low energy Optimisation and significance of ATP analysis 35 sonication (LES) treatment of a GAC sample (2 g wet weight) in 50 cm3 autoclaved tap water. Subsequently, the suspended bacteria were separated from the GAC particles and kept in autoclaved tap water for 4 hours at 20°C to allow adaptation to the new environment. A biofilm was grown on pieces (20 cm2) of plasticized PVC (PVC-P) incubated for 10 days at 15 °C in 600 cm3 of tap water inoculated with 1 cm3 of filtered river water (1.2 µm membrane filter) and enriched with nitrogen and phosphorus to prevent growth limitation (Sontheimer, Crittenden 1988).

2.5 Ultrasonic Treatment

A total of two to five gram of wet filter material was added to 50 cm3 of autoclaved tap water (pH 8.4±0.2) in a 100 cm3 screw-capped borosilicate flask. Sonication was applied for two minutes. A volume of five cm3 of the obtained suspension was collected and kept on ice for examination. The surplus liquid was decanted, 50 cm3 autoclaved tap water was added and sonication was repeated. This procedure was repeated several times depending on the objective of the experiment. Low-energy sonication (LES) was applied with a Branson sonication unit 5050 at a constant frequency of 43 kHz and 180 W power output. Samples contained in 50 cm3 of autoclaved tap water were placed in the sonication chamber with 10 L of demineralised water. High-energy sonication (HES) was applied with a Sonifier II W-250 at a constant frequency of 20 kHz and an adjustable power output. Ultrasound was applied to the sample via a titanium microtip (D = 5 mm), with power inputs ranging from 10 to 40 W.

2.6 ATP Measurement

The measurement of ATP is based on the production of light in the luciferine-luciferase assay. ATP was released from suspended cells with nucleotide-releasing buffer (NRB, Celsis). The intensity of the emitted light was measured in a luminometer (Celsis AdvanceTM) calibrated with solutions of free ATP (Celsis) in autoclaved tap water following the procedure as given by the manufacturer. The detection limit was 1 ng ATP L-1 of sample. Standard additions of free ATP dissolved in autoclaved tap water were used for recovery experiments. The influence of the adsorptive properties of GAC on the ATP measurement was tested by addition of free ATP to suspensions of virgin and preloaded GAC in autoclaved tap water (0.04 - 0.1 g GAC cm-3, dry weight). The concentrations of added ATP were of 2, 100, 600 and 3000 ng L-1 respectively.

Autoclaved tap water with similar concentrations of free ATP was used as control. Recovery of added ATP was calculated from the measured ATP concentrations in the controls and in samples, which were corrected for the concentration of autochthonous ATP.

2.7 Total Direct Cell Count (TDC)

A defined volume of the microbial suspension as obtained by sonication of GAC samples was filtered through a 0.22 µm polycarbonate filter, stained with acridine orange as described by36 Microbial activity in GAC filters in drinking water treatment [15] and observed with epifluorescence microscopy.

2.8 Heterotrophic Plate Counts (HPC)

Volumes of 0.05 cm3 of appropriate decimal dilutions of microbial suspensions obtained by sonication of GAC, sand or anthracite in autoclaved tap water were spread in triplicate over the surface of R2A agar (Oxoid Ltd.) plates, which were incubated during 10 days at 25° C.

2.9 Statistical Analysis

The statistical significance of the effects of ultrasonic treatment on microbial ATP and cultivability was determined by a t-test after correction of the data for the values measured in the controls.

3 3. RESULTS AND DISCUSSION

3.1 Interaction Between Microbial Activity and Adsorption

Biodegradability of Organic Compounds

For the purposes of discussing the interactions between biological and physicochemical processes occurring on the GAC surface, it is convenient to divide organics into the following four categories:

1. nonadsorbable, nonbiodegradable
2. nonadsorbable, biodegradable
3. adsorbable, nonbiodegradable
4. adsorbable, biodegradable

While simple, this classification scheme is ambiguous because the term nonbiodegradable is not easily defined. Instead, Alexander has adopted the adjective recalcitrant to describe substances that persist for extended periods under all environmental conditions thus far tested. Recalcitrant organics include those that are degraded slowly or not at all in nature. Occasionally, some organics that are usually biodegradable are recalcitrant due to specific environmental causes. Alexander outlined the conditions required for biodegradation and listed 15 possible mechanisms to explain recalcitrance. While some of these may be overcome by the environment in a GAC bed, others, such as accessibility of substrate and low concentration of substrate, may not. Attempts have been made to relate biodegradability to chemical structure. However, general rules are difficult to propose. Of all of the organic chemicals that are potentially harmful to health, only benzene was included in an extensive review of biodegradability studies made by Ludzack and Ettinger, who concluded that benzene was resistant to biodegradation. Helfgott et al. (1977) prepared a relative index for organics that are tested. Included were some organics with potential health effects, i.e., benzene, chloroform, DDT, and vinyl chloride. They considered all four compounds to be resistant to biodegradation, but benzene has subsequently been shown to be biodegradable (Weber 1970).

Pitter evaluated 123 organics in the aliphatic, cycloaliphatic, and aromatic classes. Of these, 21 organics were described as "biologically hard to decompose," and all were in the aromatic class. Included among these organics were dinitrobenzene's and phenols, phenyldiamines, trichlorophenol, nitroanilines, and naphthylamines. In all of the studies that were found in the literature, the experimental conditions were not those to be expected in treatment of drinking water. Most important, the concentrations of organics were in milligrams per liter rather than in micrograms per liter, and the concentrations of biomass were more consistent with biological waste treatment. The organics that are most likely to be attacked microbially fall in the nonadsorbable, biodegradable category. In treatment with GAC, these organics could be removed by the biofilm surrounding the carbon granules without affecting the adsorption process. The absorbable, biodegradable category of organics is of more interest because of the strong interaction expected between biological and adsorption processes. This sets GAC apart from sand or other inert media. Various researchers describe the removal mechanism as biodegradation in the biofilm, followed by adsorption of remaining substrate in the internal pore structure. Ying and Weber have adopted the term bio sorption to describe this process. The resulting mathematical model, which has been applied to concentrations of organics as high as those found in wastewater, indicates that adsorption is responsible for removal of biodegradable organics during the initial period of GAC operation. Later, the biofilm develops enough so that biodegradation predominates. This provides for continued, steady-state removals and can explain extended bed life, particularly in wastewater treatment applications of GAC.

In water treatment, the removal of adsorbable, biodegradable organics by microbial activity, rather than by adsorption, enhances the opportunity for the GAC bed to remove the adsorbable, nonbiodegradable organics. This latter category is generally of more concern because it contains many of the synthetic organics that may be suspected carcinogens.

3.2 Microbial Activity on GAC in Wastewater Treatment

From pilot plant studies at Pomona, California, determined that adsorption capacity could not be completely exhausted even after very long service times. More detailed investigations of microbial degradation on GAC were provided by Weber et al. (1970, 1972) in other pilot plant studies that addressed the benefits of microbial activity more directly. Comparison studies of GAC and nonsorptive, bituminous coal clearly showed that more TOC removal and microbial growth occurred on GAC. This suggested that microbial activity was in some way interrelated with the adsorption process. Proposed that the rate of microbial activity in a GAC bed should be even higher than that in a conventional biological treatment process because the substrate adsorbed on the activated carbon is much higher than in the liquid (Weber 1970).

4 CONCLUSIONS AND RECOMMENDATIONS

1. No direct evidence has been given for removal of specific organics of potential harm to health by microbial activity. Only one study implied that precursors to THM formation were removed by microbial action. An indirect benefit of microbial activity may be the lengthening of GAC service time by removing organics that would otherwise occupy adsorption sites; but this in itself does not imply more effective removal of organics of health concern.

2. While operation of GAC beds for 6 months to 2 yr without regeneration has shown that microbial activity removes organics, as measured by group parameters such as TOC, potassium permanganate demand, COD, and UV absorbance, the efficiency of biodegradation is less than that of adsorption, i.e., biodegradation to remove approximately 1.5 mg/liter TOC requires about 30 min of GAC contact (Benedek, 1977; Eberhardt et al., 1974; Jekel, 1977), while adsorption only requires approximately 10 to 15 min.

3. Pilot and full-scale tests of preozonization have shown that organics are efficiently removed through the total system, including the flocculation and adsorption steps; however, the ability of microbial action to aid in removal of specific organics of potential harm to health has not been reported. Thus far, one study, at the EPA Cincinnati pilot plant (U.S. Environmental Protection Agency, 1978c), implied that precursors to THM formation were somehow altered by ozonization in such a manner that their removal was made more effective.

4. The role of ozone in promoting microbial activity remains unclear. Conversion of organics to more biodegradable forms and addition of oxygen to the biofilm are two possibilities. On the negative side, evidence shows that ozone may decrease the adsorbability of some organics.

5. While prechlorination does not stop microbial growth on GAC, there is some evidence to suggest that it results in formation of organics that are much more resistant to biodegradation on the GAC surface.
6. The bacteria that have been reported to grow on GAC beds are considered nonpathogenic. Increases in bacteria in the effluent may be controlled by backwashing.
7. Microorganisms can generate a variety of highly potent, low-molecular-weight toxicants in culture. Studies of models of several environments and some natural systems indicate that toxicants may be, or are indeed, formed. Such products may also be produced on GAC. However, there is a lack of information regarding the identification of organics in the effluent of GAC beds. Thus, while no evidence exists, the possibility that microbial end products of concern enter the finished water cannot be ruled out.
8. The endotoxin content of water that has been filtered through GAC is either not increased or not significantly increased. The measured levels are very low and should pose no risk.

Thus far, evidence of microbial activity in GAC beds has largely been indirect. Of special importance is the identification of organics that are biodegraded. If these are not potentially harmful to health, then more investigation is needed to determine whether microbial action has actually extended or improved the adsorption of other organics that are potentially harmful to health. This may result if fewer of the internal adsorption sites are occupied because of biodegradation at the surface. Related to this concern about extended removal of organics is the unresolved question of bioregeneration. More careful investigation is needed to determine if adsorbed organics, which are more resistant to biodegradation, can be acted upon by microbes. The role of ozonization in promoting microbial action needs much further elucidation. This would involve studies to determine the changes in biodegradability and adsorbability that are brought about by ozonization of specific organics that could affect public health. There are similar concerns regarding the practice of prechlorination. If the use of GAC is to become widespread, research is necessary to identify the factors that are responsible for the initiation of microbial growth on the GAC surface and the microbiological generation of organic compounds on the carbon. Attention should be given not only to the compounds that are present and retained in the bed, but also to those that are released and could appear in the effluent. Research should be directed toward compounds generated from innocuous, natural precursors, as well as toward those that may be formed as a result of microbial action on synthetic industrial waste, agricultural pesticides, household effluents, and compounds that are formed as a result of chlorination of water. The investigations should identify the specific organic molecules thus formed and not simply the quantity of organic carbon that is emitted from GAC.

Acknowledgements

The author would like to fully thank all who contributed to conduct this work and supported it.

References

Benedek, A. 1977. The effect of ozone on activated carbon adsorption—a mechanistic analysis of water treatment data. Presented at IOI Symposium on Advanced Ozone Technology. International Ozone Institute, Toronto, Ontario. 26 pp.

Clifford F. 1983. Thermal regeneration of PAC in biological sludge mixture water research, Vol.17, NO.9, PP.1125-1138.

Crittenden JC, Weber WJ. 1978. Predictive model for design on fixed-bed adsorbers: parameter estimation and model development, Journal of Environ mental Engineering, ASCE, Vol.106, NO.EE4.

Eberhardt, M., S. Madsen, and H. Sontheimer. 1974. Untersuchungen zur Verwendung biologisch arbeitender Acktivkohlefilter bei der Trinkwasseraufbereitung. Heft 7, Engler-Bunte-Instit, der Universitaet Karlsruhe, Federal Republic of Germany. EPA translation TR-77-503 (Investigations of the use of biologically effective activated carbon filters in the processing of drinking water.) 48 pp.

Femogarcia MA, Utrina RJ. 1988. Adsorption by products, Journal of Carbon, Vol.26, NO.3, PP.365-373.

Hassler JW. 1963. Activated Carbon Chemical Publishing Company Edition, New York, USA, PP.320-326.
Hutton DG, Cheremision N. 1978. Carbon adsorption hand book combined powdered activated carbon biological treatment, Ann Arbor Scince, Michigan, USA, PP.589-448.
Hutchison D. 1990. A microbal regeneration Process for GAC: 1- Process modeling, water research, Vol.24, No.10, PP.1209-1215.
Jekel, M. 1977. Biological treatment of surface waters in activated carbon filters. Presented at the meeting of the Water Research Center (England), KIWA (Netherlands) and EBI (Federal Republic of Germany), Engler-Bunte Institute, Karlsruhe University, Federal Republic of Germany.
Mckinny, Roos E. 2004. Environmental pollution control microbiology (1) (2).
Nawwar SS. 1989. Removal of dyes from effluent of using low agriculture by protection, total environment Using low agriculture by protection, total environment, Vol.79, NO.3, PP.271-279.
Sarai, Darshan. 2006. Water treatment made simple for Operators, PP.256-79.
Simsek M. 1970. Active carbon, Elsevier publicsherInc, Amstrdam, Netherlands.
Smethurst G. 1979. Basic Water treatment: for application world – wide, PP.15-20.
Snoeyink VK. 1987. Effect of temperature time and biomass on wet air regeneration of carbon, Journal of WPCF, Vol.59, NO.3, PP.139-144.
Sontheimer A, Crittenden A. 1988. "Activated carbon for water treatment", DWGM, forschungsstelle FRC. English edition, Germany.
Weber WJ. 1970. Physico chemical processes for water and wastewater treatment, Jahan Wiley Inter science, New York, USA

Building Tomorrow's Society
Bâtir la Société de Demain

Fredericton, Canada
June 13 to 16, 2018 / *13 au 16 juin, 2018*

ULTRAVIOLET LIGHT EMITTING DIODES (UV LEDS) FOR BIOFILM CONTROL

Stephanie Gora[1,2], Kyle Rauch[1], Carolina Ontiveros[1], Graham Gagnon[1]
[1]Dalhousie University, Halifax, Nova Scotia
[2]stephanie.gora@dal.ca

Abstract: Biofilm growth on treatment equipment and biofilm formation in drinking water distribution systems are major ongoing challenges in the drinking water industry. Control mechanisms include flushing, chemically intensive cleaning procedures, chlorination, and the removal of water matrix components that encourage the growth of biofilm-forming bacteria (e.g. natural organic matter). These procedures can be onerous and shorten the useful lifetime of the equipment and there is demand for novel disinfection technologies that can provide fast and effective disinfection without damaging equipment or infrastructure. Germicidal ultraviolet violet (UV) light is one potential option for these applications, and UV light emitting diodes (UV LEDs), which are smaller and more energy efficient than standard mercury UV lamps, are particularly promising.

In this study, *Pseudomonas aeruginosa* biofilms were grown on plastic and stainless steel coupons and exposed to UV LEDs emitting light at 265 nm for times ranging from 1 to 60 minutes. The biomass remaining on the coupons was then suspended in sterile phosphate buffered solution, which was then plated on tryptic soy agar to determine colony forming units and analyzed for adenosine triphosphate (ATP) concentration, a quick and effective measure of biological activity. Preliminary results indicate that 1.5-2 log inactivation can be achieved at a UV dose of 8.1 mJ/cm^2). This dose is higher than those reported for pure cultures of planktonic bacteria (Beck et al., 2017; Rattanakul and Oguma, 2018) in bench scale experiments but below that required to inactivate *Pseudomonas aeruginosa* in biofilms in real life medical applications (Bak et al., 2010).

The results of this study confirm that UV LED irradiation is a viable technology for surface disinfection. These findings have implications for the drinking water industry and for other industries that struggle to control biofilm growth. Future experiments will compare UV LED disinfection to standard biofilm mitigation strategies for water treatment equipment and water distribution infrastructure.

1 INTRODUCTION

UV emitting mercury lamps are commonly used to disinfect municipal water and wastewater in Canada. These lamps are large and power intensive, making them difficult to use in point of entry (POE) and point of use (POU) drinking water applications. UV LEDs, in contrast, are small, can be manufactured to emit nearly any wavelength of light, and are quickly becoming less expensive and more energy efficient (Ibrahim et al., 2014). Researchers in Canada, the United States, and Japan are spearheading research into the application of UV LEDs for drinking water disinfection. Beck et al. (2017) demonstrated that UV LEDs could be used to inactivate a wide range of common human pathogens and that the wavelength dependence and energy efficiency of UV LED disinfection varies from one organism to another. Researchers at the University of Tokyo have demonstrated that significant levels of disinfection can be achieved in flow through UV LED units that may eventually be appropriate for POU drinking water treatment applications (Oguma et al., 2016, 2013).A standard method for the characterization of UV LED experiment apparatus has recently been published out of the University of British Columbia (Kheyrandish et al., 2017), and promises to increase the number of studies and improve the quality of UV LED research in the drinking water field.

Biofilms consist of microorganisms and extracellular polymeric substances attached to a surface. They are common in drinking water distribution systems and premise plumbing (Falkinham et al., 2015), and can provide a safe harbour for opportunistic human pathogens (Lau and Ashbolt, 2009). A recent study by Rattanakul and Oguma.(2018) showed that UV LEDs were effective for the inactivation of planktonic (free-floating) pure cultures of *Pseudomonas aeruginosa*, a common biofilm forming bacteria, and *Legionella pneumophilia*, a well known opportunistic pathogen. At a wavelength of 265 nm, a 2-log inactivation of *Pseudomonas aeruginosa* was achieved with a UV dose of 4 mJ/cm^2. Only 2 mJ/cm^2 of UV light was required to achieve this level of inactivation for *Legionella pneumophilia*.

Only a small number of studies have explored the use of UV light, UV LED or otherwise, for the inactivation of biofilm-bound bacteria on real surfaces. The studies that do exist suggest that mixed bacterial cultures and real clinical surfaces are more difficult to treat with UV light than biofilms grown from pure bacterial cultures under controlled lab conditions. A pair of studies conducted on medical catheters by Bak et al. illustrate this eloquently. In the first study (Bak et al., 2009), the researchers collected used catheters from a hospital, established the types of bacteria present in the biofilms attached to their inner surfaces, and then attempted to inactivate these bacterial communities with conventional UVC lamps. Species detected in the catheters included *Escherichia coli*, *Pseudomonas aeruginosa*, and *Enterococcus faecalis*. A dose of 1,500 mJ/cm^2 was required to achieve an overall 4 log kill of the bacteria present on the catheters. The researchers also noted that bacteria in thicker biofilms were more difficult to inactivate, likely due to the protective effects of EPS and light scattering cause by particulates in the biofilm matrix. In the second study (Bak et al., 2010), thin biofilms of *Pseudomonas aeruginosa* were grown in clean catheters in the laboratory and exposed to UVC LEDs. Approximately 8 mJ/cm^2 was required to achieve 4-log reduction of the bacteria present in these thin, single culture biofilms, which is just under 200 times less than the dose of UV light required to achieve this level of inactivation in multi-culture, mature biofilms from used catheters. The improved kill was attributed to the characteristics of the biofilms (thin, pure culture) as well as the increased efficiency achieved using UV LEDs, which can be applied nearly in-situ, compared to conventional UV lamps. These studies, though conducted in a medical setting rather than a civil engineering context, underscore the challenges of using UV light to inactivate biofilm-bound microorganisms, but also the promise that UV LEDs hold for biofilm mitigation.

2 METHODS

The collimated beam of light emitted by the UV LEDs at 265 nm was characterized using a methodology originally proposed by Bolton et al.(2003) and a spectrometer calibrated to measure in the UV range (Ocean Optics USB4000. The average intensity and petri factor of the light emitted by the LED apparatus were determined at five different distances from the edge of the collimator.

Pseudomonas aeruginosa biofilms were grown on polycarbonate or stainless steel coupons in a CDC biofilm growth reactor and exposed to UV LEDs emitting light at 265 nm. The biomass remaining on the coupons after treatment was suspended in sterile phosphate buffered solution, which was then plated on tryptic soy agar to determine colony forming units and analyzed for adenosine triphosphate (ATP) concentration, a quick and effective measure of biological activity, using LuminUltra's QGA method conducted on a bacterial sample cultured from the viable bacteria remaining on the coupon surface after UV exposure.

3 RESULTS

3.1 3.1 Characterization of UV LED Apparatus

The intensity of the beam across the illuminated surface at a distance of 50 mm from the collimator is depicted in Figure 1.

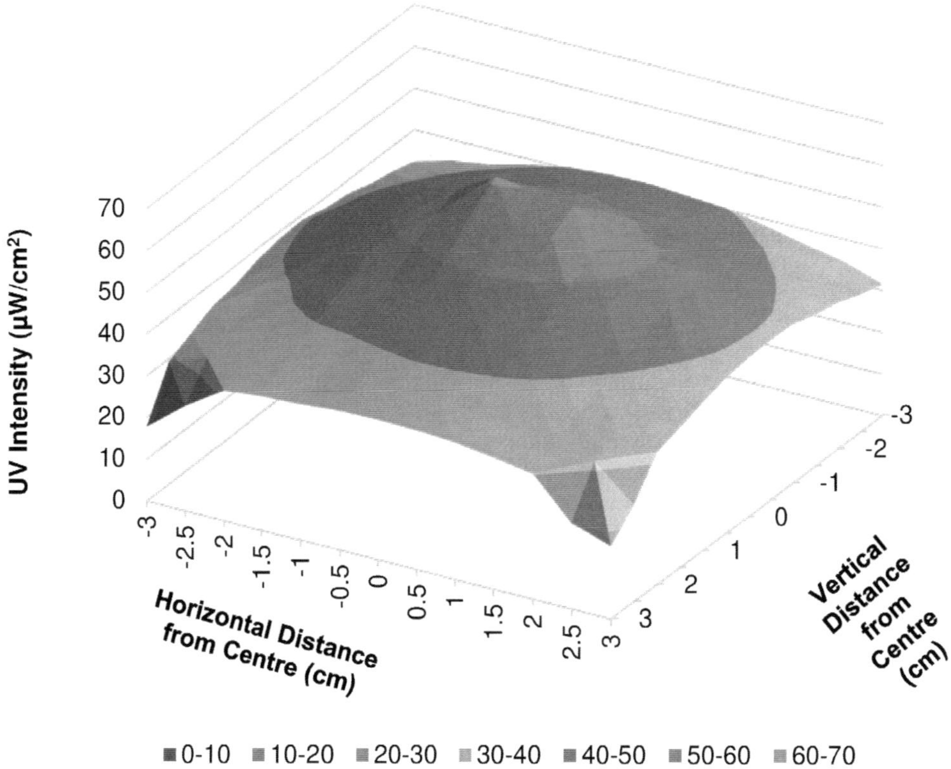

Figure 1: Intensity of UV light (265 nm) at different locations across the illuminated surface in the collimated beam apparatus at a distance of 50 mm from the collimator

The intensity distribution in Figure 1 shows that the intensity of the UV light beam varied substantially across the illuminated area (total area = 36 cm^2) and was higher in the centre than at the outer edges of the beam.

The average intensity and petri factor of the collimated beam issuing from the 265 nm LEDs in the collimated beam apparatus along with the time required to achieve a UV dose of 8 mJ/cm^2, the UV dose reported to provide 4-log reduction of *Pseudomonas aeruginosa* by previous researchers (Bak et al., 2010), at different distances from the collimator (Figure 2). Note that the average intensity and petri factor values in Figure 2 were calculated for the circular area required to illuminate the coupons (9 cm^2) rather than over the entire beam area.

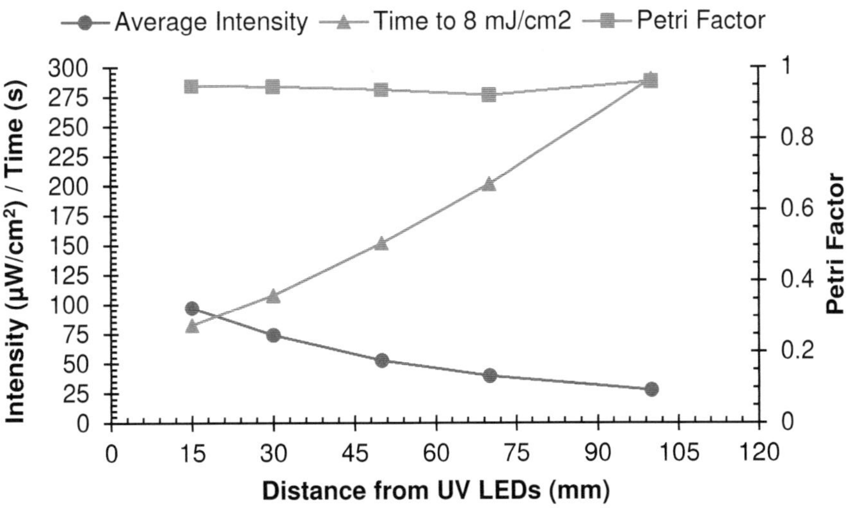

Figure 2: Intensity and uniformity (petri factor) of the light provided by the UV LED apparatus at different distances from the bottom of the collimator and the time required to achieve a UV dose of 8 mJ/cm^2 at each distance

As expected, the average intensity of the beam increased as the distance between the spectrometer sensor and the UV LEDs decreased. The petri factor was high (> 0.9) indicating that the intensity of the light beam was nearly uniform across the 3 cm diameter illuminated area irrespective of the distance between the spectrometer sensor and the UV LEDs, which means that the three replicate coupons received a uniform dose of light at each height/intensity.

3.2 Inactivation of Pseudomonas Aeruginosa – Preliminary Results

The preliminary results of the inactivation experiments confirm that biofilm-bound bacteria can be inactivated by UV LEDs. As shown in Figure 3, 1.9 ± 0.4 log inactivation was achieved at a UV dose of 8.1 mJ/cm^2. This result is in line with recent results published by the Oguma research group at the University of Tokyo (Rattanakul and Oguma, 2018), which showed 2 log reduction of planktonic (free floating) *Pseudomonas aeruginosa* at approximately 4 mJ/cm^2 under otherwise similar treatment conditions. The slightly higher UV dose required to achieve inactivation in the current study is relative to the results published by Rattanakul and Oguma likely reflects the fact that the biofilm-bound bacteria in our study were protected from UV irradiation by the layers of EPS in the biofilm, unlike the planktonic bacteria used in their study, which had no such protection.

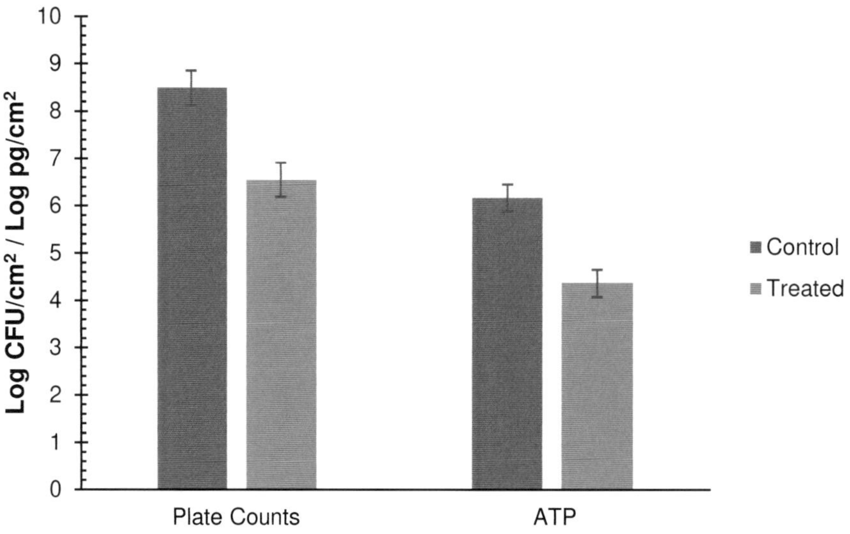

Figure 3: Log surface concentration of *Pseudomonas aeruginosa* and ATP in biofilms exposed to 8.1 mJ/cm^2 of UV light at a distance of 70 mm from the bottom of the collimator (n = 3, error bars represent the 95% confidence interval on the mean)

The UV dose required to achieve 2 log reduction of *Pseudomonas aeruginosa* in this study was similar to that applied by Bak et al. in their study with the same microorganism grown on the inner surfaces of catheters (Bak et al., 2010) but 200 times lower than that required to inactivate indigenous bacterial communities on real catheters (Bak et al., 2009). Although the Bak et al. studies were conducted in a medical setting, they do clearly show that indigenous bacterial communities in wild biofilms are likely to be more difficult to inactivate than bacteria in biofilms grown from pure cultures under controlled laboratory conditions. Given that the bacterial communities present on real drinking water infrastructure components will likely be complex, it is probable that higher UV doses will be required to achieve meaningful levels of bacterial inactivation in biofilms in real drinking water applications. The ATP results mirrored the plate count results, indicating that this parameter was a good measure of bacterial inactivation by UV LED irradiation.

3.3 Effect of Intensity on Inactivation

Establishing the effect of intensity on the inactivation of biofilm-bound *Pseudomonas aeruginosa* was one of the goals of the current project. A previous study completed Sommer et al. (1998) examined the effect UV intensity had on inactivation of *E. coli*, MS2, ψX174, B40-8, and *B. subtillis*. Their study concluded that an additional 1 log reduction in *E. coli* was achieved with the high intensity (0.2 mW/cm^2) versus the low intensity (0.002 mW/cm^2), and no difference in the inactivation was found for the other microorganisms studied.

3.3.1 Plate Counts

As in earlier experiments, polycarbonate coupons coated with *Pseudomonas aeruginosa* biofilms were exposed to the 265 nm light in the UV LEDs apparatus. The intensity of the light hitting the biofilm-covered coupons was varied by placing the coupons at different distances from the collimator as summarized in Table 1. The exposure time was varied to the ensure that all samples were exposed to the same UV dose (8.1 mJ/cm^2).

Table 1: Relationship between sample placement and light intensity at 265 nm in UV LED apparatus

Distance from Collimator (mm)	Intensity (mW/cm^2)	Exposure Time (min)
15	0.097	1.4
30	0.074	1.8
50	0.053	2.6
70	0.040	3.4

A one-way ANOVA was used to determine whether there were any significant differences between the biofilms exposed to different intensities of UV LED light. All of the treatment levels were significantly different from the control at a 95% confidence level. There was less obvious variability between the treatments, and the only statistically significant difference was between 0.0527 mW/cm^2 and 0.0970 mW/cm^2. When the samples prepared at 0.0397 mW/cm^2 (height = 70 mm) were removed from the analysis, however, there was a linear relationship between concentration and intensity (Figure 4).

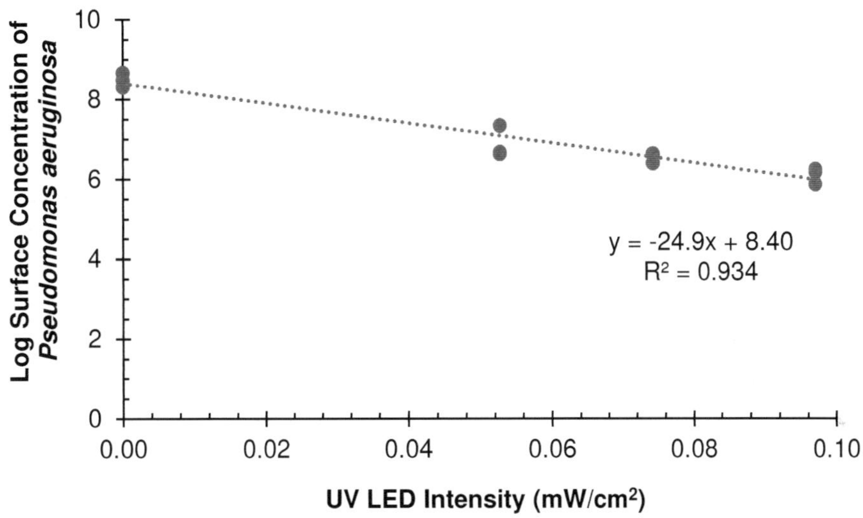

Figure 4: Relationship between intensity and log surface concentration of *Pseudomonas aeruginosa* after exposure to 8.1 mJ/cm^2 of 265 nm UV light at different intensities (n = 3)

3.3.2 ATP Results

As with the plate count results, the ATP content of the biofilms recovered from the control and treated coupons was analyzed with a one-way ANOVA at the 95% confidence level. Although the biofilms recovered from all of the treated coupons contained significantly less ATP than those recovered from the control biofilms, there were no significant differences between biofilm coated coupons exposed to different intensities of UV light.

3.4 Effect of Coupon Material

Experiments are ongoing to examine the impact of coupon material (polycarbonate vs. stainless steel) on biofilm development and UV LED treatment efficacy.

4 CONCLUSIONS

The UV dose required to achieve 2 log inactivation of *Pseudomonas aeruginosa* in this study was in line with previous studies on pure cultures, confirming that the methods employed in this study were congruent with previous research in the field. The UV dose required in the current study was, however, lower than the UV doses required to achieve this level of inactivation in a study that employed mixed bacterial cultures on real clinical surfaces, and it is expected that similarly higher UV doses will be required in real drinking water applications as well.

The effects, if any, of UV intensity on *Pseudomonas aeruginosa* inactivation were difficult to elucidate because of the small range of intensities explored in this study. This limited range was imposed by the configuration of the apparatus used in this study, specifically the length of the collimator. Nonetheless, the plate counts from this part of the study suggest a positive relationship between intensity and inactivation and justify further investigation into this phenomenon.

Acknowledgements

This work was financially supported by NSERC (Engage Grant) and BlueLight Analytics. The authors would like to acknowledge the contributions of Chris Felix, Leeza Hack, and Bronwyn Woolfrey of BlueLight Analytics and Amina Stoddart at the Centre for Water Resources Studies at Dalhousie University.

References

Bak, J., Ladefoged, S.D., Tvede, M., Begovic, T., Gregersen, A., 2010. Disinfection of Pseudomonas aeruginosa biofilm contaminated tube lumens with ultraviolet C light emitting diodes. Biofouling 26, 31–38. doi:10.1080/08927010903191353

Bak, J., Ladefoged, S.D., Tvede, M., Begovic, T., Gregersen, A., 2009. Dose requirements for UVC disinfection of catheter biofilms. Biofouling 25, 289–296. doi:10.1080/08927010802716623

Beck, S.E., Ryu, H., Boczek, L.A., Cashdollar, J.L., Jeanis, K.M., Rosenblum, J.S., Lawal, O.R., Linden, K.G., 2017. Evaluating UV-C LED disinfection performance and investigating potential dual-wavelength synergy. Water Res. 109, 207–216. doi:10.1016/j.watres.2016.11.024

Bolton, J.R., Linden, K.G., Asce, M., 2003. Standardization of Methods for Fluence: UV Dose Determination in Bench-Scale UV Experiments. J. Environ. Eng. 129, 209–215.

Falkinham, J., Pruden, A., Edwards, M., 2015. Opportunistic Premise Plumbing Pathogens: Increasingly Important Pathogens in Drinking Water. Pathogens 4, 373–386. doi:10.3390/pathogens4020373

Ibrahim, M.A.S., Macadam, J., Autin, O., Jefferson, B., 2014. Evaluating the impact of LED bulb development on the economic viability of ultraviolet technology for disinfection. Environ. Technol. (United Kingdom) 35, 400–406. doi:10.1080/09593330.2013.829858

Kheyrandish, A., Mohseni, M., Taghipour, F., 2017. Development of a method for the characterization and operation of UV-LED for water treatment. Water Res. 122, 570–579. doi:10.1016/j.watres.2017.06.015

Lau, H.Y., Ashbolt, N.J., 2009. The role of biofilms and protozoa in legionella pathogenesis: Implications for drinking water. J. Appl. Microbiol. 107, 368–378. doi:10.1111/j.1365-2672.2009.04208.x

Oguma, K., Kita, R., Sakai, H., Murakami, M., Takizawa, S., 2013. Application of UV light emitting diodes to batch and flow-through water disinfection systems. Desalination 328, 24–30. doi:10.1016/j.desal.2013.08.014

Oguma, K., Kita, R., Takizawa, S., 2016. Effects of Arrangement of UV Light-Emitting Diodes on the Inactivation Efficiency of Microorganisms in Water. Photochem. Photobiol. 92, 314–317. doi:10.1111/php.12571

Rattanakul, S., Oguma, K., 2018. Inactivation kinetics and efficiencies of UV-LEDs against Pseudomonas aeruginosa, Legionella pneumophila, and surrogate microorganisms. Water Res. 130, 31–37. doi:10.1016/j.watres.2017.11.047

Building Tomorrow's Society
Bâtir la Société de Demain

Fredericton, Canada
June 13 to 16, 2018 / *13 au 16 juin, 2018*

VOLATILE FATTY ACID PRODUCTION FROM POTATO WASTE USING THERMAL HYDROLYSIS PROCESS DURING MESOPHILIC ANAEROBIC DIGESTION

Bhargav, Mohit[1,2], and Singh, Kripa[1]
1 University of New Brunswick, Canada
2 mohit.bhargav@unb.ca

Abstract: Approximately 20,000 hectares of potatoes are cultivated every year in New Brunswick of which 25% is diverted towards the fresh market, 19% towards seeding and the remaining 56% is destined for industrial processes to manufacture several potato products which in turn generates an abundant waste stream rich in starch. Considering the applicability of potato waste as a substrate for anaerobic digestion, this study investigates the effectiveness of degree of solubilization due to thermal pre-treatment on volatile fatty acids (VFAs) production under mesophilic anaerobic digestion (35°C). This study is carried out with heat shocked mixed culture and quantitatively establishing operating conditions (Organic Loading Rate [OLR] and Hydraulic Retention Time [HRT]) which helps in predicting the product spectrum for VFAs production. The experimental runs were dedicated to produce maximum VFA concentrations (dependent variable) considering potato waste as a feedstock at multiple OLRs (6 and 9 kg/m^3.day) and multiple HRTs (0.5, 1, 2 days), keeping pH constant at 7. Experimental runs are dedicated to quantifying the independent variables influencing VFA production rates by changing one variable at a time and evaluating their effects on acetic acid yield. This study will assist in determining the feasibility of pretreatment technology to produce volatile fatty acids using mesophilic anaerobic digester. Potato starch from a local potato processing industry was thermally hydrolyzed in a stirred and externally heated Parr reactor. Experimental results for thermal pre-treatment indicate that raising the pre-treatment temperature from 100°C to 140°C led to an increase in the ratio of ((SolubleCOD, SCOD)/(TotalCOD, TCOD)) from 63-67% to 87-92% along with volatile suspended solids reduction of 59-65% and 85-89% respectively due to solubilization. At an OLR of 9 kg/m^3.day and HRT of 2 days, maximum VFA yield and maximum acetic acid yield of 0.66 $mgCOD_{volatilefattyacids}/mgCOD_{influent}$ and 0.29 $mgCOD_{aceticacid}/mgCOD_{influent}$ were reported from potato waste that was thermally hydrolyzed at 100°C, making these operating conditions ideal for maximum concentrations of VFAs.

1 INTRODUCTION

Resource recovery is an attractive and sustainable waste management approach aimed at extracting valuable resources from waste. Because of increasing oil prices, this process has gained tremendous appreciation over the years, and research is focused on maximizing selective extraction of value-added products using anaerobic digestion (AD). AD has the potential to provide a valuable carbon sources from low-value organic waste streams, which in turn makes harvesting bio-products from the wastewater treatment process more economical.

Approximately 20,000 hectares of potatoes are cultivated every year in New Brunswick of which 25% is diverted towards the fresh market, 19% towards seeding and the remaining 56% is destined for industrial processes [New Brunswick Potatoes 2017] to manufacture several potato products which in turn generates an abundant waste stream rich in starch. This potato waste poses significant disposal challenges for the industries such as bulk volume, logistics and disposal costs. The typical disposal method for potato waste is moving it to landfills, where it will degrade with due time, but it will create several problems like leachate production [Shu et al. 2006], and emission of greenhouse gases like methane and carbon dioxide [Abichou et al. 2006] which contributes towards global warming. These problems are primarily because of high organic matter (predominantly carbohydrates) and moisture content present in the waste. This organic matter makes potato waste an ideal substrate for anaerobic digestion leading to a well-established process of biogas production [Lin et al. 1982]. Biogas containing methane has a significant positive impact on the environment by utilizing this low-value organic waste to produce either heat or electricity and minimizing our dependency on non-renewable natural resources.

Generation of value-added products from AD follows four sequential steps: i.e., hydrolysis, acidogenesis, acetogenesis, and methanogenesis [McCarty 1964]. The first step, i.e., hydrolysis of organic matter, can often be a rate-limiting step considering the complex physical and chemical structure of the substrate [Izumi et al. 2010, Ma et al. 2011, Rafique et al. 2010]. Pre-treatment methods are employed before the anaerobic digestion for enhanced resource recovery from organic waste [Mossier et al. 2005, Hendricks et al. 2009, Ariunbaatar et al. 2014]. Pre-treatments results in disintegration of cell membranes which lead to partial or full solubilization of organic compounds [Haug et al. 1978], making the organic matter readily fermentable, and this results in increased hydrolysis rate. As the microbial culture present in the fermenter has less concentration of complex molecules to hydrolyze, this is a direct measure of enhanced anaerobic biodegradability. Methane production rate and substrate utilization rate were found inversely proportional to particle size [Esposito et al. 2011, Kim et al. 2000]. Thermal, chemical, and thermo-chemical pre-treatment methods have been studied widely for enhanced AD for several substrates [Kim et al. 2002, Valo et al. 2004, Tanaka et al. 2002, Carrère et al. 2009]. To further enhance resource recovery, two-phase systems are preferred for acidogenic and methanogenic activity which improve process control and enhance process stability [McCarty 1964].

Despite the attractiveness of methane production using AD, the economics of utilizing biogas is often questioned based on the market value. In addition to that, detailed knowledge of the acid phase of the two-phase AD is still required as it has the potential to provide several other bioproducts like hydrogen [Oh et al. 2003, Ginkel et al. 2005, Zhang et al. 2003], which produces volatile fatty acids as a byproduct. This has initiated the development of processes that yield products like volatile fatty acids (VFAs) consisting of carboxylic acids like acetic acid, which is an intermediate product formed during AD. Production of acetic acid remains a lucrative option considering its application as a chemical reagent in the field of chemistry.

To obtain maximum acetate production from polysaccharide breakdown during methanogenesis, low hydrogen partial pressure must be maintained which is accomplished by scavenging homoacetogens and hydrogentropic methanogens [McCarty 1964]. Inhibition of these homoacetogens and hydrogentropic methanogens by heat shocking the mixed culture should provide volatile fatty acids and hydrogen as the product, instead of methane. Obtaining VFAs as the final product has always been challenging as it is difficult to answer which VFAs and in what proportions are obtained. Although several studies have indicated acetic acid production during dark fermentation for hydrogen production [Oh et al. 2003, Su et al. 2009], where conversion of long chain fatty acids can be a rate limiting step for production of acetate, as interspecies hydrogen transfer is affected by heat shocking as it inhibits homoacetogens/hydrogentropic methanogens which increases the partial pressure of hydrogen (Speece 1996), and eventually limiting the conversion of propionic acid to acetic acid. In such a case, it needs to be understood that there are some inherent biological limitations when producing other bioproducts other than methane as some of the influent chemical oxygen demand (COD) goes towards hydrogen generation and only certain amount of volatile fatty acids can be produced.

A significant knowledge gap exists to determine the effect of solubilization due to thermal pretreatment on the production of VFAs from heat-shocked mixed cultures. This alternate process requires further research and development to provide a substitute for the conventional process of commercial VFAs production, which follows chemical routes. Hence, this study aims to investigate the effect of solubilization due to thermal pre-treatment at different temperatures on VFAs production to obtain maximum concentration of

VFAs. In addition to that, the effect of different operating conditions like hydraulic retention time (HRT) and organic loading rate (OLR) was also analyzed for VFAs production.

2 MATERIALS

2.1 Substrate and Inoculum

Potato waste from a local potato processing industry was used in this study for anaerobic digestion. Anaerobic seed sludge was obtained from ADI® Systems pilot reactor used for the treatment of sugar and starch-based wastewater and was used as inoculum. Wastewater composition and seed sludge characteristics are provided in Table 1 and Table 2 respectively. All samples were stored at 2°C-4°C.

Table 1: Characteristics of wastewater

Parameters (g/L)	Wastewater
Total Solids	15.008 ± 0.32
Total Suspended Solids	14.258 ± 0.28
Volatile Suspended Solids	13.255 ± 0.14
Total COD (TCOD)	14.983 ± 0.29
Soluble COD (SCOD)	0.1228 ± 0.03
pH	7.87 ± 0.31
Alkalinity (g-$CaCO_3$/L)*	5.885 ± 0.35

Average ± standard deviation

*After addition of sodium bicarbonate

Table 2: Sludge characteristics

Parameters (g/L)	Values
Mixed Liquor Suspended Solids (MLSS)	27.154 ± 0.42
Mixed Liquor Volatile Suspended Solids (MLVSS)	25.905 ± 0.35

Average ± standard deviation

2.2 Experimental set-up

2.2.1 Reactor Operation

2.2.1.1 Parr Reactor (4560 Series, Parr Instrument Company, Illinois, USA) (for thermal pre-treatment)

A Parr Reactor (total volume 600 mL and working volume 400 mL) was employed in the study and operated in batch mode. It was used to treat starch solution at two different pre-treatment temperatures (100°C and 140°C) to enhance solubilization. The temperature was controlled using an external heater. All the experiments were conducted at the same agitator rpm and time, i.e., 150 rpm and 30 minutes, respectively.

2.2.1.2 Anaerobic Respirometer System (Challenge AER-208, Arkansas, USA)

Reactors (total volume 310mL and working volume 250mL) were employed in the study and operated in batch mode. Two identical reactors each for starch (no thermal pre-treatment, NO_THP), starch (after thermal pre-treatment at 100°C, THP_100) and starch (after thermal pre-treatment at 140°C, THP_140) were used. Each reactor was stirred at the constant speed of about 250 rpm. The temperature was controlled at 35°C using a water bath connected to a water recirculator. Nutrients were supplied along with substrate and to minimize pH variation, sodium bicarbonate ($NaHCO_3$) was used as pH buffering reagent to maintain the pH between 6.8-7.2. Reactors were flushed with nitrogen gas to create anaerobic conditions.

The effect of acidogenesis of potato starch was investigated by operating the reactor at OLRs of 6, 9, 12 kg/m^3.day, while the effect of HRT was investigated by operating the reactor at HRTs of 0.5, 1, 2 days.

2.3 Analytical Method

All analysis was carried out in accordance with Standard Methods [APHA, AWWA, WEF 2005]. All COD and solids analysis were triplicated unless stated otherwise and conducted before and after every experimental run. All COD analysis was measured using the titrimetric method. COD analysis after thermal pre-treatment was carried out after centrifugation at 4750 rpm for 30 minutes and then, filtering the solution through 1.2µm glass fiber filter and 0.45µm membrane filter for filtered COD and soluble COD measurements respectively. VFA samples, before being analyzed using gas chromatography, were centrifuged at 4750 RPM for 30 minutes followed by filtering the sample solution through a 0.45µm membrane filter and then adding 5% v/v phosphoric acid. VFAs were analyzed with Varian CP-3800 Gas Chromatograph. It is equipped with a wall coated open tubular fused silica column (0.32 mm inner diameter) with a wall thickness of 1.2µm and a flame ionization detector (FID). The temperature of the injection port is initially set to 80°C, which rises to a maximum of 180°C, while the FID temperature of the oven is set to 300°C. Helium (He) is used as a carrier gas as a flow rate of 5 mL/min.

3 RESULTS AND DISCUSSION

3.1 Selection of thermal pre-treatment temperature

The degree of COD solubilization can be studied at several different temperatures. However, it has been reported that pre-treatment temperature is directly proportional to the degree of solubilization [Haug et al. 1978]. 24% increase of methane production was reported if kitchen waste was thermally treated at 120°C [Ma et al. 2011], whereas thermal pretreatment of fruit and vegetable waste at 175°C provided 11.7% decrease in methane production [Liu et al. 2012]. Treatments at a higher temperature (above 150°C) showed increased solubilization but no increase in methane production because of the Maillard reaction (Jiang et al. 2012). Therefore, it can be said that enhanced biogas production was mostly due to solubilization of particulates and any temperature higher than 145°C-150°C might lower the performance of anaerobic digester due to the formation of melanoidins [Dwyer et al. 2008]. This was the reason why pre-treatment temperatures of 100°C and 140°C were chosen.

3.2 Effect of thermal pre-treatment at 100°C and 140°C

Wastewater sample was thermally hydrolyzed in Parr reactor. Table 3 compares the result for thermally hydrolyzed starch at different pre-treatment temperatures. Thermal pre-treatments results in partial or full solubilization of organic compounds [Haug et al. 1978], making the organic matter readily fermentable, and this results in increased hydrolysis rate. In case of starch, thermal pre-treatment leads to granular swelling and particulate solubilization which resulted in total suspended solids reduction of 59-65% and 85-89% at 100°C and 140°C respectively, in comparison to starch (NO_THP). Since a non-buffered solution was used, pH dropped down to approximately 6.25 and 5.38 from 7.91 for thermally hydrolyzed starch at 100°C and 140°C respectively. Drop in pH was also reported during gelatinization [Marchal et al. 1999]. The degree of solubilization was measured as the ratio of ((SolubleCOD, SCOD)/(TotalCOD, TCOD)). There is loss of some water as the solution is allowed to cool down at room temperature after thermal pre-treatment, hence TCOD values after thermal pre-treatment are higher for starch(THP_100) and starch(THP_140).

Table 3: Characteristics of wastewater after thermal pre-treatment at different temperature

Characteristics	TCOD (g/L)	FCOD (g/L)	SCOD (g/L)	TS (g/L)	TSS (g/L)	VSS (g/L)	pH -
NO_THP	14.96 ± 0.61	0.123 ± 0.53	0.090 ± 0.04	15.74 ± 0.59	14.59 ± 0.98	13.68 ± 0.26	7.91 ± 0.17
THP_100	15.01 ± 0.69	12.09 ± 0.14	8.02 ± 0.19	15.45 ± 0.84	5.82 ± 0.12	5.15 ± 0.16	6.25 ± 0.25
THP_140	15.07 ± 0.16	13.98 ± 0.85	13.12 ± 0.07	15.19 ± 0.43	2.51 ± 0.29	2.09 ± 0.94	5.38 ± 0.62

Average ± standard deviation

3.3 Effect of Organic Loading Rate and Hydraulic Retention Time

The effect of OLR on VFAs production was conducted at mesophilic conditions, i.e., 35°C, pH of 7.0 was maintained using $NaHCO_3$ and HRT of 0.5, 1, 2 days. Individual reactors were assumed to be at a steady state when VFA concentration in the product stream, i.e., effluent reached a constant value. Gas composition, i.e., the ratio of methane to carbon dioxide remained almost constant with values ranging from 0.05 to 0.1 during fermentation process for starch with NO_THP, THP_100, THP_140.

3.3.1 Starch with no thermal pre-treatment (NO_THP)

Figure 1 shows the variation in concentration of VFAs at OLRs of 6, 9 and 12 kg-COD/m^3.day with HRTs of 0.5, 1, 2 days. For above mentioned operating conditions, the concentration of individual organic acids increased significantly reaching a steady state value after 10-14 days. GC analysis for VFA classification revealed the presence of several organic acids like acetic acid (AA), propionic acid (PA), and butyric acid (BA). Table 4 shows the concentration of individual VFAs at the end of digestion process under different operating conditions. Maximum VFA yield of 0.651 was obtained at an OLR of 9 kg/m^3.day and an HRT of 2 days. Since it is always difficult to predict the product spectrum, analysis of individual VFAs was necessary as the distribution of fermentation products may reveal relevant information regarding the operational conditions and signify which set of conditions promote the production of a particular organic acid.

Irrespective of operating conditions, acetic acid was the dominant product followed by butyric acid and then propionic acid. At an OLR of 6 kg/m3.day and HRT of 2 days, acetic acid and butyric acid together combined to cover approximately 65-72% of the product spectrum. Application of higher HRT proved beneficial as the concentration of VFAs increased with increasing HRT. Alkalinity i.e. buffering capacity for the anaerobic sludge was maintained by dosing sodium bicarbonate.

The concentration of VFAs produced was significantly affected at high OLR, especially at 12 kg/m^3.day (data not shown). Acetic acid remained the dominant product, followed by butyric acid, and together they combined only approximately 38% of the product spectrum. Propionic acid concentration was significantly higher in comparison to OLR of 9 kg/m^3.day. Often, production of isobutyric acid, isovaleric acid, and valeric acid was also noted; but often their concentrations were low, and hence were not quantified by GC for every injection. Reactor operation at this OLR was highly unstable as substrate loading rate was high, and so was the viscosity of fermentation broth.

Table 4: Effect of different OLRs and HRTs on VFAs production with starch (NO_THP)

	Organic Loading Rate (kg/m^3.day)					
	6			9		
HRT (d)	0.5	1	2	0.5	1	2
VFAs (g/L)	1.48 ± 0.62	3.00 ± 0.32	4.87 ± 0.48	2.28 ± 1.12	4.83 ± 0.71	7.13 ± 0.67
AA (g/L)	0.68 ± 0.25	1.28 ± 0.13	2.45 ± 0.26	1.06 ± 0.53	2.55 ± 0.41	3.85 ± 0.30
PA (g/L)	0.20 ± 0.08	0.65 ± 0.15	0.99 ± 0.13	0.54 ± 0.22	0.93 ± 0.13	1.47 ± 0.16
BA (g/L)	0.60 ± 0.28	1.07 ± 0.09	1.42 ± 0.13	0.69 ± 0.40	1.36 ± 0.20	1.81 ± 0.25
AA yield	0.048	0.092	0.174	0.075	0.181	0.274
VFA yield	0.14	0.286	0.447	0.207	0.442	0.651

Average ± standard deviation
AA yield = Acetic Acid yield = ($COD_{aceticacid}/COD_{influent}$); VFA yield = ($COD_{volatilefattyacids}/COD_{influent}$)

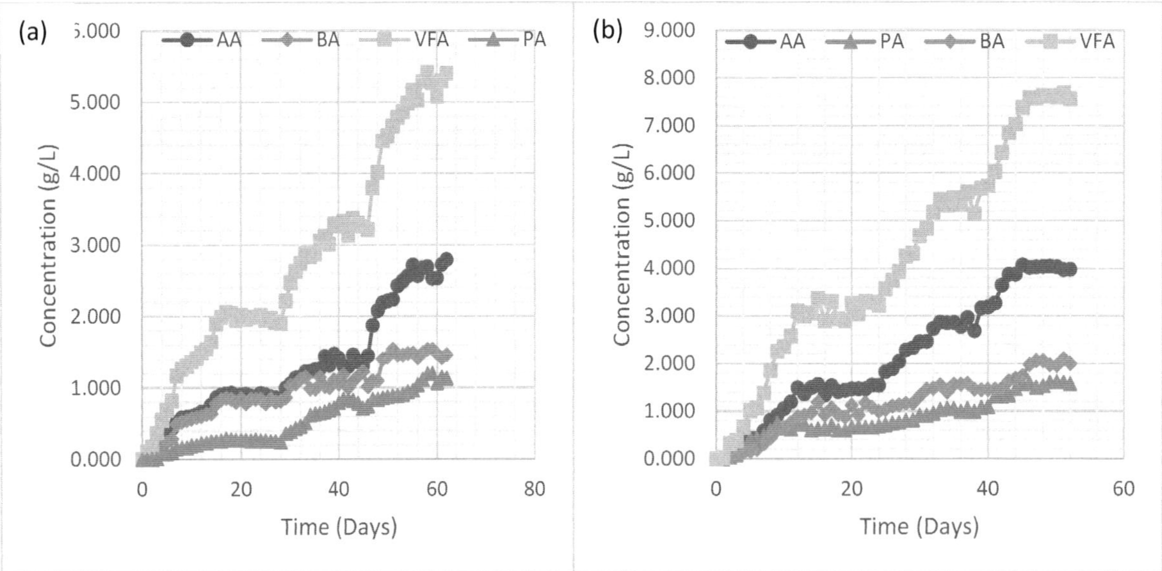

Figure 1: Concentration profiles for VFAs at OLR of a) 6 kg/m^3.day and b) 9 kg/m^3.day

3.3.2 Starch with thermal pre-treatment at 100°C (THP_100)

To analyze the effect of COD solubilization on VFA production, starch was thermally hydrolyzed at 100°C, and Table 2 provides characteristics of starch after pre-treatment. COD solubilization helped in achieving steady state conditions faster than starch that was not thermally pre-treated, which is evident from Figs 2 showing the variation on the concentration of individual VFAs at OLRs of 6 kg/m^3.day and 9 kg/m^3.day with HRTs of 0.5, 1, 2 days. Table 3 shows the concentration of individual VFAs at the end of the anaerobic digestion process.

Irrespective of operating conditions, acetic acid remains the dominant product as was reported before with starch (NO_THP) followed by either propionic acid or butyric acid. At an OLR of 9 kg/m^3.day and HRT of 2 days, maximum VFA yield, and acetic acid yield was found to be 0.66 and 0.29. To maintain the reactor operating pH near neutral, $NaHCO_3$ was used to maintain alkalinity.

OLR of 12 kg/m^3.day (data not shown) significantly reduced reactor performance which may be attributed to the combination of high soluble COD and high substrate loading rate at the beginning of the study. Similar observations were reported by Kim et al. 2003. Concentrations of propionic acid and butyric acid decreased to as low as 0.01 g/L and 0.06 g/L respectively. Because of reactor foaming, the study was stopped after ten days.

Table 5: Effect of different OLRs and HRTs on VFAs production with starch (THP_100)

	Organic Loading Rate (kg/m^3.day)					
	6			9		
HRT (d)	0.5	1	2	0.5	1	2
VFAs (g/L)	1.86 ± 0.82	3.78 ± 0.35	5.13 ± 0.91	3.25 ± 1.21	5.43 ± 0.87	7.40 ± 0.8
AA (g/L)	0.93 ± 0.41	1.64 ± 0.80	2.4 ± 0.6	1.73 ± 0.59	2.96 ± 0.57	4.07 ± 0.52
PA (g/L)	0.38 ± 0.29	1.02 ± 0.07	1.3 ± 0.16	0.83 ± 0.35	1.1 ± 0.24	1.64 ± 0.18
BA (g/L)	0.55 ± 0.18	1.11 ± 0.20	1.45 ± 0.11	0.69 ± 0.26	1.36 ± 0.13	1.70 ± 0.16
AA yield	0.106	0.11	0.17	0.12	0.22	0.29
VFA yield	0.17	0.35	0.475	0.29	0.48	0.66

Average ± standard deviation
AA yield = Acetic Acid yield = ($COD_{aceticacid}/COD_{influent}$); VFA yield = ($COD_{volatilefattyacids}/COD_{influent}$)

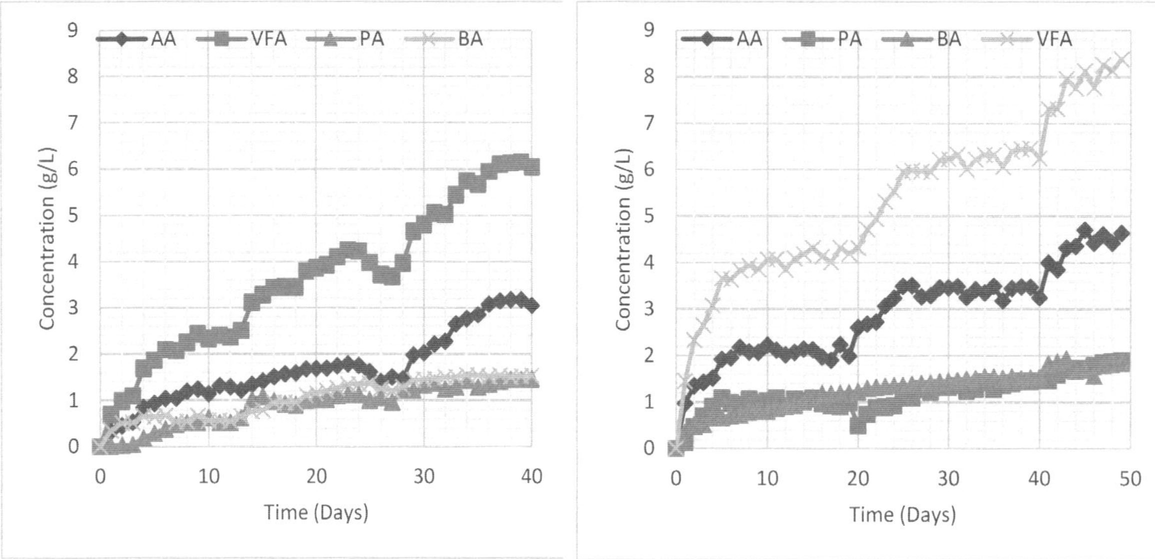

(a) (b)

Figure 2: Concentration profiles for VFAs at OLR of a) 6 kg/m^3.day and b) 9 kg/m^3.day

3.3.3 Starch with thermal pre-treatment at 140°C (THP_140)

Table 2 provides characteristics of starch after pre-treatment at 140°C. Maximum solubilization was achievable as the ratio of SCOD/TCOD was approximately 85-89%. Figure 3 shows the variation in concentration of VFAs at OLRs of 6 kg/m^3.day and 9 kg/m^3.day with HRTs of 0.5, 1, 2 days. Operational instability persisted for 12 kg/m^3.day irrespective of HRT. For other OLRs, i.e., 6 and 9 kg/m^3.day, the concentration of individual organic acids increased significantly at the beginning of each experimental run but was always less than the concentration achieved with for experimental runs with starch (NO_THP) and starch (THP_100°C). This may be attributed to the formation of complex particles which are not readily degraded by the microbial culture. GC analysis for VFA speciation revealed the presence of acetic acid, propionic acid, and butyric acid. Table 5 shows the concentration of individual VFAs at the end of the anaerobic digestion process. Acetic acid and propionic acid were the dominant products, while minimal quantities of butyric acid were formed during the entire run.

Table 5: Effect of different OLRs and HRTs on VFAs production with starch (THP_140)

	Organic Loading Rate (kg/m³.day)					
	6			9		
HRT (d)	0.5	1	2	0.5	1	2
VFAs (g/L)	0.66 ± 0.50	1.71 ± 0.16	2.8 ± 0.44	1.14 ± 0.42	2.71 ± 0.41	4.40 ± 0.42
AA (g/L)	0.38 ± 0.28	1.03 ± 0.07	1.28 ± 0.16	0.68 ± 0.25	1.3 ± 0.14	1.9 ± 0.19
PA (g/L)	0.28 ± 0.21	0.69 ± 0.1	1.06 ± 0.11	0.45 ± 0.18	1.16 ± 0.17	1.52 ± 0.04
BA (g/L)	0	0	0.45 ± 0.19	0.0	0.25 ± 0.1	0.93 ± 0.22
AA yield	0.027	0.073	0.09	0.05	0.09	0.129
VFA yield	0.055	0.142	0.254	0.09	0.23	0.395

Average ± standard deviation
AA yield = Acetic Acid yield = $(COD_{aceticacid}/COD_{influent})$; VFA yield = $(COD_{volatilefattyacids}/COD_{influent})$

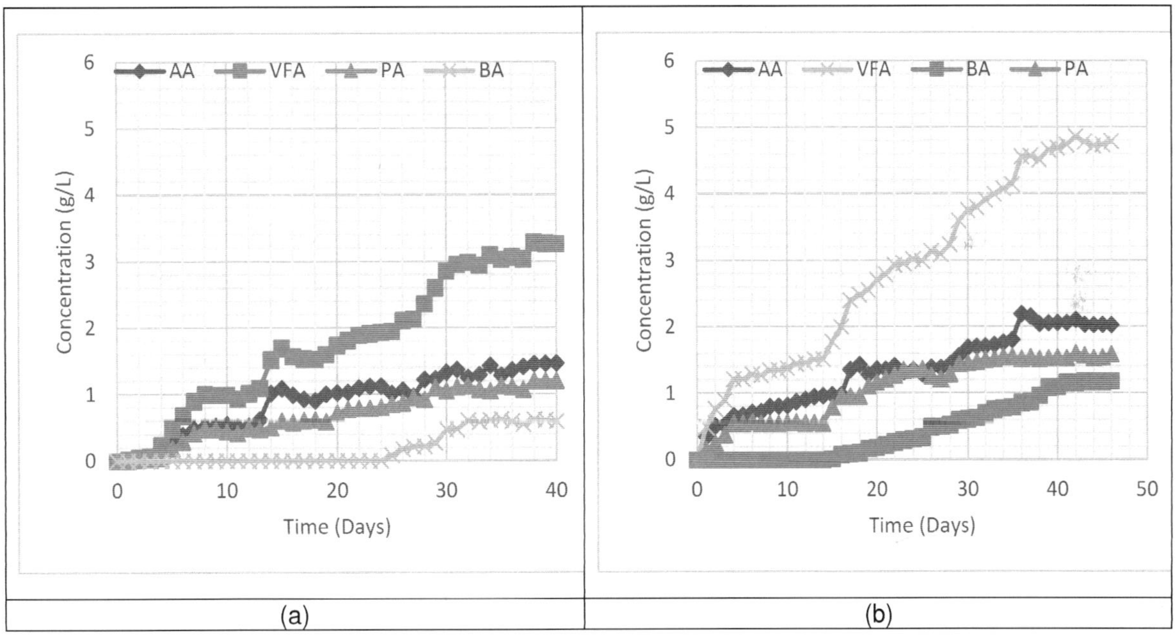

Figure 3: Concentration profiles for VFAs at OLR of a) 6 kg/m³.day and b) 9 kg/m³.day

For all the above experimental runs, the concentration of VFAs increased with increasing OLR and increasing HRT, but the VFAs yield and acetic acid yield remained the approximately the same for starch (NO_THP) and starch (THP_100). Yields for individual VFAs decreased significantly for starch (THP_140), which as suggested before can be attributed to overcooking of the starch sample.

4 CONCLUSIONS

The maximum VFAs yield of 0.66 and acetic acid yield 0.29 was observed for starch (THP_100), at operating conditions of 9 kg/m^3.day and HRT of 2 days. In addition to that, reactors reached the steady-state conditions faster which are a result of solubilized starch, that is readily fermentable by the microbial culture. Above this pre-treatment temperature, the concentration of VFAs decreased resulting in a simultaneous decrease in VFAs yield. This implies that thermal pre-treatment at a temperature above 100°C led to increased solubilization but had no positive effects on VFAs production. This is a result of decreased biodegradability of the substrate due to the possible formation of complex and toxic compounds. Since the yields for VFAs and acetic for starch (NO_THP) and starch (THP_100) are approximately the same, this implies that this is the maximum concentration of VFAs that can be obtained from this potato waste.

Acknowledgement

The authors would like to acknowledge the funding agencies including Natural Science and Engineering Research Council (NSERC) and New Brunswick Innovation Foundation (NBIF). Dr. Dennis Connor is acknowledged for his technical support.

References

Abichou, T., J. Chanton, D. Powelson, J. Fleiger, S. Escoriaza, Y. Lei, and J. Stern. 2006. "Methane Flux and Oxidation at Two Types of Intermediate Landfill Covers." Waste Management 26 (11): 1305-1312. doi:10.1016/j.wasman.2005.11.016

Ariunbaatar, J., A. Panico, G. Esposito, F. Pirozzi, P.N.L. Lens. (2014) "Pretreatment methods to enhance anaerobic digestion of organic solid waste." Applied Energy 123: 143-156.

Carrère, H., B. Sialve, and N. Bernet. 2009. "Improving Pig Manure Conversion into Biogas by Thermal and Thermo-Chemical Pretreatments." Bioresource Technology 100 (15): 3690-3694. doi:10.1016/j.biortech.2009.01.015

Dwyer, J., D. Starrenburg, S. Tait, K. Barr, D. J. Batstone, and P. Lant. 2008. "Decreasing Activated Sludge Thermal Hydrolysis Temperature Reduces Product Colour, without Decreasing Degradability." Water Research 42 (18): 4699-4709. doi:10.1016/j.watres.2008.08.019.

Esposito, G., L. Frunzo, A. Panico, and F. Pirozzi. 2011. "Modelling the Effect of the OLR and OFMSW Particle Size on the Performances of an Anaerobic Co-Digestion Reactor." Process Biochemistry 46 (2): 557-565. doi:10.1016/j.procbio.2010.10.010

Haug, R. T., D. C. Stuckey, J. M. Gossett, and P. L. McCarty. 1978. "Effect of Thermal Pretreatment on Digestibility and Dewaterability of Organic Sludges." Journal of the Water Pollution Control Federation 50 (1): 73-85.

Hendriks, A. T. W. M. and G. Zeeman. 2009. "Pretreatments to Enhance the Digestibility of Lignocellulosic Biomass." Bioresource Technology 100 (1): 10-18. doi:10.1016/j.biortech.2008.05.027.

Izumi, K., Y. -K Okishio, N. Nagao, C. Niwa, S. Yamamoto, and T. Toda. 2010. "Effects of Particle Size on Anaerobic Digestion of Food Waste." International Biodeterioration and Biodegradation 64 (7): 601-608. doi:10.1016/j.ibiod.2010.06.013

Jiang, Y., L. Marang, J. Tamis, M. C. M. van Loosdrecht, H. Dijkman, and R. Kleerebezem. 2012. "Waste to Resource: Converting Paper Mill Wastewater to Bioplastic." Water Research 46 (17): 5517-5530. doi:10.1016/j.watres.2012.07.028.

Kim, I. S., D. H. Kim, and S. -H Hyun. 2000. Effect of Particle Size and Sodium Ion Concentration on Anaerobic Thermophilic Food Waste Digestion. Water Science and Technology. Vol. 41.

Kim, J, C. Park, T. Kim, M. Lee, S. Kim, SW Kim, J. Lee. 2002. "Effects of various pretreatment for enhance anaerobic digestion with waste activated sludge." Journal of Bioscience and Bioengineering 95(3): 271-275.

Kim, M., C. Y. Gomec, Y. Ahn, and R. E. Speece. 2003. "Hydrolysis and Acidogenesis of Particulate Organic Material in Mesophilic and Thermophilic Anaerobic Digestion." Environmental Technology (United Kingdom) 24 (9): 1183-1190. doi:10.1080/09593330309385659.

Lin, K. C., R. C. Landine, and S. Bliss. 1982. "Temperature Effect on the Anaerobic Treatment of Potato-Processing Wastewater." Canadian Journal of Civil Engineering 9 (3): 549-557. doi:10.1139/l82-064.

Liu, X., W. Wang, X. Gao, Y. Zhou, and R. Shen. 2012. "Effect of Thermal Pretreatment on the Physical and Chemical Properties of Municipal Biomass Waste." Waste Management 32 (2): 249-255. doi:10.1016/j.wasman.2011.09.027.

Ma, J., T. H. Duong, M. Smits, W. Verstraete, and M. Carballa. 2011. "Enhanced Biomethanation of Kitchen Waste by Different Pre-Treatments." Bioresource Technology 102 (2): 592-599. doi:10.1016/j.biortech.2010.07.122.

Marchal, L. M., J. Jonkers, G. Th Franke, C. D. De Gooijer, and J. Tramper. 1999. "The Effect of Process Conditions on the α-Amylolytic Hydrolysis of Amylopectin Potato Starch: An Experimental Design Approach." Biotechnology and Bioengineering 62 (3): 348-357. doi:10.1002/(SICI)1097-0290(19990205)62:3<348::AID-BIT11>3.0.CO;2-F.

McCarty, P.L., 1964. Anaerobic waste treatment fundamentals. Public Works 95 (9), 107-112.

Mosier, N., C. Wyman, B. Dale, R. Elander, Y. Y. Lee, M. Holtzapple, and M. Ladisch. 2005. "Features of Promising Technologies for Pretreatment of Lignocellulosic Biomass." Bioresource Technology 96 (6): 673-686. doi:10.1016/j.biortech.2004.06.025.

New Brunswick Potatoes. c2017. Agriculture, Aquaculture and Fisheries; [accessed 2017 June 25]. http://www2.gnb.ca/content/gnb/en/departments/10/agriculture/content/crops/potatoes.html

Oh, S. -E, S. Van Ginkel, and B. E. Logan. 2003. "The Relative Effectiveness of pH Control and Heat Treatment for Enhancing Biohydrogen Gas Production." Environmental Science and Technology 37 (22): 5186-5190. doi:10.1021/es034291y.

Rafique, R., T. G. Poulsen, A. -S Nizami, Z. -U -Z Asam, J. D. Murphy, and G. Kiely. 2010. "Effect of Thermal, Chemical and Thermo-Chemical Pre-Treatments to Enhance Methane Production." Energy 35 (12): 4556-4561. doi:10.1016/j.energy.2010.07.011.

Shu, H. -Y, H. -J Fan, M. -C Chang, and W. -P Hsieh. 2006. "Treatment of MSW Landfill Leachate by a Thin Gap Annular UV/H 2O2 Photoreactor with Multi-UV Lamps." Journal of Hazardous Materials 129 (1-3): 73-79. doi:10.1016/j.jhazmat.2005.08.009.

Speece. R.E. 1996. Anaerobic biotechnology for Industrial Wastewaters. Nashville, TN: Archae Press.

Standard Methods for the examination of water and wastewater. APHA, AWWA, WEF 2005.

Su, H., J. Cheng, J. Zhou, W. Song, and K. Cen. 2009. "Improving Hydrogen Production from Cassava Starch by Combination of Dark and Photo Fermentation." International Journal of Hydrogen Energy 34 (4): 1780-1786. doi:10.1016/j.ijhydene.2008.12.045.

Tanaka, S., K. Kamiyanm. 2002. "Thermochemical pretreatment in the anaerobic digestion of waste activated sludge." Journal of Water Science and Technology 46(10): 173-179

Valo, A., H. Carrère, and J. P. Delgenès. 2004. "Thermal, Chemical and Thermo-Chemical Pre-Treatment of Waste Activated Sludge for Anaerobic Digestion." Journal of Chemical Technology and Biotechnology 79 (11): 1197-1203. doi:10.1002/jctb.1106.

Van Ginkel, S. W., S. -E Oh, and B. E. Logan. 2005. "Biohydrogen Gas Production from Food Processing and Domestic Wastewaters." International Journal of Hydrogen Energy 30 (15): 1535-1542. doi:10.1016/j.ijhydene.2004.09.017

Zhang, T., H. Liu, and H. H. P. Fang. 2003. "Biohydrogen Production from Starch in Wastewater Under Thermophilic Condition." Journal of Environmental Management 69 (2): 149-156. doi:10.1016/S0301-4797(03)00141-5.

Building Tomorrow's Society
Bâtir la Société de Demain

Fredericton, Canada
June 13 to 16, 2018 / *13 au 16 juin, 2018*

WATER RECOVERY FROM HIGH STRENGTH BREWERY WASTEWATER USING INTEGRATED ELECTRO-CHEMICAL AND REVERSE OSMOSIS PROCESS

Anwar, Nawrin[1]; Sultana, Sharmin[1]; Choudhury, Mahbuboor R.[1]; Chen, Tiantian[1]; Rahaman, M. Saifur[1,2]
[1] Building Civil and Environmental Engineering, Concordia University, Montreal, QC, Canada
[2] saifur.rahaman@encs.concordia.ca

Abstract: Freshwater scarcity has become a major challenge in today's world, driving the development of new technologies for freshwater production. Advanced treatment of wastewater is a potential source of water recovery for reuse in industrial sectors as it can both deal with water usage, the driver of the water shortage, while reducing the amount of wastewater discharged. Beer is the fifth most consumed beverage in the world and around 70% of water used in the brewing industry is discarded as wastewater; a solution which contains high concentrations of sugar, ethanol, soluble starch, total suspended solids, and volatile fatty acids, requiring treatment prior to its disposal. The high cost of wastewater disposal has led brewers to looking for a sustainable solution where water recovery for reuse within industrial premises is possible. This study aims to investigate the potential of integrated electro-chemical and reverse osmosis (RO) processes for the treatment of brewery wastewater for water recovery and possible reuse. An electro-chemical process in a multi-plated electrolysis cell, with boron doped diamond (BDD) anode was used in this study to reduce a substantial amount of chemical oxygen demand (COD) and total suspended solids from the brewery wastewater. The effluent was then further treated using RO to enhanced water quality for possible reuse in the brewing industry. Prior to all experiments, brewery wastewater was filtrated through 0.2 μm filtration units to remove suspended materials for reduction of RO membrane fouling propensity. The COD of both the feed and permeate were measured to evaluate COD removal efficiency of the process. Implementation of an electrochemical process substantially removed organics from the brewery wastewater, leading to enhanced RO permeate flux—36% higher as compared to the RO process alone. Preliminary results reveal the integrated electro-chemical and reverse osmosis process has enormous potential to be used in recovering water from brewery wastewater.

1 INTRODUCTION

Industrial and domestic wastewater discharges into the environment have become serious global concerns due to their adverse impacts on nature. The elevated levels of solids, organic compounds, and nutrients present in wastewaters make them hazardous due to effect such as eutrophication (Parawira et al. 2005; Simate et al. 2012). Brewery industries have high economic values globally as beer is the fifth most consumed beverage in the world. However, for beer production, a significant amount of water is consumed by the brewing industry, generating highly polluted effluent with 70% of the consumed water being discharged (Valta et al. 2015). The brewing industry generates effluents containing both chemical and microbial constituents including sugars, soluble starch, ethanol, volatile fatty acids (VFAs), total suspended solids, and a low concentration of heavy metals (Simate et al. 2011; Doubla et al. 2007). Therefore, proper treatment of brewery wastewater effluent is important prior to disposal, in order to comply with regulations, as well as to protect the environment, aquatic life and human health. The increasing cost of intake water and wastewater disposal is leading brewers to look for a sustainable solution to be located inside the industrial facility with the potential to generate reusable water for production or non-production purpose. Conventional brewery wastewater treatment includes physical, chemical, and biological processes. Biological processes require high energy input, high maintenance, and high retention times in terms of oxygen input, moisture control, adequate temperature for bacterial activation, and specific forms of bacteria for high removal efficiencies, that often lead to process complexities and inefficiencies (Doubla et al. 2007; Feng et al. 2008; Wang et al. 2007; Chen et al. 2016). In this study, the effect of an integrated Electro-Oxidation and Reverse Osmosis process for treatment of high strength brewery wastewater for possible water recovery was evaluated. Integration of an emerging Electro Oxidation-Electro Fenton (EO-EF) process with a Reverse Osmosis (RO) improved the flux rate of the reverse osmosis membrane as well as enhanced the removal of chemical oxygen demand (COD), total nitrogen (TN), and total phosphorous (TP).

1.1 Overview of Integrated Process

In this study, Electro-Chemical Process (Electro Oxidation and Electro Fenton) was integrated with a Reverse Osmosis Process to evaluate the water recovery performance from brewery wastewater.

1.1.1 Electro-Chemical (Electro Oxidation and Electro Fenton) Process

Electrochemical Advanced Oxidation Processes (EAOPs) combine advanced oxidation processes with electrochemistry, allowing removal of organic pollutants from high strength wastewater (Sultana et al. 2018). The basis of this technology is in-situ electrogenerated hydroxyl radicals (•OH) (Santana-Martínez et al. 2016). EAOPs-based Fenton's reaction is an emerging technology for water and wastewater reclamation, and it has been efficient in treating pharmaceuticals such as caffeine (Ganzenko et al. 2015), antibiotics (Dirany et al. 2010; El-Ghenymy et al. 2014), along with photographic processing wastewater (Bensalah et al. 2013). The fundamentals of Electro Fenton (EF) technology is the continuous electrical generation of H_2O_2 at a suitable cathode surface saturated with oxygen or air (Eq. 1) along with the addition of iron to the sample solution, as a catalyst, to produce hydroxyl radicals in the bulk solution via the Fenton reaction (Eq. 2). In this process, soluble Fe^{3+}, from the Fenton reaction, can be reduced to Fe^{2+} (Eq. 3) (Zhang et al. 2017). Hydroxyl radicals are formed at the surface of high oxidation potential anodes (Eq.4), such as SnO_2, PbO_2, or Boron-Doped Diamond (BDD), as an intermediate from water oxidation of O_2. Further, organics in an electrolytic cell can be degraded by these hydroxyl radicals (Barrera-Díaz et al. 2014). These •OH radicals react with most organics, up to full/partial mineralization, without any selectivity, (i.e., convert them into CO_2, water, and inorganic ions) (El-Ghenymy et al. 2014; Jiang and Zhang 2007).The reaction between an organic pollutant (R) and •OH is shown in Eq.5 (Ganzenko et al. 2015; El-Ghenymy et al. 2014; Dirany et al. 2010; Bensalah et al. 2013).

[1] $O_2 + 2H^+ + 2e^- \rightarrow H_2O_2$

[2] $Fe^{2+} + H_2O_2 \rightarrow Fe^{3+} + \bullet OH + OH^-$

[3] $Fe^{3+} + e^- \rightarrow Fe^{2+}$

[4] $M + H_2O \rightarrow M(\bullet OH) + H^+ + e^-$

[5] $R + \bullet OH \rightarrow$ Products

1.1.2 Reverse Osmosis Process

The Reverse Osmosis (RO) process is an efficient method for water recovery from wastewaters. RO uses semi-permeable membranes for selective mass transport (Greenlee et al. 2009). RO has exhibited enhanced performance in the treatment of semi conductor effluent (Ndiaye et al. 2005), metal mining processing wastewater (Ricci et al. 2015), pharmaceutical wastewater (Radjenovic et al. 2008), etc. Commercially available thin film composite (TFC) membranes have been successfully integrated for RO

(Elimelech and Phillip 2011). However, membrane fouling, caused by organics and biological substances, hinders the widespread application of RO (Rana and Matsuura 2010).

2 MATERIALS AND METHODS

2.1 Brewery Wastewater Characteristics

Brewery wastewater samples were collected from Brasserie Labatt (Lasalle, Quebec) after fermentation. The samples were stored at temperatures of 2-8°C. Prior to conducting the experiments, the samples were brought out of the refrigerator and maintained at room temperatures until the sample temperature reached room temperature. Table 1 represents the physical and chemical characteristics of the triplicate brewery wastewater samples from Labatt.

Table 1: Physical and Chemical characteristics of Labatt Brewery Wastewater

Parameters	Values
pH	6.91±0.38
Conductivity, μS	552±59
TS, mg/L	913±5
TSS, mg/L	148±15
TDS, mg/L	765±20
VTSS, mg/L	135±9
FTSS, mg/L	13±7
COD, mg/L	1860±70
TOC, mg/L	574±32
TN, mg/L	74±10
TP, mg/L	6.71±3.5
TFe, ppb	66±12

2.2 Electro Chemical Set-Up

Figure 1 shows the experimental setup for the electro-oxidation process where a multi-plated electrode configuration containing two anodes and three cathodes in a monopolar parallel connection were used at room temperature (22±2°C) in an open, cylindrical glass cell. Boron Doped Diamond (BDD) electrodes from Fraunhofer USA, Inc. were used as anodes and graphite electrodes were used as cathodes. The effective surface areas of the electrodes were 25 cm^2. The electrodes were placed in a vertical alignment. Sample volume for each experiment was 450 mL with initial pH adjusted to 3. A DC power supply (Agilent Technologies) was used for all experiments with a current density of 16 mA/cm^2. Experiments were conducted for 6 hours. The BDD electrodes were subjected to an auto cleaning procedure by immersing them in an acid solution (1M H_2SO_4) for 30-min after each experiment to remove adsorbed molecules at the electrode surface and were then rinsed with DI water until a neutral pH was attained. The graphite cathodes were alternatively soaked for 30 minutes in an alkaline solution (1M NaOH) and an acidic solution (1M H_2SO_4) for cleaning, and then rinsed with DI water until it reached neutral pH. Prior to each analysis, all samples were mixed using a vortex mixer.

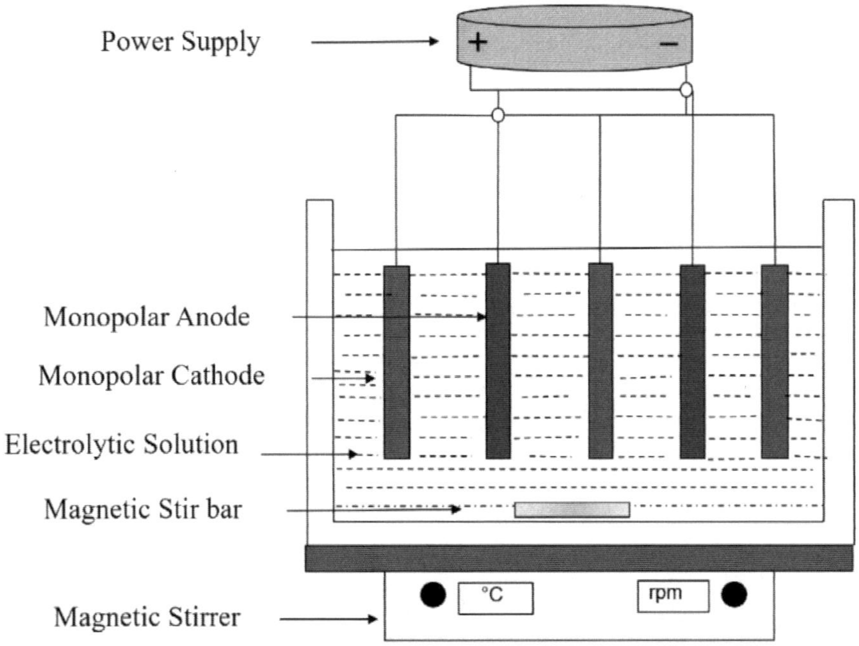

Figure 1: An open cylindrical electrolysis cell containing 450ml of sample at initial pH 3, room temperature, current density 16 mA/cm^2, electrolysis time 6h, DC current source

2.3 Reverse Osmosis Set-Up

Figure 2 shows the cross-flow mode Reverse Osmosis (RO) experimental set-up. The RO process was operated at 400 psi pressure with a 0.5 LPM flow rate and at 20°C temperature. Toray membranes (UTC-82V) were used for the RO process. The membranes were initially compacted with DI water for 15 hrs to stabilize the permeate flux following which the treated water of the electro oxidation process was used and the resulting permeate was collected in a container.

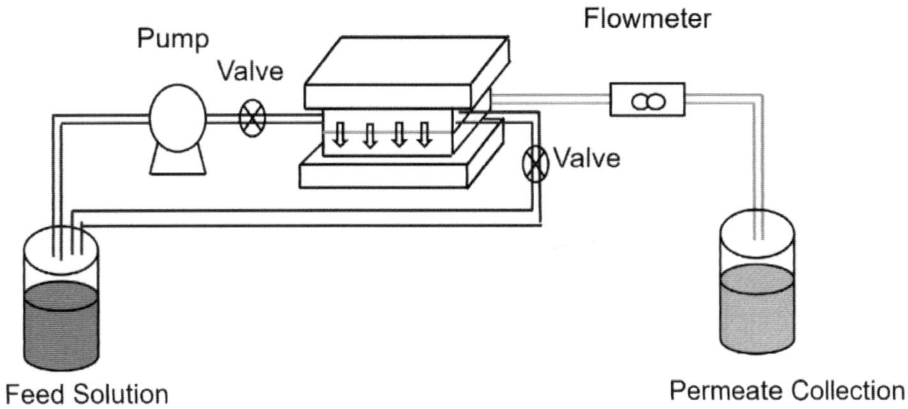

Figure 2: Schematic of Reverse Osmosis (RO) experimental setup in cross flow mode

2.4 Evaluation Technique

A pH meter (Oakton instruments, 310 series, Vernon Hills, IL, USA), turbidity meter (Orion, Thermos scientific), and conductivity meter (Oakton instruments, Vernon Hills, IL, USA) were respectively used for measurements of pH, turbidity, and conductivity. Also, reactor digestion methods of Hach kits, along with a spectrophotometer (DR2800, Hach) for absorbance measurements, were used to measure chemical oxygen demand (COD), total nitrogen (TN), and total phosphorous (TP).

3 RESULTS AND DISCUSSION

To evaluate the efficiency of the integration of the EO-RO process, flux measurements were initially performed. Also, to asses the removal efficiency of the integrated process, COD, TP, and TN measurements were conducted.

3.1 Enhanced Membrane Flux with Integrated Process

Flux measurements of the Reverse Osmosis process for brewery wastewater both with and without electro oxidation treatment (Figure 3) shows that integration of the electro-chemical process with RO was able to achieve high flux as compared to that of the RO process alone. Also, flux of the integrated process is 36% higher compared to RO process. Low flux reduction of the integrated process is an indication of reduced RO membrane fouling.

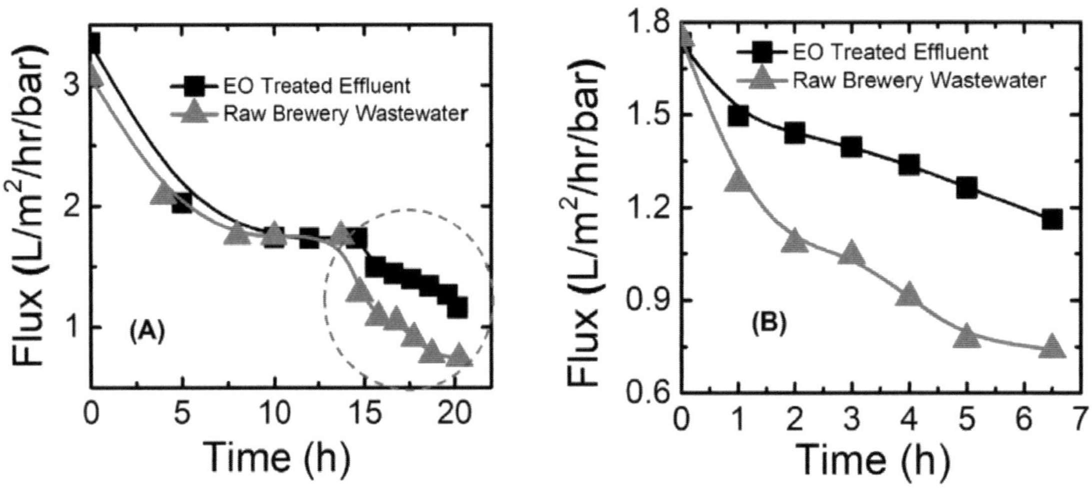

Figure 3: (A) Flux measurements of RO process with and without electrochemical treatment, and (B) enlarge view of circled portion in (A)

3.2 High Organic Removal with Integrated Process

Integration of the electro-chemical process increased the removal efficiency of organics. Figure 4 represents the comparison between the removal efficiency of RO alone and the integrated process. Results showed that incorporation of the electro-chemical process prior to RO achieved 93% COD removal efficiency, whereas 88% was achieved with RO alone. The in situ generation of hydroxyl radical in the electrochemical process rapidly oxidized the organic matter present in the wastewater and high organic removal resulted in reduced load in RO membrane with low RO membrane fouling and higher RO flux .

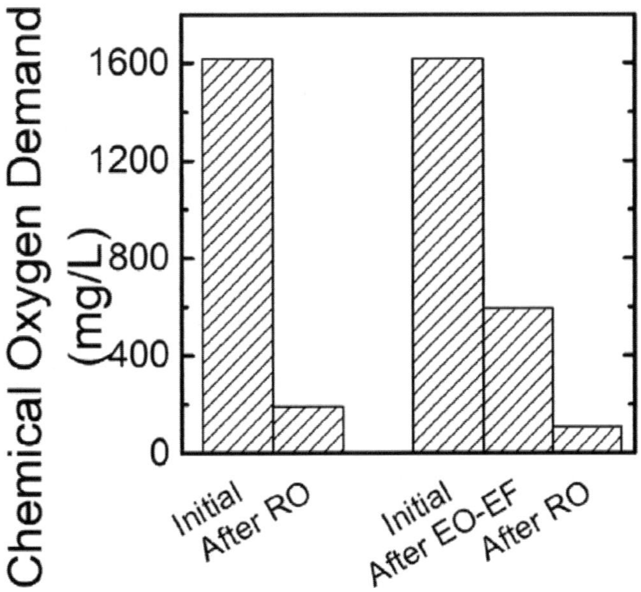

Figure 4: Chemical Oxygen Demand (COD) reduction in brewery wastewater using RO processes on raw wastewater and electrooxidation-pretreated wastewater respectively

3.3 Enhanced Nutrient Removal with Integrated Process

The integration of electrochemical process and reverse osmosis process achieved a 95% and 93% removal efficiency for TP and TN whereas a 95% and 81% removal efficiency were achieved by RO for TP and TN, respectively. Results show that integration of electrochemical and reverse osmosis processes resulted in a 13% increase in total nitrogen removal where this process did not affect phosphorous removal.

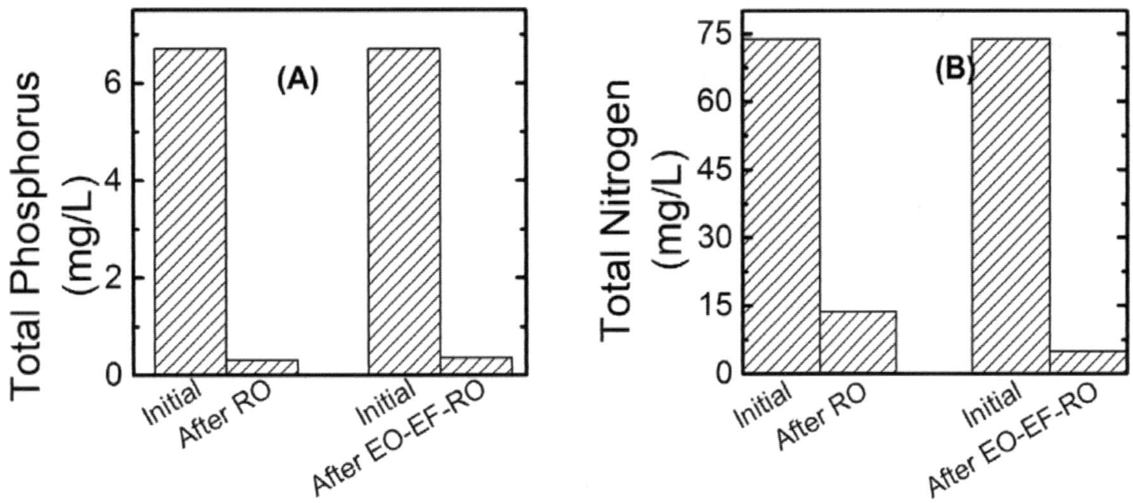

Figure 5: Removal of Total Phosphorus (A) and Total Nitrogen (B) from brewery wastewater using RO process for raw wastewater and electrooxidation pretreated wastewater respectively

4 CONCLUSIONS

Electrochemical processes have the potential to be integrated with the reverse osmosis process for brewery wastewater treatment, resulting in high-quality product water for possible reuse in the industry for both production or non-production purposes. Brewery wastewater contains trace amounts of iron, which enhances the electrochemical process by introducing Electro Fenton along with Electro Oxidation without any addition of excess iron. The integrated process achieved a 36% higher high flux rate along with high COD and TN removal efficiency. However, the formation of the passivation layer on the electrodes leads to further questions, requiring thorough investigation to obtain possible maximum reuse. Further detailed analysis of RO membrane fouling would also provide an insight into how membrane fouling is altered by incorporation of the electrochemical process. Overall, this study provides an understanding of the effects of integrating electrochemical processes with membrane processes for water recovery and possible reuse.

Acknowledgements

The authors gratefully acknowledge the Ministre de l'Agriculture, des Pêcheries et de l'Alimentation, Québec (MAPAQ) and the Natural Sciences and Engineering Research Council (NSERC) of Canada for providing funding supports for this project. We thank Claude Danis and Julien Beaulieu Sabourin (from Labatt) for collecting and shipping the brewery process water for this project.

References

Barrera-Díaz, C., P. Cañizares, F. J. Fernández, R. Natividad, and M.A. Rodrigob. 2014. "Electrochemical Advanced Oxidation Processes: An Overview of the Current Applications to Actual Industrial Effluents." *Journal of Mexican Chemical Society* 58 (3): 256-275.

Bensalah, Nasr, Ahmed Bedoui, Shankararaman Chellam, and Ahmed Abdel-Wahab. 2013. "Electro-Fenton Treatment of Photographic Processing Wastewater." *Clean - Soil, Air, Water* 41 (7): 635–44. doi:10.1002/clen.201200521.

Chen, Han, Sheng Chang, Qingbin Guo, Youngseck Hong, and Ping Wu. 2016. "Brewery Wastewater Treatment Using an Anaerobic Membrane Bioreactor." *Biochemical Engineering Journal* 105. Elsevier B.V.: 321–31. doi:10.1016/j.bej.2015.10.006.

Dirany, Ahmad, Ignasi Sirés, Nihal Oturan, and Mehmet A. Oturan. 2010. "Electrochemical Abatement of the Antibiotic Sulfamethoxazole from Water." *Chemosphere* 81 (5). Elsevier Ltd: 594–602. doi:10.1016/j.chemosphere.2010.08.032.

Doubla, A., S. Laminsi, S. Nzali, E. Njoyim, J. Kamsu-Kom, and J. L. Brisset. 2007. "Organic Pollutants Abatement and Biodecontamination of Brewery Effluents by a Non-Thermal Quenched Plasma at Atmospheric Pressure." *Chemosphere* 69 (2): 332–37. doi:10.1016/j.chemosphere.2007.04.007.

El-Ghenymy, Abdellatif, Rosa María Rodríguez, Enric Brillas, Nihal Oturan, and Mehmet A. Oturan. 2014. "Electro-Fenton Degradation of the Antibiotic Sulfanilamide with Pt/carbon-Felt and BDD/carbon-Felt Cells. Kinetics, Reaction Intermediates, and Toxicity Assessment." *Environmental Science and Pollution Research* 21 (14): 8368–78. doi:10.1007/s11356-014-2773-3.

Elimelech, Menachem, and William A. Phillip. 2011. "The Future of Seawater Desalination: Energy, Technology, and the Environment." *Science* 333 (6043): 712–17.

Feng, Yujie, Xin Wang, Bruce E. Logan, and He Lee. 2008. "Brewery Wastewater Treatment Using Air-Cathode Microbial Fuel Cells." *Applied Microbiology and Biotechnology* 78 (5): 873–80. doi:10.1007/s00253-008-1360-2.

Ganzenko, Oleksandra, Nihal Oturan, David Huguenot, Eric D. Van Hullebusch, Giovanni Esposito, and Mehmet A. Oturan. 2015. "Removal of Psychoactive Pharmaceutical Caffeine from Water by Electro-Fenton Process Using BDD Anode: Effects of Operating Parameters on Removal Efficiency." *Separation and Purification Technology* 156. Elsevier B.V.: 987–95. doi:10.1016/j.seppur.2015.09.055.

Greenlee, Lauren F., Desmond F. Lawler, Benny D. Freeman, Benoit Marrot, and Philippe Moulin. 2009. "Reverse Osmosis Desalination: Water Sources, Technology, and Today's Challenges." *Water Research* 43 (9): 2317–48.

Jiang, Cheng-chun, and Jia-fa Zhang. 2007. "Progress and Prospect in Electro-Fenton Process for Wastewater Treatment." *Journal of Zhejiang University-SCIENCE A* 8 (7): 1118–25. doi:10.1631/jzus.2007.A1118.

Ndiaye, P.I., P. Moulin, L. Dominguez, J. C. Millet, and F. Charbit. 2005. "Removal of Fluoride from Electronic Industrial Effluentby RO Membrane Separation." *Desalination* 173 (1): 25–32.

Parawira, W., I. Kudita, M. G. Nyandoroh, and R. Zvauya. 2005. "A Study of Industrial Anaerobic Treatment of Opaque Beer Brewery Wastewater in a Tropical Climate Using a Full-Scale UASB Reactor Seeded with Activated Sludge." *Process Biochemistry* 40 (2): 593–99. doi:10.1016/j.procbio.2004.01.036.

Radjenovic, J., M. Petrovic, F. Ventura, and D. Barcelo. 2008. "Rejection of Pharmaceuticals in Nanofiltration and Reverse Osmosis Membrane Drinking Water Treatment." *Water Research* 42 (14): 3601–10.

Rana, D., and T. Matsuura. 2010. "Surface Modification for Antifouling Membrane." *Chem. Rev.* 110: 2448–71.

Ricci, Bárbara C., Carolina D. Ferreira, Alice O. Aguiar, and Miriam C.S. Amaral. 2015. "Integration of Nanofiltration and Reverse Osmosis for Metal Separation and Sulfuric Acid Recovery from Gold Mining Effluent." *Separation and Purification Technology* 154: 11–21.

Santana-Martínez, Germán, Gabriela Roa-Morales, Eduardo Martin Del Campo, Rubí Romero, Bernardo A. Frontana-Uribe, and Reyna Natividad. 2016. "Electro-Fenton and Electro-Fenton-like with in Situ Electrogeneration of H2O2and Catalyst Applied to 4-Chlorophenol Mineralization." *Electrochimica Acta* 195: 246–56. doi:10.1016/j.electacta.2016.02.093.

Simate, Geoffrey S., John Cluett, Sunny E. Iyuke, Evans T. Musapatika, Sehliselo Ndlovu, Lubinda F. Walubita, and Allex E. Alvarez. 2011. "The Treatment of Brewery Wastewater for Reuse: State of the Art." *Desalination* 273 (2–3). Elsevier B.V.: 235–47. doi:10.1016/j.desal.2011.02.035.

Simate, Geoffrey S., Sunny E. Iyuke, Sehliselo Ndlovu, and Mike Heydenrych. 2012. "The Heterogeneous Coagulation and Flocculation of Brewery Wastewater Using Carbon Nanotubes." *Water Research* 46 (4). Elsevier Ltd: 1185–97. doi:10.1016/j.watres.2011.12.023.

Sultana, Sharmin, Mahbubbor Rahman Choudhury,Ahmed Refaat Bakr, Nawrin Anwar, and Md. Saifur Rahaman, 2018. "The Heterogeneous Coagulation and Flocculation of Brewery Wastewater Using Carbon Nanotubes." *Journal of Applied Electrochemisty* 48 (5): 519-528. doi.org/10.1007/s10800-018-1185-3.

Valta, Katerina, Tatjana Kosanovic, Dimitris Malamis, and Maria Moustakas, Konstantinos Loizidou. 2015. "Overview of Water Usage and Wastewater Management in the Food and Beverage Industry." *Desalination and Water Treatment* 53 (12): 3335–47.

Wang, Shu Guang, Xian Wei Liu, Wen Xin Gong, Bao Yu Gao, Dong Hua Zhang, and Han Qing Yu. 2007. "Aerobic Granulation with Brewery Wastewater in a Sequencing Batch Reactor." *Bioresource Technology* 98 (11): 2142–47. doi:10.1016/j.biortech.2006.08.018.

Zhang, Yan, Mingming Gao, Shu Guang Wang, Weizhi Zhou, Yuanhua Sang, and Xin Hua Wang. 2017. "Integrated Electro-Fenton Process Enabled by a Rotating Fe3O4/gas Diffusion Cathode for Simultaneous Generation and Activation of H2O2." *Electrochimica Acta* 231. Elsevier Ltd: 694–704. doi:10.1016/j.electacta.2017.02.091.

AUTHOR INDEX